Vector Optimization

Series Editor:

Johannes Jahn
University of Erlangen-Nürnberg
Department of Mathematics
Cauerstr. 11
81058 Erlangen
Germany
johannes.jahn@fau.de

More information about this series at
http://www.springer.com/series/8175

Vector Optimization

The series in Vector Optimization contains publications in various fields of optimization with vector-valued objective functions, such as multiobjective optimization, multi criteria decision making, set optimization, vector-valued game theory and border areas to financial mathematics, biosystems, semidefinite programming and multiobjective control theory. Studies of continuous, discrete, combinatorial and stochastic multiobjective models in interesting fields of operations research are also included. The series covers mathematical theory, methods and applications in economics and engineering. These publications being written in English are primarily monographs and multiple author works containing current advances in these fields.

Sorin-Mihai Grad

Vector Optimization and Monotone Operators via Convex Duality

Recent Advances

Springer

Sorin-Mihai Grad
Faculty of Mathematics
TU Chemnitz
Chemnitz
Germany

ISSN 1867-8971 ISSN 1867-898X (electronic)
ISBN 978-3-319-08899-0 ISBN 978-3-319-08900-3 (eBook)
DOI 10.1007/978-3-319-08900-3
Springer Cham Heidelberg New York Dordrecht London

Library of Congress Control Number: 2014948065

Printed on acid-free paper

Springer is part of Springer Science+Business Media (www.springer.com)

To Carmen

Preface

Convex analysis plays an increasingly important role in different areas of mathematics and its applications, especially in optimization. One of its main tools is represented by the conjugate functions that, employed in the duality theory, gave birth to the conjugate duality and, in a larger sense, to convex duality.

The importance of the duality approach in the optimization theory comes mainly from the facts that it provides a lower bound for the objective values of a given minimization problem and, on the other hand, it leads to necessary and sufficient optimality conditions. These can be used for determining the optimal solutions of the primal problem, for instance by employing them in generating different algorithmic approaches for solving optimization problems.

The step from scalar to vector optimization problems can be made, from the point of view of duality, in several ways, depending on the desired outcomes. In this book, several duality approaches for vector optimization problems are investigated and, to some extent, compared. Moreover, we deal with different solution concepts for vector optimization problems, too.

One can generalize the framework even further, for instance by considering set-valued multifunctions. Among the problems where they appear, an increasing interest is attracted by the ones involving monotone operators, especially since new methods for approaching them by means of convex analysis were developed. Following this path, we provide in the book several results concerning different properties of the monotone operators, too.

This book started as the habilitation thesis of the author [107], where he collected the majority of his contributions to the fields of vector optimization and monotone operators during the last decade, most of them obtained by means of duality theory for convex optimization problems. Several results, remarks, comments and explanations were added to the original thesis version in order to make the book more complete and reader-friendly.

The book is structured into seven chapters. The first one is dedicated to an introduction and to several necessary preliminaries meant to make this work as self-contained as possible. In the second chapter, we deal with scalar optimization problems. After assigning conjugate dual problems to them, we characterize their stable ε-duality gap via inclusions involving epigraphs and subdifferentials, respectively. These characterizations are further used to rediscover, in case $\varepsilon = 0$ and when the involved functions are endowed with convexity and topological properties, some recent closedness type regularity conditions from the literature and the formulae they guarantee.

Within the third chapter, different minimality concepts for sets are considered and investigated. Moreover, the corresponding minimality sets are compared. Several of these minimality concepts are presented in more general frameworks than considered so far in the literature and it is shown that some of their properties remain valid in the new setting, too, in some cases under additional weak hypotheses. These minimality notions lead to solution concepts for vector optimization problems, the so-called efficiency concepts.

A vector duality approach via a general scalarization for vector optimization problems is introduced in the fourth chapter. Different scalarization functions considered in the literature are then employed as special cases of the general scalarization, leading to various vector dual problems for a general vector optimization problem. The latter contains as special cases the classes of both constrained and unconstrained vector optimization problems and vector duals and corresponding duality statements are derived in each case.

Extending the classical Wolfe and Mond-Weir duality concepts for both scalar and vector general nondifferentiable optimization problems is the aim of the fifth chapter. Considered so far in the literature mainly for optimization problems involving differentiable functions, both these duality concepts lead to dual problems that contain the corresponding optimality conditions as constraints.

In the sixth chapter, new vector dual problems are assigned to the classical linear vector optimization problem in finitely dimensional spaces with respect to both efficient and weakly efficient solutions and then they are extended to infinitely dimensional spaces, too. It is shown that these new vector duals encounter no trouble in some special cases where some of their older counterparts from the literature failed to achieve strong duality. Comparisons between the known vector duals and the new ones are delivered, too. Moreover, we propose a new vector dual problem to a semidefinite vector optimization problems, too.

The last chapter presents results involving monotone operators obtained mainly by means of convex analysis, via conjugate functions and duality formulae. First we deal with Brézis-Haraux type approximations for the range of a sum of monotone operators, then we provide characterizations involving representative functions for the surjectivity of a sum of monotone operators, from which closedness type regularity conditions for various interesting special cases can be derived. We also show how approaching the maximal monotonicity of the bifunctions via representative functions leads to generalizing known results and to positively answering to some recently posed conjectures.

I would like to express my thanks to Gert Wanka for his continuous support and for giving me the opportunity first to study and then to do my research at

Chemnitz University of Technology. I am very grateful to my mentor and friend Radu Ioan Boţ for his guidance and many useful discussions during all the years of working together and for his advices regarding this book. I would also like to thank Ernö Robert Csetnek for his valuable comments concerning some of the results presented here and for carefully reading a preliminary version of the manuscript. Thanks are also due to the former and current members of the research group Approximation Theory within the Faculty of Mathematics of the Chemnitz University of Technology for discussions and remarks regarding different parts of this book and for contributing to the friendly and inspiring working atmosphere we all enjoyed and benefited from. I am grateful to my coauthors and to the anonymous reviewers of the papers integrated within this book who contributed with comments and remarks to their improvement, as well as to the fellow mathematicians for the valuable feed-back provided at different conferences where I presented parts of this manuscript. The financial support from DFG (German Research Foundation) via projects WA 922/1-3 and WA 922/8-1 that made possible some of the research presented in this book is thankfully acknowledged, too. Last but not least, I am grateful to my family and especially to my better half Carmen for unconditioned love, patience and support.

For updates and errata the reader is referred to the author's website

http://www.tu-chemnitz.de/~gsor

Chemnitz, Germany
May 2014

Sorin-Mihai Grad

List of Symbols and Notations

Sets and Elements

$\mathbb{R}$	The set of the real numbers
$\mathbb{R}^k_{+/-}$	The nonnegative (nonpozitive) orthant in $\mathbb{R}^k$
$\mathbb{Q}$	The set of the rational numbers
$\mathbb{N}$	The set of the nonnegative integers
X^*	Topological dual space of X
X^{**}	Topological bidual space of X
$\omega(X, X^*)$	Weak topology on X induced by X^*
$\omega(X^*, X)$	Weak* topology on X^* induced by X
$L^\perp$	Orthogonal subspace to the subspace L
K^*	Topological dual cone K^* of the cone K
K^0	The set $\{x \in K : \langle x^*, x\rangle > 0 \ \forall x^* \in K^* \setminus \{0\}\}$
K^{*0}	The set $\{x^* \in K^* : \langle x^*, x\rangle > 0 \ \forall x \in K \setminus \{0\}\}$
$\hat{K}$	The cone qi $K \cup \{0\}$, where K is a convex cone
$\tilde{K}$	The cone qri $K \cup \{0\}$, where K is a convex cone
$\ell(K)$	Linearity space of the cone K
$A^\dagger$	Moore-Penrose pseudo-inverse of the matrix A
A_{ij}	Entry lying on both the i-th row and j-th column of the matrix $A \in \mathbb{R}^{n\times m}$, $i = 1, \ldots, n$, $j = 1, \ldots, m$
$\mathscr{S}^k$	Set of the symmetric $k \times k$ matrices
$\mathscr{S}^k_+$	Set of the symmetric positive semidefinite $k \times k$ matrices
$\hat{\mathscr{S}}^k_+$	Set of the symmetric positive definite $k \times k$ matrices
e	The vector $(1, \ldots, 1)^T$
Δ_{X^m}	The set $\{(x, \ldots, x) \in X^m : x \in X\}$
$N_U(x)$	Normal cone to the set U at x
$N^\varepsilon_U(x)$	ε-normal set to the set U at x
$T_U(x)$	Bouligand tangent cone to the set U at x
0^+U	Recession cone of $U \subseteq \mathbb{R}^n$
U^o	Polar set of $U \subseteq \mathbb{R}^n$

$F_U(x^*)$	Face of $U \subseteq X$ exposed by $x^* \in X^*$
$\operatorname{lin} U$	Linear hull of the set U
$\operatorname{aff} U$	Affine hull of the set U
$\operatorname{co} U$	Convex hull of the set U
$\overline{\operatorname{co}} U$	Closed convex hull of the set U
$\operatorname{cone} U$	Conical hull of the set U
$\operatorname{coneco} U$	Convex conical hull of the set U
$\operatorname{cl} U$	Closure of the set U
$\operatorname{bd} U$	Boundary of the set U
$\operatorname{int} U$	Interior of the set U
$\operatorname{core} U$	Algebraic interior of the set U
$\operatorname{qi} U$	Quasi interior of the set U
$\operatorname{qri} U$	Quasi relative interior of the set U
$\operatorname{sqri} U$	Strong quasi relative interior of the set U
$\operatorname{ri} U$	Relative interior of the set $U \subseteq \mathbb{R}^n$
$\dim U$	Dimension of the set U
$\mathscr{S}_{\dots}$	Set of strongly cone-increasing scalarization functions
$\mathscr{T}_{\dots}$	Set of strictly cone-increasing scalarization functions
$\mathscr{F}_T$	Fitzpatrick family of the monotone operator T

Functions and Operators

$\langle x^*, x \rangle$	The value of $x^* \in X^*$ at $x \in X$
c	Coupling function defined on $X \times X^*$
$\\|\cdot\\|_{(2)}$	(Euclidean) norm on a normed space
$\\|\cdot\\|_*$	Dual norm
id_X	Identity function on X
$\operatorname{Pr}_X U$	Projection of the set $U \subseteq X \times Y$ on X
P_D	(Orthogonal) projection onto the set D
$\mathscr{L}(X, Y)$	Set of linear continuous mappings from X to Y
A^*	Adjoint mapping of $A \in \mathscr{L}(X, Y)$
$\operatorname{Im} A$	Image set of $A \in \mathscr{L}(X, Y)$
$A^{-1}(W)$	Counter image of a set $W \subseteq Y$ through $A \in \mathscr{L}(X, Y)$
$\ker A$	Kernel of $A \in \mathscr{L}(X, Y)$
$s_{\dots}$	Scalarization function
$f_1 \Box \dots \Box f_m$	Infimal convolution of the functions $f_i, i = 1, \dots, m$
$f \Box_2 g$	Bivariate convolution of the functions f and g
δ_U	Indicator function of the set U
σ_U	Support function of the set U
δ_U^v	Vector indicator function of the set U
γ_U	Gauge of the set U
$\operatorname{dom} f$	(Effective) domain of the (vector) function f
$\operatorname{epi} f$	Epigraph of the function f

$\operatorname{epi}_K h$	K-epigraph of the vector function h
(v^*h)	The function $\langle v^*, h\rangle$, where h is a vector function and $v^* \in K^*$
Af	Infimal function of the function f through $A \in \mathscr{L}(X,Y)$
$h(\cdot)$	Scalar infimal value function
$\operatorname{co} f$	Convex hull of the function f
$\bar{f}$	Lower semicontinuous hull of the function f
$\overline{\operatorname{co}} f$	Lower semicontinuous convex hull of the function f
f^*	Conjugate function of the function f
f_S^*	Conjugate function of the function f with respect to the set S
$\partial f(x)$	Subdifferential of the function f at $x \in X$
$\partial_\varepsilon f(x)$	ε-subdifferential of the function f at $x \in X$
$\nabla f(x)$	Gradient of the function f at $x \in X$
$f^\top$	Transpose of the function f
$\hat{h}$	The function $\hat{h}(x,y) = h(x,-y)$, where h is a given function
Δ_U	Oriented distance function to the set U
$\operatorname{Tr} A$	Trace of the matrix A
$T : X \rightrightarrows X^*$	Monotone operator defined on X with (set) values in X^*
$D(T)$	Domain of the monotone operator T
$R(T)$	Range of the monotone operator T
$G(T)$	Graph of the monotone operator T
$\overline{T}$	Monotone closure of the monotone operator T
φ_T	Fitzpatrick function of the monotone operator T
h_T	Representative function of the monotone operator T
ψ_T	The function $\overline{\operatorname{co}}(c + \delta_{G(T)})$
$\mathscr{J}$	Duality map on a Banach space
$F : C \times C \to \mathbb{R}$	Bifunction defined on C
A^F, ${}^F\!A$	Diagonal subdifferential operators attached to a bifunction F

Partial Orderings

$\leqq_K$	Partial ordering induced by the convex cone K
$x \leq_K y$	$x \leqq_K y$ and $x \neq y$
$x <_K y$	$y - x \in \operatorname{qi} K$ (or $y - x \in \operatorname{core} K$ or $y - x \in \operatorname{int} K$), where K is a convex cone with $\operatorname{qi} K \neq \emptyset$ ($\operatorname{core} K \neq \emptyset$ or $\operatorname{int} K \neq \emptyset$)
$x <^r_K y$	$y - x \in \operatorname{qri} K$ (or $y - x \in \operatorname{ri} K$), where K is a convex cone with $\operatorname{qri} K \neq \emptyset$ ($\operatorname{ri} K \neq \emptyset$)
$\preceqq_k$	Partial ordering induced by the cone $\mathscr{S}^k_+$
$\preceq_k$	$x \preceqq_k y$ and $x \neq y$
∞_K	Greatest element with respect to the ordering cone K attached to a space
$V^\bullet$	The space V to which the element ∞_K is added

Minimality Notions (with Respect to the Cone $K \subseteq V$)

$\mathrm{Min}(M, K)$	Set of minimal elements of the set M
$\mathrm{Max}(M, K)$	Set of maximal elements of the set M
$\mathrm{WMin}(M, K)$	Set of weakly minimal elements of the set M
$\mathrm{WMax}(M, K)$	Set of weakly maximal elements of the set M
$\mathrm{RMin}(M, K)$	Set of relatively minimal elements of the set M
$\mathrm{RMax}(M, K)$	Set of relatively maximal elements of the set M
$\mathrm{PMin}_{\ldots}(M, K)$	Generic notation for sets of properly minimal elements of the set M
$\mathscr{E}(PV\ldots)$	Efficiency set of the problem $(PV\ldots)$
$\mathscr{P}\mathscr{E}_{\ldots}(PV\ldots)$	Proper efficiency set of the problem $(PV\ldots)$
$\mathscr{W}\mathscr{E}(PV\ldots)$	Weak efficiency set of the problem $(PV\ldots)$

Generic Notations

$(P^{\cdots})$	Primal optimization problem
$(PV^{\cdots})$	Primal vector optimization problem
$v(P^{\cdots})$	Optimal objective value of the problem $(P^{\cdots})$
$(D^{\cdots})$	Dual optimization problem
$(DV^{\cdots})$	Dual vector optimization problem
$\mathscr{A}^{\cdots}$	Feasible set of a primal vector problem
$\mathscr{B}^{\cdots}$	Feasible set of a dual vector problem
$h^{\cdots}$	Objective function of a dual vector problem
$\Phi^{\cdots}$	Perturbation function
$\Phi_v^{\cdots}$	Vector perturbation function
$(RC^{\cdots})$	Regularity condition for scalar optimization problems
$(RCV^{\cdots})$	Regularity condition for vector optimization problems
$(RCM^{\cdots})$	Regularity condition for monotone operators
$(RCS^{\cdots})$	Regularity condition for sets

Contents

Chapter 1
Introduction and Preliminaries

1.1 Introduction

In this book we present some recent advances on duality for vector optimization problems and on monotone operators obtained by means of conjugate duality and other methods of the convex analysis. This is not intended to be a systematic presentation of the state-of-the-art in these fields, but an extended collection of the results obtained by the author in these research areas by making use of tools like conjugate functions, duality and convexity. His more than a decade long interest in vector optimization was materialized so far in the papers [31, 32, 34, 37, 51, 108–110, 201]. Moreover, he is the coauthor of the book [48], where the state-of-the-art on duality for vector optimization at the moment was presented. On the other hand, his research results on monotone operators and related topics can be found in [30, 33, 35, 40, 42, 44].

Some of the results on duality for vector optimization problems gathered in this book have arisen from investigations begun while preparing [48] and, in a way, the part dedicated to vector optimization of this work can be seen as a continuation and development of the mentioned book into several research directions, presented there only in a limited manner. Unless otherwise specified, the content of this book, excepting most of the preliminaries, is represented by the contributions of the author (together with his co-authors) to the research fields of vector optimization and monotone operators. Besides this introductory part, the work contains six other chapters, each of them evolving around a main theme and beginning with a short overview on the available literature together with the motivations behind the investigations that follow.

Chapter 2. This chapter is dedicated to scalar optimization problems, more precisely to characterizing via epigraph and subdifferential inclusions the situation of ε-duality gap, i.e. when the difference between the optimal objective value of a primal minimization problem and the one of its dual problem is less than ε. We

S.-M. Grad, *Vector Optimization and Monotone Operators via Convex Duality*,
Vector Optimization, DOI 10.1007/978-3-319-08900-3_1

deliver such characterizations in the most general framework, when the involved functions are taken only proper. Endowing them with convexity and topological properties, we obtain other useful equivalences, from which one can derive when $\varepsilon = 0$ closedness type regularity conditions, useful for strong duality or different conjugate formulae. After presenting these investigations for general optimization problems, we deal with both constrained and unconstrained optimization problems, showing how the mentioned results can be specialized for them, too. This chapter is based on the author's contributions [13, 14, 28, 45], moreover ideas and techniques from [46, 47, 49, 50] being also involved.

Chapter 3. In the beginning of the third chapter different proper minimality concepts for sets regarding the partial ordering induced by a convex cone that is not necessarily pointed are introduced and analyzed. Inclusion relations between the proper minimality sets of a given set in various senses are provided, too. Then the ordering cone is taken to be pointed and weak conditions that guarantee characterizations via linear scalarization of some of the considered proper minimality notions are delivered. On the other hand, we consider properly minimal elements of a set defined by means of a general scalarization function, too. Then the concept of weak minimality is extended to the case where the ordering cone has a nonempty quasi interior, showing that some properties, including the ones regarding the linear scalarization, of the classical weakly minimal points with respect to a cone with nonempty interior are inherited. Similar investigations are made for relatively minimal elements, that are defined with respect to ordering cones that have a nonempty quasi-relative interior. This chapter is based on the author's articles [109, 110], containing also several previously unpublished results.

Chapter 4. Given a general vector optimization problem, the properly minimal elements in the sense of a general scalarization of its image set lead to corresponding properly efficient solutions. Depending on the monotonicity properties of the scalarization function we differentiate between two classes of such properly efficient solutions. With respect to them several vector dual problems are attached to the primal vector optimization problem. We investigate these vector dual problems and we deliver weak and strong duality statements concerning them and the vector primal problem, as well as the corresponding necessary and sufficient optimality conditions. Several scalarization functions considered in the literature are employed in our general scheme, leading to different vector duals to the original general vector optimization problem. Afterwards we particularize the primal problem to be first constrained, then unconstrained, and vector duals are derived from the general scheme in each case. This chapter is based on the author's pieces [31, 37], together with some previously unpublished statements.

Chapter 5. This chapter presents detailed investigations on Wolfe and Mond-Weir type duality for both scalar and vector convex nonsmooth optimization problems. Both these duality concepts were considered in the literature mainly for constrained optimization problems involving differentiable (generalized) convex functions. We propose a duality approach via perturbations for a general scalar optimization problem which leads in the particular case of a primal constrained differentiable optimization problem to the classical Wolfe and Mond-Weir duals of the latter, respectively. By employing different perturbation functions, we deliver several Wolfe and Mond-Weir type dual problems to both constrained and unconstrained optimization problems. Moreover, we show that our duality approach is open towards optimization problems involving generalized convex functions, too. The approach is then extended to vector optimization problems by following two directions. Some of the Wolfe and Mond-Weir type vector dual problems of classical type to a general vector optimization problem turn out to rediscover in case of a constrained differentiable primal their counterparts from the literature. On the other hand, by making use of an idea of constructing vector dual problems considered mainly in convex vector optimization, we propose other Wolfe and Mond-Weir type vector dual problems. These alternative vector duals have larger image sets than their classical counterparts and we investigate some connections between these two classes of Wolfe and Mond-Weir type vector dual problems. Several examples are provided in order to show that some of the obtained inclusions can be sometimes strictly fulfilled and to prove that the Wolfe type duals attached to a constrained optimization problem via different perturbation functions act quite differently than their conjugate or Mond-Weir type counterparts. This chapter is based on the author's papers [29, 32, 108], employing moreover results and ideas from [39, 41].

Chapter 6. In the sixth chapter we deal with two important particular vector optimization problems, namely linear and semidefinite ones. We begin by revisiting the vector duality for the classical linear vector optimization problem in finitely dimensional spaces. We propose a new vector dual to it, for which weak, strong and converse duality are proven, comparing it moreover with its counterparts from the literature. Then we extend our investigations to infinitely dimensional spaces, showing that the vector dual we proposed can be generalized to that framework, too, maintaining most of its properties. Other vector duals, like the ones of Wolfe and Mond-Weir type from Chap. 5, are considered in this setting, too. Moreover, we deal with the mentioned linear vector optimization problems with respect to weakly efficient solutions, too. Last but not least, we propose a similar duality approach for a vector optimization problem consisting in vector minimizing with respect to the corresponding semidefinite cone a matrix function subject to semidefinite inequality constraints. This chapter is based on the author's works [34, 51], including moreover some previously unpublished material.

Chapter 7. Within the last chapter of this book there are presented some recent results involving monotone operators that are obtained mainly by techniques and tools belonging to convex analysis. After some preliminaries on monotone operators and their approach by means of convex analysis, we deliver Brézis-Haraux type approximations for the range of the sum of a monotone operator with another one composed with a linear mapping. We note the differences between what happens in general Banach spaces and how these results are modified when the involved spaces are moreover reflexive. Among the special cases of our main result we provide corrections and generalizations of earlier results from the literature. Afterwards we characterize via closedness type conditions involving representative functions the surjectivity of the sum of two maximally monotone operators. From them we derive regularity conditions for different results concerning ranges of sums of both general and particular maximally monotone operators, that are shown to be weaker than their counterparts from the literature. Last but not least, we introduce a way of approaching the maximal monotonicity of bifunctions via representative functions that allowed us to extend from reflexive to general Banach spaces different recent results from the literature and, moreover, to provide affirmative answers to some recently posed conjectures. This chapter is based on the author's publications [30,33,35,38,40,42,44].

1.2 Preliminaries

In order to make the book as self-contained as possible, some preliminary notions and results are needed. Most of them belong to the folklore of convex analysis or optimization and literature sources are indicated only for the not so widely known ones. The presentation is based on books like [21, 48, 127, 128, 140, 178, 221] and some of the references therein, unless otherwise specified.

1.2.1 Sets

Within this work we shall work with both finitely and infinitely dimensional real topological vector spaces. For a positive integer $k \in \mathbb{N}$, by $\mathbb{R}^k$ we denote the k dimensional real Euclidean space. All the vectors in $\mathbb{R}^k$ are column vectors, an upper index "$^\top$" being used to transpose them into row vectors. By $\mathbb{R}^k_+$ we denote the *nonnegative orthant* in $\mathbb{R}^k$, while $\mathbb{R}^k_-$ is the corresponding *nonpositive orthant.* Moreover, we denote $e = (1, \dots, 1)^\top \in \mathbb{R}^k$. The space of all $k \times n$ real matrices is denoted by $\mathbb{R}^{k\times n}$, while the subspace of $k \times k$ symmetric matrices is denoted by $\mathscr{S}^k$. The set of the symmetric positive semidefinite $k \times k$ matrices is $\mathscr{S}^k_+$ and its interior, the set of the symmetric positive definite $k \times k$ matrices is $\hat{\mathscr{S}}^k_+$. The entries of a matrix $A \in \mathbb{R}^{k\times k}$ will be denoted by A_{ij}, $i, j = 1, \dots, k$, while its *trace* by $\operatorname{Tr} A$ and its *transpose* by $A^\top$.

If X is a Hausdorff locally convex space, its topological *dual space* is denoted by X^*, while X^{**}, the *bidual space* of X, is the dual of the latter. When X is a normed space, we denote its norm by $\|\cdot\|$, its dual norm, i.e. the norm of its dual space, by $\|\cdot\|_*$ and one can identify X with its image under its canonical injection into a subspace of X^{**}. The space X^* can be endowed with different topologies, for instance the weak* one, denoted by $\omega(X^*, X)$, that will be considered for all the dual spaces everywhere in this book except for Chap. 7, where also topologies that do not necessarily rended X as dual space to X^* are considered. Note that in Chap. 7 whenever the topology of X^* is not mentioned the strong one is understood. The natural topology on $\mathbb{R}$ is denoted by $\mathscr{R}$. By $\langle x^*, x\rangle = x^*(x)$ we denote the value at $x \in X$ of the linear continuous functional $x^* \in X^*$. Consider also the *coupling function* $c : X \times X^* \to \overline{\mathbb{R}} = \mathbb{R} \cup \{\pm\infty\}$, $c(x, x^*) = \langle x^*, x\rangle$ for $(x, x^*) \in X \times X^*$.

A *cone* $K \subseteq X$ is a nonempty subset of X which fulfills $\alpha K \subseteq K$ for all $\alpha \geq 0$ and it is said to be *nontrivial* if it does not coincide with either $\{0\}$ or X. A convex cone $K \subseteq X$ induces on X the partial ordering "$\leqq_K$" defined by $x \leqq_K y \Leftrightarrow y - x \in K$ when $x, y \in X$. Moreover, if $x \leqq_K y$ and $x \neq y$ we write $x \leq_K y$. When $K = \mathbb{R}^k_+$ these cone inequality notations are simplified to "$\leqq$" and "$\leq$", respectively, while if $X = \mathscr{S}^k$ and $K = \mathscr{S}^k_+$ they become "$\preceqq_k$" and "$\preceq_k$", respectively. To X can be then attached a greatest element ∞_K with respect to "$\leqq_K$" which does not belong to X and let be $X^\bullet = X \cup \{\infty_K\}$. Then for any $x \in X$ one has $x \leq_K \infty_K$ and we consider on $X^\bullet$ the operations $x + \infty_K = \infty_K + x = \infty_K$ for all $x \in X$ and $\alpha \cdot \infty_K = \infty_K$ for all $\alpha \geq 0$. The *dual cone* of K is $K^* = \{x^* \in X^* : \langle x^*, x\rangle \geq 0 \ \forall x \in K\}$. The dual cone of K^* is the *bidual cone* of K, being denoted by K^{**}, and it coincides with K when the latter is convex and closed and the topology considered on X^* renders X as its dual space. By convention, $\langle x^*, \infty_K\rangle = +\infty$ for all $x^* \in K^*$. The *linearity space* of a convex cone $K \subseteq X$ is $\ell(K) = K \cap (-K)$. When $\ell(K) = \{0\}$ K is said to be *pointed.*

Given a subset U of X, by $\operatorname{cl} U$, $\operatorname{int} U$, $\operatorname{core} U$, $\operatorname{lin} U$, $\operatorname{aff} U$, $\operatorname{co} U$, $\overline{\operatorname{co}} U$, $\operatorname{cone} U$, $\operatorname{coneco} U$, $\operatorname{bd} U$ and $\dim U$ we denote its *closure, interior, algebraic interior (core), linear hull, affine hull, convex hull, closed convex hull, conical hull, convex conical hull, boundary* and *dimension*, respectively. Moreover, if U is convex, by

$$\operatorname{sqri} U = \big\{x \in U : \operatorname{cone}(U - x) \text{ is a closed linear subspace}\big\}$$

we denote its *strong quasi-relative interior*,

$$\operatorname{qri} U = \big\{x \in U : \operatorname{cl} \operatorname{cone}(U - x) \text{ is a linear subspace of } X\big\}$$

is its *quasi-relative interior*, while the *quasi interior* of U is the set

$$\operatorname{qi} U = \big\{x \in U : \operatorname{cl} \operatorname{cone}(U - x) = X\big\}.$$

In case $U \subseteq \mathbb{R}^k$, $\operatorname{ri} U$ denotes the *relative interior* of U. The *indicator function* of the set U is $\delta_U : X \to \overline{\mathbb{R}}$, defined as $\delta_U(x) = 0$ if $x \in U$ and $\delta_U(x) = +\infty$ otherwise, while its *support function* $\sigma_U : X^* \to \overline{\mathbb{R}}$ is given by $\sigma_U(x^*) = \sup_{x \in U} \langle x^*, x\rangle$. The *polar set* of $U \subseteq X$ is $U^o = \{x^* \in X^* : \sigma_U(x^*) \leq 1\}$. Note that when $U \subseteq X$ is a closed convex cone it holds $U^o = -U^*$. When $U \subseteq X$ and $x^* \in X^*$, the set $F_U(x^*) = \{x \in U : \langle x^*, x\rangle = \sigma_U(x^*)\}$ is said to be the *face of U exposed by x^**.

If X is partially ordered by the convex cone $K \subseteq X$, Y is a topological vector space and $V \subseteq Y$, the *vector indicator function* of V is $\delta^v_V : Y \to X^\bullet$, which fulfills $\delta^v_V(y) = 0$ if $y \in V$ and $\delta^v_V(y) = \infty_K$ otherwise. We use also the *projection function* $\operatorname{Pr}_X : X \times Y \to X$, defined by $\operatorname{Pr}_X(x, y) = x$ for $(x, y) \in X \times Y$, the *identity function* $\mathrm{id} : X \to X$, $\mathrm{id}(x) = x$ for $x \in X$ and, for $n \in \mathbb{N}$, the notation $\Delta_{X^n} = \{(x, \ldots, x) : x \in X\} \subseteq X^n$. The *(orthogonal) projection onto* $U \subseteq X$ is the operator $P_U : X \to U$. The *normal cone* associated to the set U at $x \in U$ is

$$N_U(x) = \{x^* \in X^* : \langle x^*, y - x\rangle \leq 0 \ \forall y \in U\}$$

and, for $\varepsilon \geq 0$, the ε-*normal set* associated to U at $x \in U$ is

$$N_U^{\varepsilon}(x) = \{x^* \in X^* : \langle x^*, y - x\rangle \leq -\varepsilon \ \forall y \in U\}.$$

If $U \subseteq \mathbb{R}^k$ is a convex set, its *recession cone* is $0^+U = \{x \in \mathbb{R}^k : U + x \subseteq U\}$. A subset of $\mathbb{R}^k$ is said to be *polyhedral* if it can be expressed as the intersection of some finite collection of closed half-spaces.

To a nonempty set $U \subseteq X$ one can attach the *Bouligand tangent cone* at $x \in \operatorname{cl} U$, that is

$$T_U(x) := \Big\{y \in X : \exists (x_l)_{l \geq 1} \in U \text{ and } (\lambda_l)_{l \geq 1} > 0 \text{ such that } \lim_{l \to +\infty} x_l = x \text{ and } \lim_{l \to +\infty} \lambda_l(x_l - x) = y\Big\}.$$

Note that $T_U(x)$ is always a cone and $T_U(x) \subseteq \operatorname{cl} \operatorname{cone}(U - x)$. When U is convex, it holds $\operatorname{cone}(U - x) \subseteq T_U(x)$, which yields in this case that $\operatorname{cl} T_U(x) = \operatorname{cl} \operatorname{cone}(U - x)$. If X is metrizable, then $T_U(x)$ is closed and, thus, if U is convex one has $T_U(x) = \operatorname{cl} \operatorname{cone}(U - x)$, which has as a consequence the convexity of $T_U(x)$, for all $x \in \operatorname{cl} U$.

In vector optimization it is used also the *quasi interior of the dual cone* (also called *strong dual cone*) of the convex cone $K \subseteq X$, $K^{*0} = \{x^* \in K^* : \langle x^*, x\rangle > 0 \ \forall x \in K \setminus \{0\}\}$. As shown in [48, Proposition 2.1.1], in general it holds $\operatorname{qi} K^* \subseteq K^{*0}$, which turns into equality when K is also closed. Note that $K^{*0} \neq \emptyset$ yields, as shown in [140, Lemma 1.27], that the convex cone K is pointed. If $U \subseteq X$ is a convex set, one has (cf. [20]) $\operatorname{int} U \subseteq \operatorname{core} U \subseteq \operatorname{qi} U \subseteq \operatorname{qri} U$ and when one of the sets in this chain of inclusions is nonempty, it coincides with all its mentioned supersets (cf. [20, 221]). For $\operatorname{qi} K \neq \emptyset$ we write $x <_K y$ if $y - x \in \operatorname{qi} K$, extending the notation usually considered in the literature for the case $\operatorname{int} K \neq \emptyset$, while when $\operatorname{qri} K \neq \emptyset$ we write $x <_K^r y$ if $y - x \in \operatorname{qri} K$. For all $x \in X$ it holds $\operatorname{qri}\{x\} = \{x\}$. In case $U \subseteq \mathbb{R}^k$, we have that $\operatorname{qi} U = \operatorname{core} U = \operatorname{int} U$ and $\operatorname{qri} U = \operatorname{sqri} U = \operatorname{ri} U$. In a separable Banach space the quasi interior of any nonempty convex set not contained in a hyperplane is nonempty (cf. [148]) and the quasi-relative interior of a nonempty closed convex set is nonempty (cf. [20]), but this is no longer true in general if the space is not separable. A situation where the interior of a set and all its generalized interiors but the quasi interior and the quasi-relative interior are empty follows.

Example 1.1. Let the real Banach space $\ell^2 = \ell^2(\mathbb{N})$ of the real sequences $(x_n)_{n \in \mathbb{N}}$ that fulfill $\sum_{n=1}^{\infty} |x_n|^2 < +\infty$ be equipped with the norm $\|\cdot\| : \ell^2 \to \mathbb{R}$, $\|x\| = \left(\sum_{n=1}^{\infty} |x_n|^2\right)^{1/2}$, $x = (x_n)_{n \in \mathbb{N}} \in \ell^2$. The *positive cone* of ℓ^2 is $\ell^2_+ = \{(x_n)_{n \in \mathbb{N}} \in$

$\ell^2 : x_n \geq 0 \ \forall n \in \mathbb{N}\}$. Then $\operatorname{int} \ell^2_+ = \operatorname{core} \ell^2_+ = \operatorname{sqri} \ell^2_+ = \emptyset$, but $\operatorname{qi} \ell^2_+ = \operatorname{qri} \ell^2_+ = \{(x_n)_{n\in\mathbb{N}} \in \ell^2 : x_n > 0 \ \forall n \in \mathbb{N}\}$.

In the following we present some properties of the quasi(-relative) interior of a set needed later in our presentation. For more on this the reader is referred to [20, 21, 23–25, 67, 68, 198].

Proposition 1.1. *Let $K \subseteq X$ be a convex cone.*

(a) K is not dense if and only if $0 \notin \operatorname{qi} K$.
(b) If $\operatorname{cl} K$ is pointed, then $0 \notin \operatorname{qri} K$.
(c) One has $\operatorname{qri} K + K = \operatorname{qri} K$ and, consequently, also $\operatorname{qi} K + K = \operatorname{qi} K$.
(d) The set $\operatorname{qri} K \cup \{0\}$ is a cone, hence so is $\operatorname{qi} K \cup \{0\}$, too.

Remark 1.1. Given a convex cone $K \subseteq X$, note that if $\operatorname{cl} K$ is pointed then $\operatorname{cl} K \neq X$, but the opposite assertion is not always true. Take, for instance, $X = \mathbb{R}^2$ and $K = \mathbb{R} \times \{0\}$, then $\operatorname{cl} K = K$ is not pointed even if it does not coincide with the whole space $\mathbb{R}^2$.

Proposition 1.2. *Let U and V be convex subsets of X. Then the following statements hold*

(a) $\operatorname{qri} U + \operatorname{qri} V \subseteq \operatorname{qri}(U + V)$*;*
(b) $\operatorname{qri}(U - x) = (\operatorname{qri} U) - x$ *for all* $x \in X$*;*
(c) $\lambda \operatorname{qri} U + (1 - \lambda)U \subseteq \operatorname{qri} U$ *for all* $\lambda \in (0, 1]$ *hence* $\operatorname{qri} U$ *is a convex set;*
(d) $\operatorname{qri}(\operatorname{qri} U) = \operatorname{qri} U$*;*
(e) if $\operatorname{qri} U \neq \emptyset$ *then* $\operatorname{cl} \operatorname{qri} U = \operatorname{cl} U$ *and* $\operatorname{cl} \operatorname{cone} \operatorname{qri} U = \operatorname{cl} \operatorname{cone} U$*;*
(f) if $U \subseteq V$ *then* $\operatorname{qi} U \subseteq \operatorname{qi} V$
(g) if $u \in U$*, then* $u \in \operatorname{qi} U$ *if and only if* $N_U(u) = \{0\}$ *and* $u \in \operatorname{qri} U$ *if and only if* $N_U(u)$ *is a linear subspace.*

Proposition 1.3. *Given the convex sets $U, V \subseteq X$, one has that* $\operatorname{qi}(U + \operatorname{qi} V) = U + \operatorname{qi} V \subseteq \operatorname{qi}(U + V)$.

Proof. By Proposition 1.2(f) one gets $\operatorname{qi}(U + \operatorname{qi} V) \subseteq \operatorname{qi}(U + V)$ and obviously $\operatorname{qi}(U + \operatorname{qi} V) \subseteq U + \operatorname{qi} V$. The only implication left to be prove is $U + \operatorname{qi} V \subseteq \operatorname{qi}(U + \operatorname{qi} V)$. When $\operatorname{qi} V = \emptyset$ it is trivially fulfilled.

Let $a \in U + \operatorname{qi} V$. Then there exist $u \in U$ and $v \in \operatorname{qi} V$ such that $a = u + v$. But $\operatorname{qi} V - v \subseteq \operatorname{qi} V - v + (U - u) = U + \operatorname{qi} V - a$. From here follows that $\operatorname{cone}(\operatorname{qi} V - v) \subseteq \operatorname{cone}(U + \operatorname{qi} V - a)$ and moreover $\operatorname{cl} \operatorname{cone}(\operatorname{qi} V - v) \subseteq \operatorname{cl} \operatorname{cone}(U + \operatorname{qi} V - a)$. But in this case $\operatorname{cl} \operatorname{cone}(\operatorname{qi} V - v) = \operatorname{cl} \operatorname{cone}(\operatorname{qri} V - v)$ (because $\operatorname{qi} V$ is nonempty) and Proposition 1.2(e) yields $\operatorname{cl} \operatorname{cone}(\operatorname{qri} V - v) = \operatorname{cl} \operatorname{cone}(V - v)$. As $v \in \operatorname{qi} V$, $\operatorname{cl} \operatorname{cone}(V - v) = X$, so $\operatorname{cl} \operatorname{cone}(U + \operatorname{qi} V - a) = X$, too and, since $a \in U + \operatorname{qi} V$, it follows that $a \in \operatorname{qi}(U + \operatorname{qi} V)$. □

In the literature there exist some separation results for convex sets by mean of quasi-relative interior (see, for instance, [21, 23–25, 76, 77, 89, 90]). We will use in our investigations the following one, given in [25, Theorem 2.7] and

[21, Theorem 20.6] where it is mentioned that it extends similar results from [76,77] from normed to locally convex spaces.

Lemma 1.1. *Let U be a nonempty convex subset of X and $x \in U$. If $x \notin \operatorname{qri} U$ then there exists an $x^* \in X^* \setminus \{0\}$ such that $\langle x^*, y \rangle \le \langle x^*, x \rangle$ for all $y \in U$.*

Specializing Lemma 1.1 for the quasi interior one obtains not only an implication like there, but actually an equivalence, that can be also seen as a direct consequence of Proposition 1.2(g).

Lemma 1.2. *Let U be a nonempty convex subset of X and $x \in U$. Then $x \notin \operatorname{qi} U$ if and only if there exists an $x^* \in X^* \setminus \{0\}$ such that $\langle x^*, y \rangle \le \langle x^*, x \rangle$ for all $y \in U$.*

If $K \subseteq X$ is a closed convex cone and we endow X^* with the $\omega(X^*, X)$ topology, one can immediately show via [48, Proposition 2.1.1] that $\operatorname{qi} K = \{x \in K : \langle x^*, x \rangle > 0 \ \forall x^* \in K^* \setminus \{0\}\}$. Let us denote the set in the right-hand side of this equality by K^0. Aware that in the literature this notation was also used for the interior and polar cone of K, respectively, we opted for it due to the similarity with K^{*0}. Let us see now what relations can be identified between K^0 and $\operatorname{qri} K$ in the case of a convex cone $K \subseteq X$ that is not necessarily closed (cf. [109]).

Proposition 1.4. *Let $K \subseteq X$ be a convex cone.*

(a) It holds $K^0 \subseteq \operatorname{qri} K$.
(b) If $K^0 \neq \emptyset$, then $\operatorname{qri} K \subseteq K^0$.

Proof. (a) If $x \in K^0 \setminus \operatorname{qri} K$, then $\langle x^*, x \rangle > 0$ for all $x^* \in K^* \setminus \{0\}$ and, on the other hand, Lemma 1.1 yields the existence of an $\bar{x}^* \in X^* \setminus \{0\}$ such that $\langle \bar{x}^*, x^* \rangle \le \langle \bar{x}^*, y \rangle$ for all $y \in K$. Then $\bar{x}^* \in K^* \setminus \{0\}$ and $\langle \bar{x}^*, x^* \rangle \le 0$. But $\langle \bar{x}^*, x^* \rangle > 0$, and this contradiction yields that there exists no x as taken above.

(b) If $x \in \operatorname{qri} K \setminus K^0$ then there exists an $\bar{x}^* \in K^* \setminus \{0\}$ such that $\langle \bar{x}^*, x \rangle = 0$. Then $\langle -\bar{x}^*, y - x \rangle \le 0$ for all $y \in K$, i.e. $-\bar{x}^* \in N_K(x)$. As $x \in \operatorname{qri} K$ yields that $N_K(x)$ is a linear subspace of X^*, it follows that $\bar{x}^* \in N_K(x)$, too, i.e. $\langle \bar{x}^*, y - x \rangle \le 0$ for all $y \in K$. This yields $\langle \bar{x}^*, y \rangle = 0$ for all $y \in K$, consequently $K^0 = \emptyset$. □

Remark 1.2. If the convex cone K is also closed one has $\operatorname{qi} K = K^0$, so $K^0 \neq \emptyset$ means actually $\operatorname{qi} K \neq \emptyset$, that yields $\operatorname{qi} K = \operatorname{qri} K$. Conditions that guarantee that $K^0 \neq \emptyset$ were proposed in the literature to the best of our knowledge only for this case, for instance in [140, Theorem 3.38]. Similarly, the inclusion in Proposition 1.4(b) was previously known only under the additional hypothesis $\operatorname{cl}(K - K) = X$, which also yields $\operatorname{qi} K = \operatorname{qri} K$, as done for instance in [20, Theorem 3.10] or [219, Lemma 2.5].

Another separation statement from the literature, this time in finitely dimensional spaces, that we shall use within this work is [117, Lemma 2.2(i)].

Lemma 1.3. *Let the closed convex cone $K \subseteq \mathbb{R}^k$ and the polyhedral set $U \subseteq \mathbb{R}^k$ fulfilling $U \cap K = \{0\}$. Then there exists a $\gamma \in \mathbb{R}^k \setminus \{0\}$ such that $\gamma^\top k < 0 \le \gamma^\top u$ for all $k \in K \setminus \{0\}$ and all $u \in U$.*

In order to refine some results proven via duality by employing very general regularity conditions we have introduced in [42] the notion of a set that is closed regarding another one (see also [21, 45, 50]). However, for the investigations from Chap. 2 a more general notion is required, that was originally defined in [13] (and, for the case $Z = X \times \mathbb{R}$ in [14]).

Definition 1.1. Given $\varepsilon \geq 0$, a set $U \subseteq X \times \mathbb{R}$ is said to be $(0, \varepsilon)$-*vertically closed regarding the set* $Z \subseteq X \times \mathbb{R}$ if $(\operatorname{cl} U) \cap Z \subseteq (U \cap Z) - (0, \varepsilon)$, while when $Z = X \times \mathbb{R}$, U is called simply $(0, \varepsilon)$-*vertically closed*. Moreover, a set $U \subseteq X$ that fulfills $(\operatorname{cl} U) \cap W = U \cap W$, where $W \subseteq X$, is said to be *closed regarding the set* W.

Remark 1.3. A set $U \subseteq X \times \mathbb{R}$ is closed regarding $X \times \mathbb{R}$ if and only if it is closed. A closed set is closed regarding any subset of the space it lies in, but vice versa this does not in general. For instance, the real interval $[0, 1)$ is closed regarding the set $\{0\}$, but it is not closed in general.

Remark 1.4. The notion of an ε-*closed set* was considered in the literature in different instances that have nothing in common with our research, see for instance [3, 95], while in [184, Definition 3.2] one can find a definition for a *vertically closed set*.

1.2.2 Functions

In what follows we present some preliminary notions and results involving functions needed later in our presentation. We begin with some notions which extend the classical monotonicity to functions defined on partially ordered spaces. Let $K \subseteq X$ be a convex cone.

Definition 1.2. Let the nonempty set $U \subseteq X$ and $f : X \to \overline{\mathbb{R}}$ a given function.

(i) If $f(x) \leq f(y)$ for all $x, y \in U$ such that $x \leqq_K y$ the function f is called *K-increasing* on U.
(ii) If $f(x) < f(y)$ for all $x, y \in U$ such that $x \leq_K y$ the function f is called *strongly K-increasing* on U.
(iii) If f is K-increasing on U, $\operatorname{qi} K \neq \emptyset$ and for all $x, y \in U$ fulfilling $x <_K y$ follows $f(x) < f(y)$ the function f is called *strictly K-increasing* on U.
(iv) If f is K-increasing on U, $\operatorname{qri} K \neq \emptyset$ and for all $x, y \in U$ fulfilling $x <_K^r y$ follows $f(x) < f(y)$ the function f is called *relatively strictly K-increasing* on U.
(v) When $U = X$ we call these classes of functions simply *K-increasing*, *strongly K-increasing*, *strictly K-increasing* and *relatively strictly K-increasing*, respectively.

Remark 1.5. In Definition 1.2(iii) and (iv) we extend the notion of a strictly K-increasing on U function given so far in the literature for the case $\operatorname{int} K \neq \emptyset$

(or $\operatorname{core} K \neq \emptyset$). Note also that in case $X = \mathbb{R}$ and $K = \mathbb{R}_+$, the $\mathbb{R}_+$-increasing functions are actually the monotonically increasing ones, while the strongly, strictly and relatively strictly $\mathbb{R}_+$-increasing functions are nothing but strictly monotonically increasing functions.

Let us illustrate this definition with the following example (see [48, 110]).

Example 1.2. Let $x^* \in X^*$. If $x^* \in K^*$, then for all $x_1, x_2 \in X$ such that $x_1 \leqq_K x_2$ we have that $\langle x^*, x_2 - x_1 \rangle \geq 0$. Therefore $\langle x^*, x_1 \rangle \leq \langle x^*, x_2 \rangle$ and this means that the elements of K^* are actually K-increasing functions.

If $x^* \in K^{*0}$, then for all $x_1, x_2 \in X$ such that $x_1 \leq_K x_2$ it holds $\langle x^*, x_2 - x_1 \rangle > 0$. This means by definition that the elements of K^{*0} are strongly K-increasing functions.

If $K \subseteq X$ is moreover closed, X^* is endowed with the $\omega(X^*, X)$ topology and $\operatorname{qi} K \neq \emptyset$, then $\operatorname{qi} K = \{x \in K : \langle x^*, x \rangle > 0 \ \forall x^* \in K^* \setminus \{0\}\}$ and thus every $x^* \in K^* \setminus \{0\}$ is strictly K-increasing.

If $K^0 \neq \emptyset$, then Proposition 1.4 yields $\operatorname{qri} K = K^0$ and thus every $x^* \in K^* \setminus \{0\}$ is relatively strictly K-increasing.

Having a function $f : X \to \overline{\mathbb{R}}$ we use the classical notations for its *domain* $\operatorname{dom} f = \{x \in X : f(x) < +\infty\}$, *epigraph* $\operatorname{epi} f = \{(x, r) \in X \times \mathbb{R} : f(x) \leq r\}$, *lower semicontinuous hull* $\bar{f} : X \to \overline{\mathbb{R}}$, *convex hull* $\operatorname{co} f : X \to \overline{\mathbb{R}}$, *lower semicontinuous convex hull* $\overline{\operatorname{co}} f : X \to \overline{\mathbb{R}}$ and *conjugate function* $f^* : X^* \to \overline{\mathbb{R}}$, $f^*(x^*) = \sup\{\langle x^*, x \rangle - f(x) : x \in X\}$. If $U \subseteq X$, the conjugate function of f regarding $U \subseteq X$ is $f_U^* : X \to \overline{\mathbb{R}}$, $f_U^* = (f + \delta_U)^*$. We call f *proper* if $f(x) > -\infty$ for all $x \in X$ and $\operatorname{dom} f \neq \emptyset$. For f proper and $\varepsilon \geq 0$, if $f(x) \in \mathbb{R}$ the *(convex) ε-subdifferential* of f at x is $\partial_\varepsilon f(x) = \{x^* \in X^* : f(y) - f(x) \geq \langle x^*, y - x \rangle - \varepsilon \ \forall y \in X\}$, while if $f(x) = +\infty$ we take by convention $\partial_\varepsilon f(x) = \emptyset$. The ε-subdifferential becomes in case $\varepsilon = 0$ the classical *(convex) subdifferential* denoted for f by ∂f. Note that for $U \subseteq X$ we have for all $x \in U$ and all $\varepsilon \geq 0$ that $\partial_\varepsilon \delta_U(x) = N_U^\varepsilon(x)$. Between a function and its conjugate regarding U there is the *Young-Fenchel inequality* $f_U^*(x^*) + f(x) \geq \langle x^*, x \rangle$ for all $x \in U$ and $x^* \in X^*$. If $U = X$, this inequality is fulfilled as equality if and only if $x^* \in \partial f(x)$ and in general one has $f^*(x^*) + f(x) \leq \langle x^*, x \rangle + \varepsilon$ if and only if $x^* \in \partial_\varepsilon f(x)$. Moreover, a function is said to be *upper hemicontinuous* if it is upper semicontinuous on line segments.

Considering for each $\alpha \in \mathbb{R}$ the function $\alpha f : X \to \overline{\mathbb{R}}$, $(\alpha f)(x) = \alpha f(x)$ for $x \in X$, note that when $\alpha = 0$ we take $0 f = \delta_{\operatorname{dom} f}$. Given a linear continuous mapping $A : X \to Y$, we have its *adjoint* $A^* : Y^* \to X^*$ given by $\langle A^* y^*, x \rangle = \langle y^*, Ax \rangle$ for any $(x, y^*) \in X \times Y^*$. Its *image* is $\operatorname{Im} A = \{Ax : x \in X\}$, while the *counter image* of a set $W \subseteq Y$ through A is $A^{-1}(W) := \{x \in X : Ax \in W\}$. With $\mathscr{L}(X, Y)$ we denote the set of the linear continuous mappings $A : X \to Y$. When $f, g : X \to \overline{\mathbb{R}}$ are proper, their *infimal convolution* is $f \Box g : X \to \overline{\mathbb{R}}$, $f \Box g(a) = \inf_{x \in X}[f(x) + g(a - x)]$. It is said to be *exact* at $y \in X$ when the infimum at $a = y$ is attained, i.e. there exists an $x \in X$ such that $f \Box g(y) = f(x) + g(y - x)$. If $U, W \subseteq X$, denote by $f^\top : W \times U \to \overline{\mathbb{R}}$ the *transpose* of the function $f : U \times W \to \overline{\mathbb{R}}$, which is defined as $f^{\top}(w,u) = f(u,w)$ for all $(w,u) \in W \times U$.

A vector function $F : Y \to X^{\bullet}$ is said to be *proper* if its *domain* $\operatorname{dom} F = \{y \in Y : F(y) \in X\}$ is nonempty. It is called K-*convex* if $F(tx+(1-t)y) \leqq_K tF(x)+(1-t)F(y)$ for all $x, y \in Y$ and all $t \in [0,1]$. The vector function F is said to be K-epi-*closed* if K is closed and its K-*epigraph* $\operatorname{epi}_K F = \{(y,x) \in Y \times X : x \in F(y) + K\}$ is closed, and it is called K-*lower semicontinuous* if for every $y \in Y$, each neighborhood W of zero in X and for any $b \in X$ satisfying $b \leqq_K F(y)$, there exists a neighborhood U of y in Y such that $F(U) \subseteq b + W + X \cup \{\infty_K\}$. For $x^* \in K^*$ the function $(x^*F) : Y \to \overline{\mathbb{R}}$ is defined by $(x^*F)(y) = \langle x^*, F(y)\rangle$, $y \in Y$. If F is K-lower semicontinuous then (x^*F) is lower semicontinuous whenever $x^* \in K^* \setminus \{0\}$ and if K is closed, then every K-lower semicontinuous vector function is also K-epi-closed, but, not all K-epi-closed vector functions are K-lower semicontinuous, as the situation depicted in [47, Example 1] shows.

Remark 1.6. If the function $f : X \to \overline{\mathbb{R}}$ is K-increasing, then $\operatorname{dom} f^* \subseteq K^*$. However, the analogy with the results mentioned in Example 1.2 stops here, since even in case $X = \mathbb{R}$ one can find strictly increasing functions with their conjugates having as domain $\mathbb{R}_+$ and not $(0, +\infty)$, which coincides with both $\mathbb{R}_+ \setminus \{0\}$ and $\operatorname{int} \mathbb{R}_+$, for instance the exponential function.

For an attained infimum (supremum) instead of inf (sup) we write min (max), while the optimal objective value of the optimization problem (P) is denoted by $v(P)$.

Chapter 2
Duality for Scalar Optimization Problems

2.1 Historical Overview and Motivation

Assigning a dual problem to a given minimization problem provides, due to the weak duality, a lower bound for the objective values of the latter. Moreover, if *strong duality* can be proven, the optimal objective values of the two problems coincide and they can be determined since usually the dual problem has a simpler structure than the primal one and can be easier solved. Moreover, necessary and sufficient *optimality conditions* for the primal-dual pair of problems in discussion can be derived and these can be employed for determining the optimal solutions of the primal problem when the ones of the dual, guaranteed by the strong duality statement, were already identified. The corresponding duality theory is very well developed in the convex case and can be consulted in books like [21, 48, 127, 128, 178, 221]. Moreover, it was shown in the literature (see, for instance, [39, 41, 52, 53]) that the hypotheses on the involved functions can be weakened to different generalizations of the convexity without destroying the strong duality statements.

In order to guarantee the strong duality one usually needs besides the convexity assumptions the fulfillment of a regularity condition or constraint qualification. Different such conditions were considered in the literature, the most important classes of them being the interiority type ones (see, for instance, [21, 221]) and the closedness type ones (cf. [21, 45, 48]). Moreover, strong duality is closely related to subdifferential formulae and the mentioned sufficient conditions can be employed to ensure these, too. Besides the strong duality of interest are also the so-called *stable strong duality*, i.e. the situation when strong duality holds for any linear perturbation of the objective function of the primal problem, and its stronger version *total duality* where, additional to the strong duality also an optimal solution of the primal problem is known. Characterizations for these situations involving epigraph and subdifferential inclusions, respectively, were provided for instance in [46, 47].

In some cases, however, it can be shown only that the distance between the optimal objective values of the primal and dual problem is less than some nonnegative ε,

S.-M. Grad, *Vector Optimization and Monotone Operators via Convex Duality*, Vector Optimization, DOI 10.1007/978-3-319-08900-3_2

situation called in [13, 14] *ε-duality gap*. Starting from our investigations from these papers, where we characterized via epigraph and subdifferential inclusions the ε-duality gap for composed and constrained optimization problems, respectively, we provide in this chapter similar statements for general scalar optimization problems where the involved functions are taken first only proper. Endowing them with convexity and topological properties, we obtain other useful equivalences, from which when $\varepsilon = 0$ *closedness type regularity conditions* (cf. [56, 57, 63, 64]) are derived. These can be employed, for instance, for subdifferential formulae, as done in [28, 45, 126] or, like in [27], for providing formulae for biconjugates of combinations of functions.

After presenting these investigations for general optimization problems, we deal with both *constrained* and *unconstrained optimization problems*, showing how the mentioned results can be specialized for them, too, by means of the perturbation theory (cf. [178, 221]). In this way some of our results from [13, 14, 28, 45–47, 50] as well as different others from the literature can be obtained as special cases of the general statements presented here.

2.2 Characterizations Involving Epigraphs

We begin our investigations with a general scalar optimization problem. In order to investigate its duality properties, we embed it into a family of general perturbed scalar optimization problems, to which corresponding conjugate dual problems (cf., for instance, [21, 48, 178, 221]) are assigned. Then we characterize via epigraph inclusions the so-called stable ε-duality gap regarding the primal problem and its conjugate dual. Adding convexity and topological hypotheses to the function involved, we show that this approach leads to some closedness type regularity conditions recently considered in the literature for duality and other formulae involving convex functions and their conjugates. Afterwards we particularize the primal problem to be constrained and unconstrained, respectively, and the corresponding duality statements are derived from the general case.

2.2.1 General Perturbed Scalar Optimization Problems

Consider two Hausdorff locally convex vector spaces X and Y. Most of the results presented within this chapter hold actually in the more general framework of linear spaces, but in order to avoid juggling with the spaces we use here the mentioned setting. Let the proper function $F : X \to \overline{\mathbb{R}}$ and the general optimization problem

$$(PG) \qquad \inf_{x \in X} F(x).$$

Making use of a proper *perturbation function* $\Phi : X \times Y \to \overline{\mathbb{R}}$, fulfilling $\Phi(x, 0) = F(x)$ for all $x \in X$, a hypothesis that guarantees that $0 \in \Pr_Y \operatorname{dom} \Phi$, the problem (PG) can be rewritten as

$$(PG) \qquad \inf_{x \in X} \Phi(x, 0).$$

We call Y the *perturbation space* and its elements *perturbation variables*. Note that the way Φ is defined guarantees that $0 \in \Pr_Y (\operatorname{dom} \Phi)$. To (PG) we attach the following *conjugate dual* problem

$$(DG) \qquad \sup_{y^* \in Y^*} \{-\Phi^*(0, y^*)\},$$

and for this primal-dual pair of optimization problems *weak duality* always holds, i.e. $v(DG) \leq v(PG)$. In order to investigate the duality properties of these optimization problems, for each $x^* \in X^*$ we consider the following problem

$$(PG_{x^*}) \qquad \inf_{x \in X} \left[\Phi(x, 0) - \langle x^*, x \rangle\right],$$

obtained by linearly perturbing the objective function of (PG). Thus (PG) is embedded in the family of optimization problems $\{(PG_{x^*}) : x^* \in X^*\}$, where it coincides with (PG_0). To each problem in the mentioned family one can attach the corresponding conjugate dual problem, namely, for $x^* \in X^*$,

$$(DG_{x^*}) \qquad \sup_{y^* \in Y^*} \{-\Phi^*(x^*, y^*)\}.$$

By construction, whenever $x^* \in X^*$ one has $v(DG_{x^*}) \leq v(PG_{x^*})$, i.e. for each of these pairs of primal-dual optimization problems there is weak duality. However, of interest are the situations where the optimal objective values of the primal and its corresponding dual problem coincide or their difference lies within a given small margin.

Definition 2.1. Let $\varepsilon \geq 0$. We say that there is *ε-duality gap* for the problems (PG) and (DG) if $v(PG) - v(DG) \leq \varepsilon$. If $v(PG_{x^*}) - v(DG_{x^*}) \leq \varepsilon$ for all $x^* \in X^*$, we say that for (PG) and (DG) one has *stable ε-duality gap*.

Definition 2.2. We say that there is *strong duality* for the problems (PG) and (DG) if $v(PG) = v(DG)$ and the dual problem has an optimal solution. If $v(PG_{x^*}) = v(DG_{x^*})$ and (DG_{x^*}) has an optimal solution for all $x^* \in X^*$, we say that for (PG) and (DG) one has *stable strong duality*.

Definition 2.3. Let $\varepsilon \geq 0$. An element $x \in X$ is said to be an *ε-optimal solution* to (PG) if $0 \in \partial_\varepsilon \Phi(\cdot, 0)(x)$.

Let $\varepsilon \geq 0$. The first statement we give presents a characterization via epigraph inclusions of a situation of stable ε-duality gap for the problems (PG) and (DG).

Theorem 2.1. *Let $W \subseteq X^*$. Then it holds*

$$\operatorname{epi}(\Phi(\cdot, 0))^* \cap (W \times \mathbb{R}) \subseteq \Pr_{X^* \times \mathbb{R}} \operatorname{epi} \Phi^* \cap (W \times \mathbb{R}) - (0, \varepsilon) \qquad (2.2.1)$$

if and only if for each $x^* \in W$ there exists a $\bar{y}^* \in Y^*$ such that

$$(\Phi(\cdot,0))^*(x^*) \geq \Phi^*(x^*,\bar{y}^*) - \varepsilon. \tag{2.2.2}$$

Proof. If $x^* \in W$ such that $(\Phi(\cdot,0))^*(x^*) = +\infty$ there is nothing to prove. Let $x^* \in W$ such that $(\Phi(\cdot,0))^*(x^*) \in \mathbb{R}$. Noticing that, for $y^* \in Y^*$, one has $(x^*, y^*, (\Phi(\cdot,0))^* \ (x^*)) \in \operatorname{epi} \Phi^* - (0,0,\varepsilon)$ if and only if $\Phi^*(x^*,y^*) \leq (\Phi(\cdot,0))^*(x^*) + \varepsilon$, the desired conclusion follows. □

Remark 2.1. The inequality (2.2.2) can be rewritten as $-(\Phi(\cdot,0))^*(x^*) \leq -\Phi^*(x^*, \ y^*) + \varepsilon$. While the quantity in the left-hand side of this inequality is actually $v(PG_{x^*})$, the one in the right-hand side is not necessarily $v(DG_{x^*}) + \varepsilon$, as the supremum in (DG_{x^*}) is not shown to be attained at y^*. Though, (2.2.2) implies $v(PG_{x^*}) \leq v(DG_{x^*}) + \varepsilon$.

Remark 2.2. Using (2.2.2), the stable ε-duality gap for the problems (PG) and (DG) can be equivalently characterized through the following epigraph inclusion

$$\operatorname{epi}(\Phi(\cdot,0))^* \subseteq \operatorname{epi} \inf_{y^* \in Y^*} \Phi^*(\cdot,y^*) - (0,\varepsilon). \tag{2.2.3}$$

In case $\varepsilon = 0$ both the epigraph inclusion and the inequality considered in Theorem 2.1 turn into equalities and (2.2.2) collapses into an inequality that describes actually the stable strong duality for (PG) and (DG).

Corollary 2.1. *Let $W \subseteq X^*$. Then it holds*

$$\operatorname{epi}(\Phi(\cdot,0))^* \cap (W \times \mathbb{R}) = \operatorname{Pr}_{X^* \times \mathbb{R}} \operatorname{epi} \Phi^* \cap (W \times \mathbb{R}) \tag{2.2.4}$$

if and only if for each $x^* \in W$ there exists a $\bar{y}^* \in Y^*$ such that

$$(\Phi(\cdot,0))^*(x^*) = \Phi^*(x^*,\bar{y}^*).$$

If $W = X^*$ one obtains a condition involving epigraphs that ensures that there is stable ε-duality gap for the problems (PG) and (DG).

Corollary 2.2. *It holds*

$$\operatorname{epi}(\Phi(\cdot,0))^* \subseteq \operatorname{Pr}_{X^* \times \mathbb{R}} \operatorname{epi} \Phi^* - (0,\varepsilon)$$

if and only if for each $x^* \in X^*$ there exists a $\bar{y}^* \in Y^*$ such that

$$(\Phi(\cdot,0))^*(x^*) \geq \Phi^*(x^*,\bar{y}^*) - \varepsilon.$$

Using Theorem 2.1 one can formulate necessary and sufficient ε-optimality conditions for any primal-dual pair of optimization problems (PG_{x^*})–(DG_{x^*}) when $x^* \in W \subseteq X^*$.

Theorem 2.2. *Let $W \subseteq X^*$ and $x^* \in W$.*

(a) If $\bar{x} \in X$ is an optimal solution to (PG_{x^*}) and (2.2.1) is satisfied, then there exists a $\bar{y}^* \in Y^*$, an ε-optimal solution to (DG_{x^*}), such that one has

$$\Phi(\bar{x}, 0) + \Phi^*(x^*, \bar{y}^*) \leq \langle x^*, \bar{x}\rangle + \varepsilon \tag{2.2.5}$$

or, equivalently,

$$(x^*, \bar{y}^*) \in \partial_\varepsilon \Phi(\bar{x}, 0). \tag{2.2.6}$$

(b) Assume that $\bar{x} \in X$ and $\bar{y}^* \in Y^*$ fulfill (2.2.5) or (2.2.6). Then $\bar{x}$ is an ε-optimal solution to (PG_{x^*}), $\bar{y}^*$ is an ε-optimal solution to (DG_{x^*}) and $v(PG_{x^*}) \leq v(DG_{x^*}) + \varepsilon$.

Proof. (a) Theorem 2.1 yields, via Remark 2.1, that $\Phi(\bar{x}, 0) + \Phi^*(x^*, \bar{y}^*) \leq \langle x^*, \bar{x}\rangle + \varepsilon$. Because of the weak duality for (PG_{x^*}) and (DG_{x^*}), it follows also that $v(DG_{x^*}) \leq -\Phi^*(x^*, \bar{y}^*) + \varepsilon$, i.e. $\bar{y}$ is an ε-optimal solution to (DG_{x^*}).
(b) Assuming (2.2.5) fulfilled, it follows $\Phi(\bar{x}, 0) - \langle x^*, \bar{x}\rangle \leq \varepsilon - \Phi^*(0, \bar{y}^*)$, that implies $v(DG_{x^*}) \leq v(PG_{x^*}) \leq \Phi(\bar{x}, 0) - \langle x^*, \bar{x}\rangle \leq \varepsilon - \Phi^*(0, \bar{y}^*) \leq v(DG_{x^*}) + \varepsilon \leq v(DP_{x^*}) + \varepsilon$, from which the conclusion follows. □

Remark 2.3. If $\varepsilon = 0$, relations (2.2.5) and (2.2.6) become *optimality conditions* for (PG_{x^*}) and (DG_{x^*}), while when $W = X^*$ Theorem 2.2 delivers what may be called *stable optimality conditions* for (PG) and (DG).

Remark 2.4. Taking $x^* = 0$ in Theorem 2.2 one obtains ε-optimality conditions for the primal-dual pair of optimization problems (PG)–(DG) and, moreover, that the satisfaction of the condition

$$\operatorname{epi}(\Phi(\cdot, 0))^* \subseteq \operatorname{Pr}_{X^* \times \mathbb{R}} \operatorname{epi} \Phi^* - (0, \varepsilon)$$

guarantees that there is ε-duality gap for these problems.

Enriching the function Φ with convexity and topological properties, one can deliver another characterization of (2.2.2) by means of the notion of the $(0, \varepsilon)$-vertical closedness of the conjugate of Φ regarding a cartesian product of sets (cf. Definition 1.1).

Theorem 2.3. *Let $W \subseteq X^*$ and the function Φ be also convex and lower semicontinuous. Then the set $\operatorname{Pr}_{X^* \times \mathbb{R}} \operatorname{epi} \Phi^*$ is $(0, \varepsilon)$-vertically closed regarding the set $W \times \mathbb{R}$ in the topology $\omega(X^*, X) \times \mathscr{R}$ if and only if for each $x^* \in W$ there exists a $\bar{y}^* \in Y^*$ such that (2.2.2) holds.*

Proof. According to [49, Theorem 2.3] (see also [21, Theorem 5.2]), for Φ proper, convex and lower semicontinuous it holds $\operatorname{epi}(\Phi(\cdot, 0))^* = \operatorname{cl} \operatorname{Pr}_{X^* \times \mathbb{R}} \operatorname{epi} \Phi^*$. The assertion follows via Theorem 2.1. □

In case $\varepsilon = 0$ the assertion in Theorem 2.3 rediscovers [21, Theorem 9.1].

Corollary 2.3. *Let $W \subseteq X^*$ and the function Φ be also convex and lower semicontinuous. Then the set $\Pr_{X^* \times \mathbb{R}} \operatorname{epi} \Phi^*$ is closed regarding the set $W \times \mathbb{R}$ in the topology $\omega(X^*, X) \times \mathscr{R}$ if and only if for each $x^* \in W$ there exists a $\bar{y}^* \in Y^*$ such that*

$$(\Phi(\cdot, 0))^*(x^*) = \Phi^*(x^*, \bar{y}^*) = \min_{y^* \in Y^*} \Phi^*(x^*, y^*).$$

If $W = X^*$ Theorem 2.3 delivers the following statement.

Corollary 2.4. *Let function Φ be also convex and lower semicontinuous. Then the set $\Pr_{X^* \times \mathbb{R}} \operatorname{epi} \Phi^*$ is $(0, \varepsilon)$-vertically closed in the topology $\omega(X^*, X) \times \mathscr{R}$ if and only if for each $x^* \in X^*$ there exists a $\bar{y}^* \in Y^*$ such that* (2.2.2) *holds.*

Taking in Corollary 2.3 moreover $W = X^*$, one obtains a characterization of the stable strong duality for (PG) and (DG), rediscovering thus [48, Theorem 3.2.2] (see also [64]).

Corollary 2.5. *Let the function Φ be also convex and lower semicontinuous. Then the set $\Pr_{X^* \times \mathbb{R}} \operatorname{epi} \Phi^*$ is closed in the topology $\omega(X^*, X) \times \mathscr{R}$ if and only if for each $x^* \in X^*$ there exists a $\bar{y}^* \in Y^*$ such that*

$$(\Phi(\cdot, 0))^*(x^*) = \Phi^*(x^*, \bar{y}^*) = \min_{y^* \in Y^*} \Phi^*(x^*, y^*).$$

An important consequence of Theorem 2.3, via Corollary 2.5, is the strong duality statement for (PG) and (DG) that follows (see also [48, 63, 64, 174]).

Corollary 2.6. *Assume that Φ is convex and lower semicontinuous. If $\Pr_{X^* \times \mathbb{R}} \operatorname{epi} \Phi^*$ is a closed set in the topology $\omega(X^*, X) \times \mathscr{R}$, then $v(PG) = v(DG)$ and the dual problem (DG) has an optimal solution $\bar{y}^* \in Y^*$.*

Remark 2.5. Several regularity conditions were proposed in the literature in order to achieve strong duality for (PG) and (DG). We list in the following the most important of those considered when the function Φ is convex (cf. [21, 48, 221]), namely the one involving continuity

$$(RC_1^G) \Big| \; \exists x' \in X \text{ such that } (x', 0) \in \operatorname{dom} \Phi \text{ and } \Phi(x', \cdot) \text{ is continuous at } 0,$$

a weak interiority type one,

$$(RC_2^G) \left| \begin{array}{l} X \text{ and } Y \text{ are Fréchet spaces, } \Phi \text{ is lower semicontinuous and} \\ 0 \in \operatorname{sqri} \Pr_Y(\operatorname{dom} \Phi), \end{array} \right.$$

a generalized interiority type one which works in finitely dimensional spaces,

$$(RC_3^G) \Big| \; \dim \operatorname{lin} \Pr_Y(\operatorname{dom} \Phi) < +\infty \text{ and } 0 \in \operatorname{ri} \Pr_Y(\operatorname{dom} \Phi),$$

and finally the closedness type regularity condition already mentioned in Corollary 2.6,

$$(RC_4^G) \left| \begin{array}{l} \Phi \text{ is lower semicontinuous and } \mathrm{Pr}_{X^* \times \mathbb{R}}(\mathrm{epi}\, \Phi^*) \text{ is closed} \\ \text{in the topology } \omega(X^*, X) \times \mathscr{R}. \end{array} \right.$$

Note that all these regularity conditions ensure actually stable strong duality for (PG) and (DG).

Necessary and sufficient *optimality conditions* for (PG_{x^*}) and (DG_{x^*}), where $x^* \in W \subseteq X^*$ can be derived, too, from Theorem 2.2 via Corollary 2.5 (see also [21,48,221]).

Corollary 2.7. *Let $W \subseteq X^*$ and $x^* \in W$.*

(a) Assume that Φ is convex. Let $\bar{x} \in X$ be an optimal solution to (PG_{x^*}) and assume that one of the regularity conditions (RC_i^G), $i \in \{1,2,3,4\}$, is fulfilled. Then there exists a $\bar{y}^* \in Y^*$, an optimal solution to (DG_{x^*}), such that one has

$$\Phi(\bar{x}, 0) + \Phi^*(x^*, \bar{y}^*) = \langle x^*, \bar{x} \rangle, \tag{2.2.7}$$

or, equivalently,

$$(x^*, \bar{y}^*) \in \partial \Phi(\bar{x}, 0). \tag{2.2.8}$$

(b) Assume that $\bar{x} \in X$ and $\bar{y}^* \in Y^*$ fulfill (2.2.7) or (2.2.8). Then $\bar{x}$ is an optimal solution to (PG_{x^*}), $\bar{y}^*$ is an optimal solution to (DG_{x^*}) and $v(PG_{x^*}) = v(DG_{x^*})$.

Remark 2.6. When $W = X^*$, Corollary 2.7 delivers what may be called stable optimality conditions for (PG) and (DG). Taking there $x^* = 0$ one obtains necessary and sufficient optimality conditions for (PG) and (DG) (see also [21,48,221]). Because they will be used later in the book, we give them below for reader's convenience.

Corollary 2.8. *(a) Assume that Φ is convex. Let $\bar{x} \in X$ be an optimal solution to (PG) and assume that one of the regularity conditions (RC_i^G), $i \in \{1,2,3,4\}$, is fulfilled. Then there exists a $\bar{y}^* \in Y^*$, an optimal solution to (DG), such that one has*

$$\Phi(\bar{x}, 0) + \Phi^*(0, \bar{y}^*) = 0 \tag{2.2.9}$$

or, equivalently,

$$(0, \bar{y}^*) \in \partial \Phi(\bar{x}, 0). \tag{2.2.10}$$

(b) Assume that $\bar{x} \in X$ and $\bar{y}^* \in Y^*$ fulfill (2.2.9) or (2.2.10). Then $\bar{x}$ is an optimal solution to (PG), $\bar{y}^*$ is an optimal solution to (DG) and $v(PG) = v(DG)$.

As byproducts of the duality investigations presented in this subsection we obtain some ε-Farkas statements and results involving (η, ε)-saddle points, as follows. We begin with the *ε-Farkas type* results for (PG_{x^*}) and (DG_{x^*}), where $x^* \in W \subseteq X^*$. They extend some recent Farkas type statements from the literature that generalize the classical Farkas Lemma.

Theorem 2.4. *Let $W \subseteq X^*$.*

(a) Suppose that (2.2.1) *holds. Given $x^* \in W$, if $\Phi(x,0) - \langle x^*, x\rangle \geq \varepsilon/2$ for all $x \in X$ then there exists a $\bar{y}^* \in Y^*$ such that $\Phi^*(x^*, \bar{y}^*) \leq \varepsilon/2$.*
(b) Given $x^* \in W$, if there exists a $\bar{y}^* \in Y^*$ such that $\Phi^*(x^*, \bar{y}^*) \leq -\varepsilon/2$, then $\Phi(x,0) - \langle x^*, x\rangle \geq \varepsilon/2$ for all $x \in X$.

Proof. (a) Theorem 2.1 yields the existence of a $\bar{y}^* \in Y^*$ such that $-(\Phi(\cdot,0))^*(x^*) \leq \varepsilon - \Phi^*(x^*, \bar{y}^*)$. Then $\varepsilon/2 \leq \varepsilon - \Phi^*(x^*, \bar{y}^*)$ and the conclusion follows.
(b) Using the weak duality for (PG_{x^*}) and (DG_{x^*}) it follows that $\Phi(x,0) - \langle x^*, x\rangle \geq -\Phi^*(x^*, \bar{y}^*) \geq \varepsilon/2$. □

Using (2.2.3) as a regularity condition one can give other ε-Farkas type results for (PG_{x^*}) and (DG_{x^*}), where $x^* \in W \subseteq X^*$, which can be proven analogously to the ones in Theorem 2.4.

Theorem 2.5. *Let $W \subseteq X^*$.*

(a) Suppose that (2.2.3) *holds. Given $x^* \in W$, if $\Phi(x,0) - \langle x^*, x\rangle \geq \varepsilon/2$ for all $x \in X$ then $\inf_{y^* \in Y^*} \Phi^*(x^*, y^*) \leq \varepsilon/2$.*
(b) Given $x^* \in W$, if $\inf_{y^* \in Y^*} \Phi^*(x^*, y^*) \leq -\varepsilon/2$, then $\Phi(x,0) - \langle x^*, x\rangle \geq \varepsilon/2$ for all $x \in X$.

If $\varepsilon = 0$, the ε-Farkas type results turn into equivalences, as follows.

Corollary 2.9. *Let $W \subseteq X^*$ and suppose that* (2.2.4) *holds. Given $x^* \in W$, one has $\Phi(x,0) - \langle x^*, x\rangle \geq 0$ for all $x \in X$ if and only if there exists a $\bar{y}^* \in Y^*$ such that $\Phi^*(x^*, \bar{y}^*) \leq 0$.*

Corollary 2.10. *Let $W \subseteq X^*$ and suppose that* (2.2.3) *holds as an equality for $\varepsilon = 0$. Given $x^* \in W$, one has $\Phi(x,0) - \langle x^*, x\rangle \geq 0$ for all $x \in X$ if and only if $\inf_{y^* \in Y^*} \Phi^*(x^*, y^*) \leq 0$.*

In order to deal with statements involving (η, ε)-saddle points, let us recall now the definition of the Lagrangian function for the pair of primal-dual problems (PG_{x^*})–(DG_{x^*}), where $x^* \in X^*$.

Definition 2.4. Let $x^* \in X^*$. The function $L^{(PG_{x^*})} : X \times Y^* \to \overline{\mathbb{R}}$ defined by

$$L^{(PG_{x^*})}(x, y^*) = \inf_{y \in Y} \left[\Phi(x, y) - \langle x^*, x\rangle - \langle y^*, y\rangle\right]$$

is called the *Lagrangian function* of the pair of primal-dual problems (PG_{x^*})–(DG_{x^*}) relative to the perturbation function Φ.

Remark 2.7. One can easily see that for all $x^* \in X^*$ and all $x \in X$ it holds $L^{(PG_{x^*})}(x, y^*) = -\langle x^*, x\rangle - (\Phi(x,\cdot))^*(y^*)$ for all $y^* \in Y^*$. Thus for all $x \in X$ the function $L^{(PG_{x^*})}(x,\cdot)$ is concave and upper semicontinuous. On the other hand, assuming that Φ is convex, whenever $y^* \in Y^*$ the function $L^{(PG_{x^*})}(\cdot, y^*)$ is convex, too.

Remark 2.8. Given $x^* \in X^*$, we can reformulate the primal-dual pair of problems (PG_{x^*})–(DG_{x^*}) by means of the Lagrangian $L^{(PG_{x^*})}$, namely while (DG_{x^*}) is equivalent to $\sup_{y^* \in Y^*} \inf_{x \in X} L^{(PG_{x^*})}$, while if $\Phi(x,\cdot)$ is a convex and lower semicontinuous function taking nowhere the value $-\infty$ for all $x \in X$, (PG_{x^*}) actually means $\inf_{x\in X} \sup_{y^*\in Y^*} L^{(PG_{x^*})}$. Note that even without the additional hypotheses, $v(PG_{x^*})$ is not less than $\inf_{x\in X} \sup_{y^*\in Y^*} L^{(PG_{x^*})}$.

In the following we generalize the classical notion of a saddle point.

Definition 2.5. Let $\eta, \varepsilon \geq 0$ and $x^* \in X^*$. We say that $(\bar{x}, \bar{y}^*) \in X \times Y^*$ is an (η, ε)-*saddle point* of the Lagrangian $L^{(PG_{x^*})}$ if

$$L^{(PG_{x^*})}(\bar{x}, y^*) - \eta \leq L^{(PG_{x^*})}(\bar{x}, \bar{y}^*) \leq L^{(PG_{x^*})}(x, \bar{y}^*) + \varepsilon \;\forall (x, y^*) \in X \times Y^*.$$

Remark 2.9. The notion of an ε-saddle point of a function with two variables was already considered in the literature, see for instance [147, 199], while for (η, ε)-saddle points we refer to [13].

Slightly weakening the properness hypothesis imposed on Φ and adding to it convexity and topological assumptions, one obtains the following statement connecting the (η, ε)-saddle points of $L^{(PG_{x^*})}$ with the $(\varepsilon + \eta)$-duality gap for the problems (PG_{x^*}) and (DG_{x^*}), and the existence of some $(\varepsilon + \eta)$-optimal solutions to them.

Theorem 2.6. *Let $\eta, \varepsilon \geq 0$ and $x^* \in X^*$.*

(a) If $(\bar{x}, \bar{y}^*) \in X \times Y^*$ is an (η, ε)-saddle point of $L^{(PG_{x^*})}$ and $\Phi(\bar{x}, \cdot)$ is a convex and lower semicontinuous function taking nowhere the value $-\infty$, then $\bar{x}$ is an $(\varepsilon + \eta)$-optimal solution to (PG_{x^*}), $\bar{y}^*$ is an $(\varepsilon + \eta)$-optimal solution to (DG_{x^*}) and there is $(\varepsilon + \eta)$-duality gap for the primal-dual pair of problems $(PG_{x^*}) - (DG_{x^*})$.

(b) If $\nu \geq 0$, $\bar{x} \in X$ is an ε-optimal solution to (PG_{x^*}), $\bar{y}^* \in Y^*$ is an η-optimal solution to (DG_{x^*}) and $v(PG_{x^*}) \leq v(DG_{x^*}) + \nu$, then $(\bar{x}, \bar{y}^*) \in X \times Y^*$ is an $(\eta + \varepsilon + \nu, \eta + \varepsilon + \nu)$-saddle point of $L^{(PG_{x^*})}$.

Proof. (a) From the definition of an (η, ε)-saddle point it follows via Remark 2.8 that

$$\Phi(\bar{x}, 0) - \langle x^*, \bar{x}\rangle - \eta = \sup_{y^* \in Y^*} L^{(PG_{x^*})}(\bar{x}, y^*) - \eta \leq L^{(PG_{x^*})}$$
$$(\bar{x}, \bar{y}^*) \leq \inf_{x\in X} L^{(PG_{x^*})}(x, \bar{y}^*) + \varepsilon = \varepsilon - \Phi^*(x^*, \bar{y}^*). \quad (2.2.11)$$

Using the weak duality for the problems (PG_{x^*}) and (DG_{x^*}), (2.2.11) yields $v(DG_{x^*}) - \eta \leq \varepsilon - \Phi^*(x^*, \bar{y}^*)$ and $\Phi(\bar{x}, 0) - \langle x^*, \bar{x}\rangle - \eta \leq \varepsilon - v(PG_{x^*})$, hence $\bar{x}$ is an $(\varepsilon + \eta)$-optimal solution to (PG_{x^*}) and $\bar{y}^*$ is an $(\varepsilon + \eta)$-optimal solution to (DG_{x^*}). Relation (2.2.11) implies also that $\Phi(\bar{x}, 0) - \langle x^*, \bar{x}\rangle - \eta \leq \varepsilon - \Phi^*(x^*, \bar{y}^*)$, consequently $v(PG_{x^*}) \leq v(DG_{x^*}) + \eta + \varepsilon$.

(b) Using again Remark 2.8, one obtains that $\Phi(\bar{x}, 0) - \langle x^*, \bar{x}\rangle \geq \sup_{y^* \in Y^*} L^{(PG_{x^*})}(\bar{x}, y^*) \geq L^{(PG_{x^*})}(\bar{x}, \bar{y}^*)$ and $-\Phi^*(0, \bar{y}^*) = \inf_{x \in X} L^{(PG_{x^*})}(x, \bar{y}^*) \leq L^{(PG_{x^*})}(\bar{x}, \bar{y}^*)$. But $\bar{x}$ is an ε-optimal solution to (PG_{x^*}) and $\bar{y}^*$ is an η-optimal solution to (DG_{x^*}), consequently

$$\begin{aligned} v(DG_{x^*}) - \eta \leq -\Phi^*(0, \bar{y}^*) \leq L^{(PG_{x^*})}(\bar{x}, \bar{y}^*) &\leq \Phi(\bar{x}, 0) - \langle x^*, \bar{x}\rangle \\ &\leq v(PG_{x^*}) + \varepsilon. \end{aligned}$$

Recalling that $v(PG_{x^*}) \leq v(DG_{x^*}) + \nu$, one obtains from here

$$v(PG_{x^*}) - \eta - \nu \leq L^{(PG_{x^*})}(\bar{x}, \bar{y}^*) \leq v(DG_{x^*}) + \varepsilon + \nu,$$

followed by

$$\Phi(\bar{x}, 0) - \langle x^*, \bar{x}\rangle - \varepsilon - \eta - \nu \leq L^{(PG_{x^*})}(\bar{x}, \bar{y}^*) \leq -\Phi^*(x^*, \bar{y}^*) + \eta + \varepsilon + \nu.$$

Employing again the formulae derived above via Remark 2.8 one obtains that $(\bar{x}, \bar{y}^*) \in X \times Y^*$ is an $(\eta + \varepsilon + \nu, \eta + \varepsilon + \nu)$-saddle point of $L^{(PG_{x^*})}$. □

If one takes in Theorem 2.6 $\eta = \varepsilon = \nu = 0$, the two assertions become equivalent, rediscovering [48, Theorem 3.3.2].

Corollary 2.11. *Let $x^* \in X^*$ and assume that Φ is a convex and lower semicontinuous function taking nowhere the value $-\infty$. Then $(\bar{x}, \bar{y}^*) \in X \times Y^*$ is a saddle point of $L^{(PG_{x^*})}$ if and only if $\bar{x}$ is an optimal solution to (PG_{x^*}), $\bar{y}^*$ is an optimal solution to (DG_{x^*}) and $v(PG_{x^*}) = v(DG_{x^*})$.*

The general scalar optimization problem (PG) encompasses as special cases many types of scalar optimization problems. In the next subsections we shall write both constrained and unconstrained optimization problems as special cases of (PG) and dual problems will be assigned to them by carefully choosing the employed perturbation functions.

2.2.2 Constrained Scalar Optimization Problems

The first class of particular optimization problems for which we particularize the investigations from Sect. 2.2.1 is the one of the constrained optimization problems. Consider the nonempty set $S \subseteq X$ and let the nonempty convex cone $C \subseteq Y$ induce a partial ordering on Y. Take the proper functions $f : X \to \overline{\mathbb{R}}$ and $h : X \to Y^\bullet$, fulfilling the feasibility condition $\operatorname{dom} f \cap S \cap h^{-1}(-C) \neq \emptyset$. The primal problem we treat further is

$$(PC) \qquad \inf_{x \in \mathscr{A}} f(x),$$

where the *feasible set* of the problem (PC) is

$$\mathscr{A} = \{x \in S : h(x) \in -C\}.$$

There are different choices of the perturbation function Φ for which (PC) turns out to be a special case of (PG). In the following we consider two of them, which will lead to two different dual problems to (PC) that arise from (DG).

It is known that the *classical Lagrange dual problem* to (PC),

$$(DC^L) \qquad \sup_{z^* \in C^*} \inf_{x \in S} \left[f(x) + (z^* h)(x) \right],$$

can be obtained as a special case of (DG) by using the perturbation function

$$\Phi^L : X \times Y \to \overline{\mathbb{R}},\ \Phi^L(x,z) = \begin{cases} f(x), & \text{if } x \in S, h(x) \in z - C, \\ +\infty, & \text{otherwise,} \end{cases}$$

which is proper because f and h are proper and due to the feasibility condition, and whose conjugate is

$$(\Phi^L)^* : X^* \times Y^* \to \overline{\mathbb{R}},\ (\Phi^L)^*(x^*, z^*) = (f - (z^* h) + \delta_S)^*(x^*) + \delta_{C^*}(-z^*).$$

Let $\varepsilon \geq 0$. The first statement we give presents a characterization via epigraph inclusions of a situation of stable ε-duality gap for the problems (PC) and (DC^L) that is a special case of Theorem 2.1.

Theorem 2.7. *Let $W \subseteq X^*$. Then it holds*

$$\operatorname{epi}(f + \delta_{\mathscr{A}})^* \cap (W \times \mathbb{R}) \subseteq \bigcup_{z^* \in C^*} \operatorname{epi}(f + (z^* h))^*_S \cap (W \times \mathbb{R}) - (0, \varepsilon)$$

if and only if for each $x^* \in W$ there exists a $\bar{z}^* \in C^*$ such that

$$(f + \delta_{\mathscr{A}})^*(x^*) \geq (f + (\bar{z}^* h))^*_S(x^*) - \varepsilon.$$

Remark 2.10. Analogously one can particularize the other statements from Sect. 2.2.1 to the present framework, too, rediscovering or improving different statements from [14, 141, 142]. We mention here only that for the strong duality statement for the problems (PC) and (DC^L), which follows directly from Corollary 2.6 or Remark 2.5, besides convexity assumptions which guarantee the convexity of the perturbation function Φ^L, ensured, for instance, by taking S and f convex and h C-convex, one can employ the regularity conditions obtained by particularizing (RC_i^G), $i \in \{1,2,3,4\}$, namely (cf. [48])

$$(RC_1^L)\Big|\ \exists x' \in \operatorname{dom} f \cap S \text{ such that } h(x') \in -\operatorname{int} C,$$

which is the classical *Slater constraint qualification*,

$$(RC_2^L)\left| \begin{array}{l} X \text{ and } Y \text{ are Fréchet spaces, } S \text{ is closed, } f \text{ is lower semicontinuous,} \\ h \text{ is } C-\text{epi}-\text{closed and } 0 \in \operatorname{sqri}\big(h(\operatorname{dom} f \cap S \cap \operatorname{dom} h) + C\big), \end{array}\right.$$

$$(RC_3^L)\left| \begin{array}{l} \dim \operatorname{lin}\left(h(\operatorname{dom} f \cap S \cap \operatorname{dom} h) + C\right) < +\infty \text{ and} \\ 0 \in \operatorname{ri}\left(h(\operatorname{dom} f \cap S \cap \operatorname{dom} h) + C\right), \end{array}\right.$$

and

$$(RC_4^L)\left| \begin{array}{l} S \text{ is closed, } f \text{ is lower semicontinuous, } h \text{ is } C-\text{epi}-\text{closed} \\ \text{and } \bigcup\limits_{z^* \in C^*} \operatorname{epi}(f + (z^*h) + \delta_S)^* \text{ is closed in the topology } \omega(X^*, X) \times \mathscr{R}. \end{array}\right.$$

Another perturbation function employed to assign a conjugate dual problem to (*PC*) as a special case of (*DG*) is (cf. [21,48])

$$\Phi^{FL} : X \times X \times Y \to \overline{\mathbb{R}},\ \Phi^{FL}(x,y,z) = \begin{cases} f(x+y), & \text{if } x \in S, h(x) \in z - C, \\ +\infty, & \text{otherwise.} \end{cases}$$

It is proper as well because f and h are proper and due to the fulfillment of the mentioned feasibility condition and has as conjugate the function $(\Phi^{FL})^* : X^* \times X^* \times Y^* \to \overline{\mathbb{R}}$,

$$(\Phi^{FL})^*(x^*, y^*, z^*) = f^*(y^*) + (-(z^*h) + \delta_S)^*(x^* - y^*) + \delta_{-C^*}(z^*).$$

The dual problem it attaches to (*PC*) is the *Fenchel-Lagrange dual problem*

$$(DC^{FL}) \qquad \sup_{y^* \in X^*, z^* \in C^*} \big\{ -f^*(y^*) - ((z^*h) + \delta_S)^*(-y^*) \big\}.$$

The characterization via epigraph inclusions of a situation of stable ε-duality gap for the problems (*PC*) and (DC^{FL}) that is a special case of Theorem 2.1 follows.

Theorem 2.8. *Let $W \subseteq X^*$. Then it holds*

$$\operatorname{epi}(f + \delta_{\mathscr{A}})^* \cap (W \times \mathbb{R}) \subseteq \Big(\operatorname{epi} f^* + \bigcup_{z^* \in C^*} \operatorname{epi}(z^*h)_S^* \Big) \cap (W \times \mathbb{R}) - (0, \varepsilon)$$

if and only if for each $x^* \in W$ there exist $\bar{y}^* \in X^*$ and $\bar{z}^* \in C^*$ such that

$$(f + \delta_{\mathscr{A}})^*(x^*) \geq f^*(\bar{y}^*) + (\bar{z}^*h)_S^*(x^* - \bar{y}^*) - \varepsilon.$$

Remark 2.11. Analogously one can particularize the other statements from Sect. 2.2.1 to the present framework, too. We mention here only that for strong duality for the problems (PC) and (DC^{FL}), which follows directly from Corollary 2.6 or Remark 2.5, besides convexity assumptions which guarantee the convexity of the perturbation function Φ^{FL}, ensured, for instance, by taking S and f convex and h C-convex, one can employ the regularity conditions obtained by particularizing (RC_i^G), $i \in \{1,2,3,4\}$, namely

$$(RC_1^{FL}) \Big| \ \exists x' \in \operatorname{dom} f \cap S \text{ such that } f \text{ is continuous at } x' \text{ and } h(x') \in -\operatorname{int} C,$$

$$(RC_2^{FL}) \left| \begin{array}{l} X \text{ and } Y \text{ are Fréchet spaces, } S \text{ is closed, } f \text{ is lower semicontinuous, } h \text{ is} \\ C-\text{epi}-\text{closed and } 0 \in \operatorname{sqri}\big(\operatorname{dom} f \times C - \operatorname{epi}_{-C}(-h) \cap (S \times Y)\big), \end{array} \right.$$

$$(RC_3^{FL}) \left| \begin{array}{l} \dim \operatorname{lin}\left(\operatorname{dom} f \times C - \operatorname{epi}_{-C}(-h) \cap (S \times Z)\right) < +\infty \text{ and} \\ 0 \in \operatorname{ri}\left(\operatorname{dom} f \times C - \operatorname{epi}_{-C}(-h) \cap (S \times Z)\right), \end{array} \right.$$

and

$$(RC_4^{FL}) \left| \begin{array}{l} S \text{ is closed, } f \text{ is lower semicontinuous, } h \text{ is } C-\text{epi}-\text{closed and} \\ \operatorname{epi} f^* + \bigcup\limits_{z^* \in C^*} \operatorname{epi}((z^*h) + \delta_S)^* \text{ is closed in the topology} \omega(X^*, X) \times \mathscr{R}. \end{array} \right.$$

Other perturbation functions can be employed in order to assign conjugate dual problems to (PC) as special cases of (DG), too. For instance, using the perturbation function

$$\Phi^{EFL} : X \times X \times X \times Y \to \overline{\mathbb{R}},\ \Phi^{EFL}(x,y,t,z) = \begin{cases} f(x+y), & \text{if } x \in S, h(x+t) \in z - C, \\ +\infty, & \text{otherwise,} \end{cases}$$

that is proper as well because f and h are proper and due to the fulfillment of the mentioned feasibility condition and has as conjugate the function $(\Phi^{EFL})^* : X^* \times X^* \times X^* \times Y^* \to \overline{\mathbb{R}}$,

$$(\Phi^{EFL})^*(x^*,y^*,t^*,z^*) = f^*(y^*) + (-z^*h)^*(t^*) + \sigma_S(x^* - y^* - t^*) + \delta_{-C^*}(z^*),$$

one can attach to (PC) is the *extended Fenchel-Lagrange dual problem* (cf. [14,46])

$$(DC^{EFL}) \qquad \sup_{\substack{y^*,t^* \in X^*,\\ z^* \in C^*}} \big\{ - f^*(y^*) - (z^*h)^*(t^*) - \sigma_S(-y^* - t^*) \big\}.$$

The characterization via epigraph inclusions of a situation of stable ε-duality gap for the problems (PC) and (DC^{EFL}) that is a special case of Theorem 2.1 follows.

Theorem 2.9. *Let $W \subseteq X^*$. Then it holds*

$$\operatorname{epi}(f + \delta_{\mathscr{A}})^* \cap (W \times \mathbb{R}) \subseteq \Big(\operatorname{epi} f^* + \operatorname{epi} \sigma_S + \bigcup_{z^* \in C^*} \operatorname{epi}(z^*h)^* \Big) \cap (W \times \mathbb{R}) - (0, \varepsilon)$$

if and only if for each $x^* \in W$ *there exist* $\bar{y}^*, \bar{t}^* \in X^*$ *and* $\bar{z}^* \in C^*$ *such that*

$$(f + \delta_{\mathscr{A}})^*(x^*) \geq f^*(\bar{y}^*) + (\bar{z}^* h)^*(t^*) + \sigma_S(x^* - \bar{y}^* - \bar{t}^*) - \varepsilon.$$

However, due to the fact that Φ^{EFL} and (DC^{EFL}) have quite many variables we will not work further with them in this book, as one can notice that already Φ^{FL} leads in some cases to pretty complicated formulae. The interested reader can though derive by means of Φ^{EFL} similar statements to the ones given further for the other perturbation functions attached to (PC), some of them being available in [14, 46]. Note also that another *Fenchel-Lagrange type dual problem* can be assigned to (PC) via the perturbation function

$$\Phi^{LF} : X \times X \times Y \to \overline{\mathbb{R}},\ \Phi^{LF}(x, t, z) = \begin{cases} f(x), & \text{if } x \in S, h(x + t) \in z - C, \\ +\infty, & \text{otherwise,} \end{cases}$$

namely (cf. [14, 46])

$$(DC^{LF}) \qquad \sup_{t^* \in X^*, z^* \in C^*} \left\{ - f_S^*(-t^*) - (z^* h)^*(t^*) \right\},$$

but we will not mention it further either.

Remark 2.12. In order to give stable ε-duality statements for (PC) and the dual problems we assigned to it within this subsection one can introduce the functions (cf. [14, 141, 142]) $h^\diamond$, $h_S^\diamond : X^* \to \overline{\mathbb{R}}$, defined by $h^\diamond = \inf_{z^* \in C^*} (z^* h)^*$ and $h_S^\diamond = \inf_{z^* \in C^*} (z^* h)_S^*$, respectively. Then, for instance, the stable ε-duality gap for the problems (PC) and (DC^{FL}) is characterized through

$$\operatorname{epi}(f + \delta_{\mathscr{A}})^* \subseteq \operatorname{epi}(f^* \Box h_S^\diamond) - (0, \varepsilon).$$

Remark 2.13. Other interesting statements can be derived from the ones given in this subsection by taking $f(x) = 0$ for all $x \in X$, when relations involving the feasible set $\mathscr{A}$ and, on the other hand, the constraint function h and the constraint set S can be characterized via epigraph inclusions, as done in [14, 46, 47].

2.2.3 Unconstrained Scalar Optimization Problems

Consider now the unconstrained optimization problem

$$(PU) \qquad \inf_{x \in X} [f(x) + g(Ax)],$$

where $A : X \to Y$ is a linear continuous mapping and $f : X \to \overline{\mathbb{R}}$ and $g : Y \to \overline{\mathbb{R}}$ are proper functions fulfilling the feasibility condition $\operatorname{dom} f \cap A^{-1}(\operatorname{dom} g) \neq \emptyset$. The perturbation function considered for assigning to (PU) the classical *Fenchel dual problem*

$$(DU) \qquad \sup_{y^* \in Y^*} \{-f^*(A^*y^*) - g^*(-y^*)\},$$

is (cf. [21, 221])

$$\Phi^U : X \times Y \to \overline{\mathbb{R}},\ \Phi^U(x, y) = f(x) + g(Ax + y),$$

which is proper because f and g are proper and due to the fulfillment of the mentioned feasibility condition and has as conjugate the function

$$(\Phi^U)^* : X^* \times Y^* \to \overline{\mathbb{R}}, (\Phi^U)^*(x^*, y^*) = f^*(x^* - A^*y^*) + g^*(y^*).$$

Let $\varepsilon \geq 0$. The first statement we give presents a characterization via epigraph inclusions of a situation of stable ε-duality gap for the problems (PU) and (DU) that is a special case of Theorem 2.1. Before proceeding, let us introduce the notation $(A^* \times \mathrm{id}_{\mathbb{R}})(\mathrm{epi}\, g^*) = \{(x^*, r) \in X^* \times \mathbb{R} : \exists y^* \in Y^* \text{ such that } A^*y^* = x^* \text{ and } (y^*, r) \in \mathrm{epi}\, g^*\}$.

Theorem 2.10. *Let $W \subseteq X^*$. Then it holds*

$$\mathrm{epi}(f + g \circ A)^* \cap (W \times \mathbb{R}) \subseteq \big(\mathrm{epi}\, f^* + (A^* \times \mathrm{id}_{\mathbb{R}})(\mathrm{epi}\, g^*)\big) \cap (W \times \mathbb{R}) - (0, \varepsilon)$$

if and only if for each $x^* \in W$ there exists a $\bar{y}^* \in Y^*$ such that

$$(f + g \circ A)^*(x^*) \geq f^*(A^*\bar{y}^*) + g^*(\bar{x}^* - \bar{y}^*) - \varepsilon.$$

Remark 2.14. Analogously one can particularize the other statements from Sect. 2.2.1 to the present framework, too. We mention here only that for strong duality for the problems (PU) and (DU), which follows directly from Corollary 2.6 or Remark 2.5, besides convexity assumptions which guarantee the convexity of the perturbation function Φ^U, ensured, for instance, by taking f and g convex, one can employ the regularity conditions obtained by particularizing (RC_i^G), $i \in \{1, 2, 3, 4\}$, namely

$$(RC_1^U) \big| \ \exists x' \in \mathrm{dom}\, f \cap A^{-1}(\mathrm{dom}\, g) \text{ such that } g \text{ is continuous at} Ax',$$

$$(RC_2^U) \left| \begin{array}{l} X \text{ and } Y \text{ are Fréchet spaces, } f \text{ and } g \text{ are lower semicontinuous} \\ \text{and } 0 \in \mathrm{sqri}(\mathrm{dom}\, g - A(\mathrm{dom}\, f)), \end{array} \right.$$

$$(RC_3^U) \big| \ \dim \mathrm{lin}(\mathrm{dom}\, g - A(\mathrm{dom}\, f)) < +\infty \text{ and } \mathrm{ri}\, A(\mathrm{dom}\, f) \cap \mathrm{ri}\, \mathrm{dom}\, g \neq \emptyset,$$

and

$$(RC_4^U) \left| \begin{array}{l} f \text{ and } g \text{ are lower semicontinuous and } \mathrm{epi}\, f^* + (A^* \times \mathrm{id}_{\mathbb{R}})(\mathrm{epi}\, g^*) \\ \text{is closed in the topology } \omega(X^*, X) \times \mathscr{R}. \end{array} \right.$$

Remark 2.15. Valuable special cases of the scalar optimization problem (PU), met in the literature in various circumstances, can be obtained, for instance, by taking $X = Y$ and A to be the identity mapping on X or f to be the zero function, respectively. The vector duals assigned above to (PU) and the corresponding duality and optimality conditions statements can be directly particularized for these problems, too.

A special case of Theorem 2.10 that will be used later in our presentation follows.

Proposition 2.1. *Let the proper, convex and lower semicontinuous functions* $f, g : X \to \overline{\mathbb{R}}$ *satisfying* $\operatorname{dom} f \cap \operatorname{dom} g \neq \emptyset$ *and* $x^* \in X^*$. *Then* $f^* \Box g^*$ *is* $\omega(X^*, X)$-*lower semicontinuous at* x^* *and exact at* x^* *if and only if*

$$\inf_{x \in X} \left[f(x) + g(x) - \langle x^*, x \rangle \right] = \max_{y^* \in X^*} \{ -f^*(y^*) - g^*(y^* - x^*) \}. \tag{2.2.12}$$

Proof. Taking in Theorem 2.10 $X = Y$, A to be the identity mapping on X, $W = \{x^*\}$ and $\varepsilon = 0$, one obtains, via Corollary 2.1, that (2.2.12) is equivalent to $\operatorname{epi}(f + g)^* \cap (\{x^*\} \times \mathbb{R}) = (\operatorname{epi} f^* + \operatorname{epi} g^*) \cap (\{x^*\} \times \mathbb{R})$, i.e. there exists an $a^* \in X^*$ such that $(f+g)^*(x^*) = f^*(a^*) + g^*(x^* - a^*)$. Because $(f+g)^* = \operatorname{cl}(f^* \Box g^*)$ in the present hypotheses, this means actually that $\operatorname{cl}(f^* \Box g^*)(x^*) = f^* \Box g^*(x^*)$ and the infimal convolution is exact at x^*. □

Moreover, one can see (PC) as an unconstrained optimization problem, namely

$$(PC) \qquad \inf_{x \in X} [f(x) + \delta_{\mathscr{A}}(x)],$$

where the notations are consistent with the ones in Sect. 2.2.2. Then, taking $A := \mathrm{id}_X$, $f := f$ and $g := \delta_{\mathscr{A}}$, a *Fenchel dual problem* can be attached to (PC), namely

$$(DC^F) \qquad \sup_{y^* \in X^*} \{ -f^*(y^*) - \sigma_{\mathscr{A}}(-y^*) \}.$$

This dual problem to (PC) can be obtained directly from (DG), too, by using the perturbation function

$$\Phi^F : X \times X \to \overline{\mathbb{R}},\ \Phi^F(x, y) = \begin{cases} f(x + y), & \text{if } x \in \mathscr{A}, \\ +\infty, & \text{otherwise}, \end{cases}$$

proper because f and h are proper and due to the fulfillment of the mentioned feasibility condition and having as conjugate the function

$$(\Phi^F)^* : X^* \times X^* \to \overline{\mathbb{R}},\ (\Phi^F)^*(x^*, y^*) = \sigma_{\mathscr{A}}(x^* - y^*) + f^*(y^*).$$

Remark 2.16. One can particularize the statements from Sect. 2.2.1 for this primal-dual pair of problems, too. However, we present here only the regularity conditions which, besides convexity assumptions which guarantee the convexity of the perturbation function Φ^F, ensured, for instance, by taking S and f convex and h C-convex, guarantee the strong duality. They are obtained by particularizing (RC_i^G),

$i \in \{1, 2, 3, 4\}$, being

$$(RC_1^F) \Big| \ \exists x' \in \operatorname{dom} f \cap \mathscr{A} \text{ such that } f \text{ is continuous at } x',$$

or, alternatively,

$$(RC_1^{F'}) \Big| \ \operatorname{dom} f \cap \operatorname{int} \mathscr{A} \neq \emptyset,$$

$$(RC_2^F) \left| \begin{array}{l} X \text{ is a Fréchet space, } \mathscr{A} \text{ is closed, } f \text{ is lower semicontinuous} \\ \text{and } 0 \in \operatorname{sqri}(\operatorname{dom} f - \mathscr{A}), \end{array} \right.$$

$$(RC_3^F) \Big| \ \dim \operatorname{lin}(\operatorname{dom} f - \mathscr{A}) < +\infty \text{ and } \operatorname{ri} \operatorname{dom} f \cap \operatorname{ri} \mathscr{A} \neq \emptyset,$$

and

$$(RC_4^F) \left| \begin{array}{l} \mathscr{A} \text{ is closed, } f \text{ is lower semicontinuous and } \operatorname{epi} f^* + \operatorname{epi} \sigma_{\mathscr{A}} \\ \text{is closed in the topology } \omega(X^*, X) \times \mathscr{R}. \end{array} \right.$$

Remark 2.17. In order to ensure the convexity of the set $\mathscr{A}$ it is sufficient to take the set S convex and h to be a C-convex vector function. To guarantee that the set $\mathscr{A}$ is closed it is enough to assume that S is a closed set and h a C-epi-closed vector function.

One can extend the investigations on ε-duality regarding unconstrained problems towards problems consisting in the minimization of a sum of a function with a C-increasing function composed with a vector function over the whole space X. Let the nonempty convex cone $C \subseteq Y$ induce a partial ordering on Y, the proper function $f : X \to \overline{\mathbb{R}}$, the proper and C-increasing function $g : Y \to \overline{\mathbb{R}}$ and the proper vector function $h : X \to Y^\bullet$ fulfilling the feasibility condition $\operatorname{dom} g \cap (h(\operatorname{dom} f) + C) \neq \emptyset$. To the unconstrained optimization problem

$$(PS) \qquad \inf_{x \in X} [f(x) + g(h(x))],$$

one can attach via perturbation theory different dual problems that are special cases of (DG).

Taking the perturbation function

$$\Phi^1 : X \times Y \to \overline{\mathbb{R}}, \ \Phi^1(x, y) = f(x) + g(h(x) + y),$$

one assigns to (PS) the following conjugate dual problem

$$(DS^1) \qquad \sup_{y^* \in C^*} \big\{ -g^*(y^*) - (f + (y^*h))^*(0) \big\},$$

while by means of

$$\Phi^2 : X \times X \times Y \to \overline{\mathbb{R}}, \ \Phi^2(x, y, z) = f(x + y) + g(h(x) + z),$$

one attaches to (PS) another conjugate dual problem, namely

$$(DS^2) \qquad \sup_{\substack{y^* \in X^*, \\ z^* \in C^*}} \big\{ - g^*(z^*) - f^*(y^*) - (z^* h)^*(-y^*) \big\}.$$

One can directly adapt the general statements regarding (*PG*) and (*DG*) for (PS) and its duals by considering the perturbation functions Φ^1 and Φ^2 or, alternatively, the assertions for (*PU*) and (*DU*) can be used, when one carefully constructs two functions of two variables say, F and G, such that $(F+G)^*(\cdot,0) = (f + g \circ h)^*$, as done in [21,45]. However, we will not pursue here this path, referring the interested reader to [13] for statements similar to the ones given within this section that involve (PS) and its dual problems.

2.3 Characterizations Involving Subdifferentials

After the characterizations via epigraphs of ε-duality gap statements presented in Sect. 2.2, we provide in the following similar ones, but involving subdifferential inclusions. Again, we begin our investigations with a general scalar optimization problem embedded into a family of general perturbed scalar optimization problems and then it is particularized to be constrained and unconstrained, respectively. Adding convexity and topological hypotheses to the functions involved, characterizations via closedness type regularity conditions of the zero duality gap and total duality are obtained, too.

2.3.1 General Perturbed Scalar Optimization Problems

Consider again the framework of Sect. 2.2.1. Let $\varepsilon \geq 0$. The first statement we give presents a characterization via epigraph inclusions of a situation of stable ε-duality gap for the problems (*PG*) and (*DG*).

Theorem 2.11. *Let $x \in X$. Then*

$$\partial \Phi(\cdot,0)(x) = \bigcap_{\eta>0} \mathrm{Pr}_{X^*} \partial_{\varepsilon+\eta} \Phi(x,0) \tag{2.3.13}$$

holds if and only if for each $x^* \in \partial \Phi(\cdot,0)(x)$ one has

$$(\Phi(\cdot,0))^*(x^*) \geq \inf_{y^* \in Y^*} \Phi^*(x^*,y^*) - \varepsilon. \tag{2.3.14}$$

Proof. Note first that one always has $\partial \Phi(\cdot,0)(x) \supseteq \mathrm{Pr}_{X^*} \partial_{\varepsilon+\eta} \Phi(x,0)$ whenever $\eta > 0$. Assume that the reverse inclusion holds for any $\eta > 0$. Then $x^* \in$

$\partial\Phi(\cdot,0)(x)$ if and only if for each $\eta > 0$ there exists a $y_\eta^* \in Y^*$ such that $(x^*, y_\eta^*) \in \partial_{\varepsilon+\eta}\Phi(x,0)$. This means that $\Phi(x,0) + \Phi^*(x^*, y_\eta^*) \leq \langle x^*, x\rangle + \varepsilon + \eta$, that is equivalent to $\Phi^*(x^*, y_\eta^*) - \varepsilon \leq \langle x^*, x\rangle - \Phi(x,0) + \eta$, which yields $\inf_{y^*\in Y^*} \Phi^*(x^*, y^*) - \varepsilon \leq (\Phi(\cdot,0))^*(x^*) + \eta$. The latter inequality holds for any $\eta > 0$, so letting η tend towards 0 we obtain (2.3.14). The other implication follows by making the same steps backwards, using moreover that $x^* \in \partial\Phi(\cdot,0)(x)$ if and only if $(\Phi(\cdot,0))^*(x^*) = \langle x^*, x\rangle - \Phi(x,0)$. □

Remark 2.18. The inequality (2.3.14) can be rewritten as $v(PG_{x^*}) \leq v(DG_{x^*}) + \varepsilon$, i.e. in Theorem 2.11 we provide an equivalent characterization via subdifferential inclusions of the ε-duality gap for (PG_{x^*}) and (DG_{x^*}), when $x^* \in \partial\Phi(\cdot,0)(x)$, i.e. x is an optimal solution to the problem (PG_{x^*}).

Analogously to Theorem 2.11 one can prove the following statement.

Theorem 2.12. *Let $x \in X$ and $\nu > 0$. Then the validity of* (2.3.14) *for all $x^* \in \partial_\nu\Phi(\cdot,0)(x)$ yields*

$$\partial_\nu\Phi(\cdot,0)(x) = \bigcap_{\eta>0} \Pr_{X^*}\partial_{\varepsilon+\eta+\nu}\Phi(x,0). \tag{2.3.15}$$

Viceversa, (2.3.15) *yields for any $x^* \in \partial_\nu\Phi(\cdot,0)(x)$ that*

$$(\Phi(\cdot,0))^*(x^*) \geq \inf_{y^*\in Y^*} \Phi^*(x^*, y^*) - \varepsilon - \nu. \tag{2.3.16}$$

Remark 2.19. Let $x \in X$. In case $\varepsilon = 0$, Theorem 2.12 yields that

$$\partial_\nu\Phi(\cdot,0)(x) = \bigcap_{\eta>0} \Pr_{X^*}\partial_{\eta+\nu}\Phi(x,0) \tag{2.3.17}$$

holds for all $\nu > 0$ if and only if whenever $\mu > 0$ one has $(\Phi(\cdot,0))^*(x^*) \geq \inf_{y^*\in Y^*} \Phi^*(x^*, y^*) - \mu$ for all $x^* \in \partial_\mu\Phi(\cdot,0)(x)$. The last inequality yields $(\Phi(\cdot,0))^*\ (x^*) \geq \inf_{y^*\in Y^*} \Phi^*(x^*, y^*)$ whenever $x^* \in \cap_{\mu>0}\partial_\mu\Phi(\cdot,0)(x) = \partial\Phi(\cdot,0)(x)$, and since the opposite inequality holds in general, we obtain that if (2.3.17) is valid for all $\nu > 0$ one has $(\Phi(\cdot,0))^*(x^*) = \inf_{y^*\in Y^*} \Phi^*(x^*, y^*)$ for all $x^* \in \partial\Phi(\cdot,0)(x)$.

The last assertion in Remark 2.19 can be improved in order to become an equivalence as follows.

Theorem 2.13. *The formula* (2.3.17) *is valid for all $x \in X$ and all $\nu > 0$ if and only if for all $x^* \in X^*$ one has $(\Phi(\cdot,0))^*(x^*) = \inf_{y^*\in Y^*} \Phi^*(x^*, y^*)$.*

Proof. Let $x \in X$. If $(x,0) \notin \operatorname{dom}\Phi$, there is nothing to prove, so we consider the case $\Phi(x,0) \in \mathbb{R}$. Take now $x^* \in X^*$. If $(\Phi(\cdot,0))^*(x^*) = +\infty$ there is nothing to prove, otherwise $x^* \in \partial_\mu\Phi(\cdot,0)(x)$ for all $\mu \geq \Phi(x,0) + (\Phi(\cdot,0))^*(x^*) - \langle x^*, x\rangle$.

The validity of (2.3.17) for $\nu = \Phi(x,0) + (\Phi(\cdot,0))^*(x^*) - \langle x^*, x\rangle$ yields like in the proof of Theorem 2.11 that $\inf_{y^*\in Y^*} \Phi^*(x^*, y^*) - \nu \leq \langle x^*, x\rangle - \Phi(x,0) + \eta$

for all $\eta > 0$. Letting η tend towards 0 and replacing ν with its value, it follows $\inf_{y^* \in Y^*} \Phi^*(x^*, y^*) \leq (\Phi(\cdot, 0))^*(x^*)$, which proves the sufficiency.

To show the necessity, let $\nu > 0$ and $x^* \in \partial_\nu \Phi(\cdot, 0)(x)$. Then the hypothesis yields $\Phi(x, 0) + \inf_{y^* \in Y^*} \Phi^*(x^*, y^*) \leq \langle x^*, x\rangle + \nu$. If $\eta > 0$, there exists a $y^*_\eta \in Y^*$ such that $\Phi(x, 0) + \Phi^*(x^*, y^*_\eta) \leq \langle x^*, x\rangle + \nu + \eta$, i.e. $x^* \in \mathrm{Pr}_{X^*} \partial_{\eta+\nu} \Phi(x, 0)$. As η, x and ν were arbitrarily chosen, the conclusion follows. □

Enriching the function Φ with convexity and topological properties, one can deliver another characterization of (2.3.17) (see also [28, Theorem 3.1], where the proof is made via epigraph inclusions) as well as a consequence of Theorem 2.11.

Theorem 2.14. *Let the function Φ be also convex and lower semicontinuous. The formula* (2.3.17) *is valid for all $x \in X$ and all $\nu > 0$ if and only if the function* $\inf_{y^* \in Y^*} \Phi^*(\cdot, y^*)$ *is $\omega(X^*, X)$-lower semicontinuous.*

Proof. As shown in [174] (see also [49]), the hypotheses yield that the function $(\Phi(\cdot, 0))^*$ is actually the $\omega(X^*, X)$-lower semicontinuous hull of $\inf_{y^* \in Y^*} \Phi^*(\cdot, y^*)$. The conclusion is then a consequence of Theorem 2.13. □

Corollary 2.12. *If the function Φ is also convex and lower semicontinuous and the function* $\inf_{y^* \in Y^*} \Phi^*(\cdot, y^*)$ *is $\omega(X^*, X)$-lower semicontinuous, then for all $x \in X$ it holds*

$$\partial \Phi(\cdot, 0)(x) = \bigcap_{\eta > 0} \mathrm{Pr}_{X^*} \partial_\eta \Phi(x, 0).$$

One can prove the following characterization via subdifferential inclusions of a situation of ε-duality gap for (*PG*) and (*DG*) in the same way as done for Theorem 2.11, too.

Theorem 2.15. *Let $x \in X$. Then*

$$\partial_\varepsilon \Phi(\cdot, 0)(x) = \bigcap_{\eta > 0} \mathrm{Pr}_{X^*} \partial_{\varepsilon+\eta} \Phi(x, 0)$$

holds if and only if for each $x^* \in \partial_\varepsilon \Phi(\cdot, 0)(x)$ one has

$$\Phi(x, 0) - \langle x^*, x\rangle \leq \sup_{y^* \in Y^*} \big\{ - \Phi^*(x^*, y^*)\big\} + \varepsilon. \tag{2.3.18}$$

Remark 2.20. For an $x^* \in \partial_\varepsilon \Phi(\cdot, 0)(x)$, the right-hand side of (2.3.18) is actually $v(DG_{x^*}) + \varepsilon$, while in the left-hand side we have something that can be larger than or equal to (PG_{x^*}). Thus, (2.3.18) yields the ε-duality gap for (PG_{x^*}) and (DG_{x^*}) and in Theorem 2.15 we provide a sufficient condition based on subdifferential inclusions that guarantees it.

Besides the ε-duality gap statements for (*PG*) and (*DG*) provided above, we can formulate other characterizations via subdifferential inclusions, as follows.

Theorem 2.16. *Let $x \in X$. Then*

$$\partial \Phi(\cdot,0)(x) = \Pr_{X^*} \partial_\varepsilon \Phi(x,0) \tag{2.3.19}$$

holds if and only if for each $x^* \in \partial \Phi(\cdot,0)(x)$ there exists a $y^* \in Y^*$ such that

$$(\Phi(\cdot,0))^*(x^*) \geq \Phi^*(x^*,y^*) - \varepsilon. \tag{2.3.20}$$

Proof. One always has $\partial \Phi(\cdot,0)(x) \supseteq \Pr_{X^*} \partial_\varepsilon \Phi(x,0)$. The reverse inclusion holds if and only if for each $x^* \in \partial \Phi(\cdot,0)(x)$ there exists a $y^* \in Y^*$ such that $(x^*,y^*) \in \partial_\varepsilon \Phi(x,0)$, i.e. $\Phi(x,0) + \Phi^*(x^*,y^*) \leq \langle x^*, x\rangle + \varepsilon$. But $x^* \in \partial \Phi(\cdot,0)(x)$ if and only if $(\Phi(\cdot,0))^*(x^*) = \langle x^*, x\rangle - \Phi(x,0)$ and the desired equivalence follows. □

Theorem 2.17. *One has*

$$\partial_\nu \Phi(\cdot,0)(x) = \Pr_{X^*} \partial_\nu \Phi(x,0) \tag{2.3.21}$$

for all $x \in X$ and all $\nu > 0$ if and only if for all $x^* \in X^*$ it holds $(\Phi(\cdot,0))^*(x^*) = \min_{y^* \in Y^*} \Phi^*(x^*,y^*)$.

Theorem 2.18. *Let $x \in X$. Then*

$$\partial_\varepsilon \Phi(\cdot,0)(x) = \Pr_{X^*} \partial_\varepsilon \Phi(x,0)$$

holds if and only if for each $x^* \in \partial_\varepsilon \Phi(\cdot,0)(x)$ there exists a $y^* \in Y^*$ such that

$$\Phi(x,0) - \langle x^*, x\rangle \leq -\Phi^*(x^*,y^*) + \varepsilon.$$

Enriching the function Φ with convexity and topological properties, one can deliver another characterization of (2.3.21) that follows via Corollary 2.6 (see also [28]), by means of a closedness type regularity condition this time.

Theorem 2.19. *Let the function Φ be also convex and lower semicontinuous. The formula (2.3.21) is valid for all $x \in X$ and all $\nu > 0$ if and only if the set $\Pr_{X^* \times \mathbb{R}} \operatorname{epi} \Phi^*$ is closed in the topology $\omega(X^*,X) \times \mathscr{R}$.*

Using Theorem 2.16 and Corollary 2.6 one can show the following statement (see also [28]).

Corollary 2.13. *If the function Φ is also convex and lower semicontinuous and the set $\Pr_{X^* \times \mathbb{R}} \operatorname{epi} \Phi^*$ is closed in the topology $\omega(X^*,X) \times \mathscr{R}$, then for all $x \in X$ it holds*

$$\partial \Phi(\cdot,0)(x) = \Pr_{X^*} \partial \Phi(x,0).$$

Remark 2.21. The difference between the closedness type regularity conditions considered in Corollaries 2.12 and 2.13 can be clearer observed by comparing the way they can be equivalently written as formulae for the conjugate of $\Phi(\cdot, 0)$. The first of them consists of an infimum, thus it characterizes the stable zero duality gap for (PG) and (DG), while the other one means that the same infimum is also attained, i.e. there is stable strong duality for (PG) and (DG), being thus obviously stronger than its counterpart. An example to underline this fact can be found in [56]. The difference between these two conditions can be seen also when we equivalently characterize them as formulae for the ε-subdifferential of $\Phi(\cdot, 0)$ in Theorems 2.14 and 2.19, respectively.

The statements we provided within this subsection can be employed to deliver ε-optimality conditions for (PG) and (DG), too, as follows. First we give a statement that is a consequence of Theorem 2.12.

Theorem 2.20. *(a) Assuming that the regularity condition* (2.2.3) *is fulfilled and that $x \in X$ is an ε-optimal solution to (PG), for each $\eta > 0$ there exists a $y_\eta^* \in Y^*$ such that $(0, y_\eta^*) \in \partial_{\eta+\varepsilon}\Phi(x, 0)$, i.e. $\Phi(x, 0) + \Phi^*(0, y_\eta^*) \leq \eta + \varepsilon$. Moreover, y_η^* is an $\eta + \varepsilon$-optimal solution to (DG).*
(b) If $x \in X$ and for each $\eta > 0$ there exists a $y_\eta^* \in Y^*$ such that $(0, y_\eta^*) \in \partial_{\eta+\varepsilon}\Phi(x, 0)$, then $x \in X$ is an ε-optimal solution to (PG) and each y_η^* is an $\eta + \varepsilon$-optimal solution to (DG).

Analogously one can employ Theorem 2.18 in order to achieve ε-optimality conditions for (PG) and (DG), as follows.

Theorem 2.21. *(a) Assuming that the regularity condition* (2.3.21) *is fulfilled and that $x \in X$ is an ε-optimal solution to (PG), there exists a $y^* \in Y^*$ such that $(0, y^*) \in \partial_\varepsilon\Phi(x, 0)$, i.e. $\Phi(x, 0) + \Phi^*(0, y^*) \leq \varepsilon$. Moreover, y^* is an ε-optimal solution to (DG).*
(b) If $x \in X$ and $y^* \in Y^*$ fulfill $(0, y^*) \in \partial_\varepsilon\Phi(x, 0)$, then $x \in X$ is an ε-optimal solution to (PG) and y^* an ε-optimal solution to (DG).

Remark 2.22. The other statements given in this subsection can be employed for delivering ε-optimality conditions for (PG) and (DG), too. Taking $\varepsilon = 0$ in Theorem 2.21 or in the corresponding statements following from Theorems 2.16, 2.17 or 2.19 one rediscovers the optimality condition given in Corollary 2.8.

Now let us see what happens when the primal problem is particularized to be first constrained, then unconstrained.

2.3.2 Constrained Scalar Optimization Problems

Consider again the framework of Sect. 2.2.2 and we work with the constrained primal optimization problem (PC) and the perturbations we employed for it in order to attach dual problems to it. Let $\varepsilon \geq 0$. Using first the Lagrange perturbation function Φ^L, one obtains from Theorem 2.15 the following statement where a subdifferential inclusion characterizes a situation of ε-duality gap for (PC) and (DC^L).

Theorem 2.22. *Let $x \in X$. Then*

$$\partial_\varepsilon(f + \delta_{\mathscr{A}})(x) = \bigcap_{\eta>0} \bigcup_{z^* \subset C^*} \partial_{\varepsilon+\eta+(z^*h)(x)}(f + \delta_S + (z^*h))(x)$$

if and only if for each $x^* \in \partial_\varepsilon(f + \delta_{\mathscr{A}})(x)$ one has

$$f(x) - \langle x^*, x\rangle \leq \sup_{z^* \in C^*} \{ - (f + (\bar{z}^*h))_S^*(x^*)\} + \varepsilon.$$

Analogously one can particularize the other statements from Sect. 2.3.1 to the present framework, too. For instance, Theorem 2.18 turns into the following assertion.

Theorem 2.23. *Let $x \in X$. Then*

$$\partial_\varepsilon(f + \delta_{\mathscr{A}})(x) = \bigcup_{z^* \in C^*} \partial_{\varepsilon+(z^*h)(x)}(f + \delta_S + (z^*h))(x)$$

if and only if for each $x^* \in \partial_\varepsilon(f + \delta_{\mathscr{A}})(x)$ there exists a $z^* \in C^*$ such that

$$f(x) - \langle x^*, x\rangle \leq -(f + (\bar{z}^*h))_S^*(x^*) + \varepsilon.$$

Adding convexity and topological hypotheses to the functions and sets involved, one obtains the following consequences of Theorems 2.14 and 2.19, respectively.

Theorem 2.24. *Let S be a closed and convex set, f a convex and lower semicontinuous function and h a C-convex and C-*epi*-closed vector function. The formula*

$$\partial_\nu(f + \delta_{\mathscr{A}})(x) = \bigcap_{\eta>0} \bigcup_{z^* \in C^*} \partial_{\nu+\eta+(z^*h)(x)}(f + \delta_S + (z^*h))(x)$$

is valid for all $x \in X$ and all $\nu > 0$ if and only if the function $\inf_{z^* \in C^*}(f + (\bar{z}^*h))_S^*$ *is $\omega(X^*, X)$-lower semicontinuous.*

Theorem 2.25. *Let S be a closed and convex set, f a convex and lower semicontinuous function and h a C-convex and C-*epi*-closed vector function. The formula*

$$\partial_\nu(f + \delta_{\mathscr{A}})(x) = \bigcup_{z^* \in C^*} \partial_{\nu+(z^*h)(x)}(f + \delta_S + (z^*h))(x)$$

is valid for all $x \in X$ and all $\nu > 0$ if and only if the set $\cup_{z^* \in C^*} \operatorname{epi}(f + (z^*h) + \delta_S)^*$ *is closed in the topology $\omega(X^*, X) \times \mathscr{R}$.*

Another perturbation function employed to assign a conjugate dual problem to (PC) as a special case of (DG) is Φ^{FL}. The statements from Sect. 2.3.1 particularized above for Φ^L become in this case the following ones.

Theorem 2.26. *Let $x \in X$. Then*

$$\partial_\varepsilon(f + \delta_{\mathscr{A}})(x) = \bigcap_{\eta>0} \bigcup_{\substack{z^* \in C^*, \varepsilon_1, \varepsilon_2 \geq 0, \\ \varepsilon_1+\varepsilon_2=\varepsilon+\eta+(z^*h)(x)}} \big(\partial_{\varepsilon_1} f(x) + \partial_{\varepsilon_2}((z^*h) + \delta_S)(x)\big)$$

if and only if for each $x^* \in \partial_\varepsilon(f + \delta_{\mathscr{A}})(x)$ one has

$$f(x) - \langle x^*, x\rangle \leq \sup_{\substack{z^* \in C^*, \\ y^* \in X^*}} \big\{ - f^*(y^*) - (\bar{z}^*h)^*_S(x^* - y^*)\big\} + \varepsilon.$$

Theorem 2.27. *Let $x \in X$. Then*

$$\partial_\varepsilon(f + \delta_{\mathscr{A}})(x) = \bigcup_{\substack{z^* \in C^*, \varepsilon_1, \varepsilon_2 \geq 0, \\ \varepsilon_1+\varepsilon_2=\varepsilon+\eta+(z^*h)(x)}} \big(\partial_{\varepsilon_1} f(x) + \partial_{\varepsilon_2}((z^*h) + \delta_S)(x)\big)$$

if and only if for each $x^* \in \partial_\varepsilon(f + \delta_{\mathscr{A}})(x)$ there exist $z^* \in C^*$ and $y^* \in X^*$ such that

$$f(x) - \langle x^*, x\rangle \leq - f^*(y^*) - (z^*h)^*_S(x^* - y^*) + \varepsilon.$$

Theorem 2.28. *Let S be a closed and convex set, f a convex and lower semicontinuous function and h a C-convex and C-*epi-*closed vector function. The formula*

$$\partial_\nu(f + \delta_{\mathscr{A}})(x) = \bigcap_{\eta>0} \bigcup_{\substack{z^* \in C^*, \varepsilon_1, \varepsilon_2 \geq 0, \\ \varepsilon_1+\varepsilon_2=\nu+\eta+(z^*h)(x)}} \big(\partial_{\varepsilon_1} f(x) + \partial_{\varepsilon_2}((z^*h) + \delta_S)(x)\big)$$

is valid for all $x \in X$ and all $\nu > 0$ if and only if the function $\inf_{z^* \in C^*} f \square (\bar{z}^*h)^*_S$ *is $\omega(X^*, X)$-lower semicontinuous.*

Theorem 2.29. *Let S be a closed and convex set, f a convex and lower semicontinuous function and h a C-convex and C-*epi-*closed vector function. The formula*

$$\partial_\nu(f + \delta_{\mathscr{A}})(x) = \bigcup_{\substack{z^* \in C^*, \varepsilon_1, \varepsilon_2 \geq 0, \\ \varepsilon_1+\varepsilon_2=\nu+(z^*h)(x)}} \big(\partial_{\varepsilon_1} f(x) + \partial_{\varepsilon_2}((z^*h) + \delta_S)(x)\big)$$

is valid for all $x \in X$ and all $\nu > 0$ if and only if the set $\operatorname{epi} f^* + \cup_{z^* \in C^*} \operatorname{epi}(z^*h)^*_S$ *is closed in the topology $\omega(X^*, X) \times \mathscr{R}$.*

Analogously one can particularize the other statements from Sect. 2.3.1 to the present framework, too.

2.3.3 Unconstrained Scalar Optimization Problems

Consider now the framework of Sect. 2.2.3 and we work again with the unconstrained primal optimization problem (PU) and the perturbation function Φ^U employed in order to attach dual problems to it. Let $\varepsilon \geq 0$. From Theorem 2.15 one obtains the following statement where a subdifferential inclusion characterizes a situation of ε-duality gap for (PU) and (DU).

Theorem 2.30. *Let $x \in X$. Then*

$$\partial_\varepsilon(f + g \circ A)(x) = \bigcap_{\eta>0} \bigcup_{\substack{\varepsilon_1,\varepsilon_2\geq 0,\\ \varepsilon_1+\varepsilon_2=\varepsilon+\eta}} \big(\partial_{\varepsilon_1} f(x) + A^*\partial_{\varepsilon_2} g(Ax)\big)$$

if and only if for each $x^* \in \partial_\varepsilon(f + g \circ A)(x)$ one has

$$f(x) + g(Ax) - \langle x^*, x\rangle \leq \sup_{y^*\in X^*} \big\{ - f^*(A^*y^*) - g^*(x^* - y^*)\big\} + \varepsilon.$$

Further, Theorem 2.18 turns into the following assertion.

Theorem 2.31. *Let $x \in X$. Then*

$$\partial_\varepsilon(f + g \circ A)(x) = \bigcup_{\substack{\varepsilon_1,\varepsilon_2\geq 0,\\ \varepsilon_1+\varepsilon_2=\varepsilon}} \big(\partial_{\varepsilon_1} f(x) + A^*\partial_{\varepsilon_2} g(Ax)\big)$$

if and only if for each $x^* \in \partial_\varepsilon(f + g \circ A)(x)$ there exists a $y^* \in X^*$ such that

$$f(x) + g(Ax) - \langle x^*, x\rangle \leq -f^*(A^*y^*) - g^*(x^* - y^*) + \varepsilon.$$

Adding convexity and topological hypotheses to the functions and sets involved, one obtains the following consequences of Theorems 2.14 and 2.19, respectively.

Theorem 2.32. *Let the functions f and g be also convex and lower semicontinuous. The formula*

$$\partial_\nu(f + g \circ A)(x) = \bigcap_{\eta>0} \bigcup_{\substack{\varepsilon_1,\varepsilon_2\geq 0,\\ \varepsilon_1+\varepsilon_2=\nu+\eta}} \big(\partial_{\varepsilon_1} f(x) + A^*\partial_{\varepsilon_2} g(Ax)\big)$$

is valid for all $x \in X$ and all $\nu > 0$ if and only if the function $\inf_{y^*\in C^*} \big[f^*(A^*y^*)+ g^*(\cdot - y^*)\big]$ is $\omega(X^*, X)$-lower semicontinuous.

Theorem 2.33. *Let the functions f and g be also convex and lower semicontinuous. The formula*

$$\partial_\nu(f + g \circ A)(x) = \bigcup_{\substack{\varepsilon_1,\varepsilon_2 \geq 0,\\ \varepsilon_1+\varepsilon_2=\nu}} \big(\partial_{\varepsilon_1} f(x) + A^* \partial_{\varepsilon_2} g(Ax)\big)$$

is valid for all $x \in X$ and all $\nu > 0$ if and only if the set $\operatorname{epi} f^* + (A^* \times \operatorname{id}_{\mathbb{R}})(\operatorname{epi} g^*)$ *is closed in the topology $\omega(X^*, X) \times \mathscr{R}$.*

Analogously one can particularize the other statements from Sect. 2.3.1 to the present framework, too. We present here only what becomes Corollary 2.13 in this framework, since this statement will be needed later in our presentation (see also [21,48]).

Corollary 2.14. *If the functions f and g are also convex and lower semicontinuous and the set* $\operatorname{epi} f^* + (A^* \times \operatorname{id}_{\mathbb{R}})(\operatorname{epi} g^*)$ *is closed in the topology $\omega(X^*, X) \times \mathscr{R}$, then for all $x \in X$ one has*

$$\partial(f + g \circ A)(x) = \partial f(x) + A^* \partial g(Ax).$$

Moreover, one can see (*PC*) as an unconstrained optimization problem like in Sect. 2.2.3 and the corresponding counterparts of the statements given above can be formulated for it by particularizing these assertions or by employing the perturbation function Φ^F.

Remark 2.23. From the statements provided in Sects. 2.3.2 and 2.3.3 one can derive ε-optimality conditions for (*PC*) and (*PU*) and their dual problems, as done in the general case in Theorems 2.20 and 2.21.

Remark 2.24. Further characterizations of stable ε-duality gap and strong duality statements via epigraph and subdifferential inclusions in the vein of the ones provided within this chapter for constrained optimization problems can be found in [14], while in [13] similar assertions are delivered for unconstrained composed optimization problems (see also [43]). Moreover, in [28] we have provided equivalent characterizations of stable zero duality gap and stable strong duality via epigraph inclusions for both constrained and unconstrained, as well for composed optimization problems with the involved functions taken convex.

Chapter 3
Minimality Concepts for Sets

3.1 Historical Overview and Motivation

Solving a scalar optimization problem usually means to determine the points where the objective function attains its minimum (respectively maximum) over the feasible set, but one can also look for solutions satisfying stronger conditions, like weak sharp minima or strong minima. A similar situation can be found in vector optimization, too, where due to the increased complexity of the problems the solution concepts are more diversified. The most known and widely used of them is the *(Pareto-)minimality*, whose roots go back to the late nineteen century when Edgeworth and Pareto, respectively, introduced the first notions of optimality in multiobjective optimization. Its underlying concept can be understood as follows. Given an initial allocation of goods among a set of individuals, a change to a different allocation that makes at least one individual better off without making any other individual worse off is called a Pareto-improvement. An allocation is defined as *Pareto optimal* when no further Pareto-improvements can be made. This notion was then extended for partial orders induced by convex cones, a point being called (Pareto-)minimal to a set where it belongs if there is no other element of the set which is less than it with respect to the mentioned partial ordering.

However, the minimal elements of a set can be difficult to determine in many situations and sometimes only some of them are required, thus different other minimality concepts were considered in the literature. Some of them are weaker than the classical one, as it is the case for the *weak minimality* and its generalizations, while most of them are more restrictive than it, like the *ideal minimality*, *strong minimality* or *proper minimality*. One can go even further by considering notions like *ε-minimality* or *approximate minimality*, but they surpass the purposes of the present work.

In the literature one can find several weak minimality notions, some of them mentioned for instance in [6, 7], defined via different interiority notions for the cone that partially orders the working space. However the classical weak minimality

S.-M. Grad, *Vector Optimization and Monotone Operators via Convex Duality*, Vector Optimization, DOI 10.1007/978-3-319-08900-3_3

introduced via a cone with a nonempty interior is the most used one because of its alternative characterization via the linear scalarization. There exist several proper minimality notions, too, reviewed and compared, for instance, in [48, Section 2.4]. However, for most of them the corresponding properly minimal elements are quite difficult to identify because of the complexity of their definitions. The only one for which the properly minimal elements can be relatively easily determined is actually the most restrictive of these notions, namely the *proper minimality in the sense of linear scalarization*. The corresponding properly minimal elements are actually optimal solutions to an attached scalar optimization problem. But since the linear scalarization may fail to deliver valuable results regarding the vector optimization problems investigated for various purposes (see, for instance, [22, 81]) and, on the other hand, as an unfortunate choice of its scalarization parameters can lead to unbounded scalar optimization problems (see, for instance [15,92]), other functions with similar properties, i.e. strongly or strictly cone-monotone increasing, were employed for the same purpose in works like [31, 37, 92, 93, 97, 98, 102, 103, 139, 140, 166], giving birth to new proper and weak minimality notions, which under certain conditions coincide with the already mentioned ones. Motivated by them, we propose in Sect. 3.3.1 a general scheme for defining properly minimal elements with respect to different scalarization functions, that will be employed later in Chap. 4 for duality investigations on vector optimization problems.

We begin our investigations in Sect. 3.2 by considering several minimality notions for sets regarding the partial ordering induced by a convex cone that is not necessarily pointed. We also compare them, showing that most of the inclusions between different types of properly minimal sets given in [48, Section 2.4] for pointed ordering cones remain valid in the more general framework, too. Then we take the ordering cone to be also pointed and we deliver new weak conditions which guarantee the coincidence of several types of properly minimal elements. We also show that under mild hypotheses the scalarization properties of the classical weak minimality, which is defined when the interior of the ordering cone of the space we work in is nonempty, remain valid when we extend it by considering only the quasi interior of the mentioned cone nonempty. Similar investigations are done for relatively minimal elements, that are defined when only the quasi-relative interior of the ordering cone is known to be nonempty.

3.2 General Ordering Cones

In this section we present various minimality concepts for sets that are considered regarding the partial ordering induced by a convex cone that is not necessarily pointed. Then we compare the different minimality sets corresponding to a given set.

Although some of the following definitions can be given in more general settings, consider a Hausdorff locally convex space V partially ordered by the nontrivial convex cone $K \subseteq V$. Beginning with Sect. 3.3 we will impose the fulfillment of the condition $\operatorname{qi} K^* \neq \emptyset$, that yields $K^{*0} \neq \emptyset$ and, consequently, that the cone K is pointed. But, for the moment, we work with K possibly not pointed. Let $M \subseteq V$ be a nonempty set and we begin by recalling the definition of the minimal elements of this set that is due to Borwein [18].

Definition 3.1. An element $\bar{v} \in M$ is said to be a *minimal element of M (regarding the partial ordering induced by K)*, if from $v \leqq_K \bar{v}$, $v \in M$, follows $v \geqq_K \bar{v}$. The set of all minimal elements of M is denoted by $\mathrm{Min}(M, K)$ and it is called the *minimal set of M (regarding the partial ordering induced by K).*

Remark 3.1. From Definition 3.1 follows that if $\bar{v} \in M$ is a minimal element of M then any $\tilde{v} \in M$ such that $\tilde{v} \leqq_K \bar{v}$ is also a minimal element of M. Note further that $\bar{v} \in \mathrm{Min}(M, K)$ means that for all $v \in M$ fulfilling $v \leqq_K \bar{v}$ it is binding to have $\bar{v} - v \in \ell(K)$, which, when K is pointed turns into $\bar{v} = v$.

Remark 3.2. One has $\bar{v} \in \mathrm{Min}(M, K)$ if and only if one of the following conditions is fulfilled (cf. [48])

(i) There is no $v \in M$ such that $\bar{v} - v \in K \setminus \ell(K)$;
(ii) $(\bar{v} - K) \cap M \subseteq \bar{v} + K$;
(iii) $(-K) \cap (M - \bar{v}) \subseteq K$.

The *maximal elements* of the set M (regarding the partial ordering induced by K) are defined analogously, being actually the elements of the set $\mathrm{Max}(M, K) := \mathrm{Min}(M, -K) = -\mathrm{Min}(-M, K)$.

Let us investigate now the relations that exist between the minimal elements of M and $M + K$.

Proposition 3.1. *It holds* $\mathrm{Min}(M + K, K) \cap M \subseteq \mathrm{Min}(M, K) \subseteq \mathrm{Min}(M + \ell(K), K) = \mathrm{Min}(M + K, K)$*, and the inclusions turn into equalities when K is pointed.*

Proof. If $\bar{v} \in \mathrm{Min}(M + K, K) \cap M$, and there exists a $v \in M$ such that $v \leqq_K \bar{v}$ and $\bar{v} - v \notin \ell(K)$, noting that $v \in M + K$ one obtains a contradiction to the minimality of $\bar{v}$ in $M + K$. Consequently, $\mathrm{Min}(M + K, K) \cap M \subseteq \mathrm{Min}(M, K)$

By [48, Lemma 2.4.1] it is known that $\mathrm{Min}(M, K) \subseteq \mathrm{Min}(M + K, K)$.

If $\bar{v} \in \mathrm{Min}(M + \ell(K), K)$, then $\bar{v} \in M + \ell(K) \subseteq M + K$. Let $v \in M$ and $k \in K \setminus \ell(K)$ such that and $v + k \leqq_K \bar{v}$. Then $v \leqq_K \bar{v} - k \leqq_K \bar{v}$. As $v, \bar{v} \in M + \ell(K)$ and $\bar{v} \in \mathrm{Min}(M + \ell(K), K)$, it follows that $\bar{v} \leqq_K v$, which yields $\bar{v} \leqq_K v + k$, i.e. $\bar{v} \in \mathrm{Min}(M + K, K)$. Consequently, $\mathrm{Min}(M + \ell(K), K) \subseteq \mathrm{Min}(M + K, K)$.

Finally, for $\bar{v} \in \mathrm{Min}(M + K, K)$, if $\bar{v} \in M + \ell(K)$, then, assuming that $\bar{v} \notin \mathrm{Min}(M + \ell(K), K)$ leads like above to a contradiction to the minimality of $\bar{v}$ in $M + K$. Assuming that $\bar{v} \notin M + \ell(K)$, there exist $v \in M$ and $k \in K \setminus \ell(K)$ such that $\bar{v} = v + k$. But $v \in M + K$, too, and $v - \bar{v} = -k \in K \setminus \ell(K)$, that yields, via Remark 3.2(iii), that $\bar{v} \notin \mathrm{Min}(M + K, K)$. Consequently, $\mathrm{Min}(M + K, K) \subseteq \mathrm{Min}(M + \ell(K), K)$. □

The minimality notion introduced above can be refined by defining the so-called properly minimal elements of a set with respect to the considered ordering cone. Properly minimal elements turn out to be minimal elements with additional properties. In the literature one can find different types of properly minimal elements and we shall deal in the following with some of them, namely the ones considered for the case K pointed in [48, Section 2.4]. The only exception concerns the ones in the sense of Geoffrion (cf. [48, Definition 2.4.6]), which are defined only when K is the nonnegative orthant of an Euclidian space, hence pointed. We show how the definitions of the properly minimal elements of a set (regarding the partial ordering induced by K) can be extended for the current framework. Moreover, the inclusion relations between the sets of different properly minimal elements proven in [48, Subsection 2.4.3] can be generalized for the situation when K is not necessarily pointed, too. The proofs of these results are adapted from the ones in [48, Subsection 2.4.3], but we include them here for the convenience of the reader. Note also that here we work only with properly minimal elements, for considering properly maximal elements one needs only replace the cone K by $-K$.

The first notion of proper minimality we introduce and investigate extends for not necessarily pointed ordering cones the one due to Hurwicz (cf. [113,130]). Note that it can be considered even in the more general setting of topological vector spaces.

Definition 3.2. An element $\bar{v} \in M$ is said to be a *properly minimal element of M in the sense of Hurwicz (regarding the partial ordering induced by K)* if $\operatorname{cl}(\operatorname{coneco}((M - \bar{v}) \cup K)) \cap (-K) \subseteq K$. The set of all properly minimal elements of M in the sense of Hurwicz is denoted by $\operatorname{PMin}_{Hu}(M, K)$.

Remark 3.3. The proper minimality of $\bar{v} \in M$ in the sense of Hurwicz can be equivalently written as $0 \in \operatorname{Min}\big(\operatorname{cl}(\operatorname{coneco}((M - \bar{v}) \cup K)), K\big)$.

Remark 3.4. In Definition 3.2, in Remark 3.2(iii) and further in all the definitions of properly minimal elements that will follow in this section except the ones in the sense of linear scalarization one can replace in the right-hand side of the inclusions the cone K with $\ell(K)$, respectively K' with $\ell(K')$ in Definition 3.6.

Remark 3.5. Since $M - \bar{v} \subseteq \operatorname{cl}(\operatorname{coneco}((M - \bar{v}) \cup K))$, Remark 3.2(iii) yields that $\operatorname{PMin}_{Hu}(M, K) \subseteq \operatorname{Min}(M, K)$.

Proposition 3.2. *It holds* $\operatorname{PMin}_{Hu}(M, K) = \operatorname{PMin}_{Hu}(M + \ell(K), K) = \operatorname{PMin}_{Hu}(M + K, K)$.

Proof. Since $\operatorname{coneco}((M - \bar{v}) \cup K) = \operatorname{coneco}((M + \ell(K) - \bar{v}) \cup K) = \operatorname{coneco}((M + K - \bar{v}) \cup K)$, the conclusion follows immediately. □

To overcome the drawback of Geoffrion's classical definition of proper minimality that is limited to the ordering cone $K = \mathbb{R}^k_+$, Borwein proposed in [16] a more general notion of proper minimality, extended for not necessarily pointed ordering cones as follows.

Definition 3.3. An element $\bar{v} \in M$ is said to be a *properly minimal element of M in the sense of Borwein (regarding the partial ordering induced by K)* if $\operatorname{cl} T_{M+K}(\bar{v}) \cap (-K) \subseteq K$. The set of all properly minimal elements of M in the sense of Borwein is denoted by $\operatorname{PMin}_{Bo}(M, K)$.

Remark 3.6. The proper minimality $\bar{v} \in M$ in the sense of Borwein can be equivalently written as $0 \in \operatorname{Min}\left(\operatorname{cl} T_{M+K}(\bar{v}), K\right)$. Moreover, if V is metrizable, then the tangent cone is closed and in this situation one may omit the closure operation within Definition 3.3.

Remark 3.7. If $K = \{0\}$, [48, Remark 2.4.6(b)] yields $\operatorname{PMin}_{Bo}(M, K) \subseteq \operatorname{Min}(M, K)$. The same conclusion can be obtained when $K \neq \{0\}$, too, as follows. Assuming that $\bar{v} \in \operatorname{PMin}_{Bo}(M, K) \setminus \operatorname{Min}(M, K)$, there exists a $v \in M$ such that $\bar{v} - v \in K \setminus \ell(K)$. For $l \geq 1$, taking $v_l = \bar{v} + (1/l)(v - \bar{v})$, one gets $v_l = v + ((l-1)/l)(\bar{v} - v) \in M + K$. As $\lim_{l \to +\infty} v_l = \bar{v}$ and $\lim_{l \to +\infty} l(v_l - \bar{v}) = v - \bar{v}$, it follows $v - \bar{v} \in T_{M+K}(\bar{v})$. Then $v - \bar{v} \in T_{M+K}(\bar{v}) \cap (-K \setminus \ell(K)) \subseteq \operatorname{cl} T_{M+K}(\bar{v}) \cap (-K)$, but, since $v - \bar{v} \in K \setminus \ell(K)$, this contradicts the proper minimality of $\bar{v}$.

Proposition 3.3. *It holds* $\operatorname{PMin}_{Bo}(M, K) \subseteq \operatorname{PMin}_{Bo}(M + \ell(K), K) = \operatorname{PMin}_{Bo}(M + K, K)$.

Proof. It always holds $M + K \subseteq M + K + \ell(K) \subseteq (M + K) + K$, and the convexity of K guarantees that $M + K = M + K + \ell(K) = (M + K) + K$. Therefore, $T_{M+K}(\bar{v}) = T_{M+\ell(K)+K}(\bar{v}) = T_{M+K+K}(\bar{v})$ for all $\bar{v} \in M + K$. Taking into consideration that $M \subseteq M + \ell(K) \subseteq M + K$, it follows that $\operatorname{PMin}_{Bo}(M, K) \subseteq \operatorname{PMin}_{Bo}(M + \ell(K), K) \subseteq \operatorname{PMin}_{Bo}(M + K, K)$.

Take $\bar{v} \in \operatorname{PMin}_{Bo}(M + K, K)$. If $\bar{v} \in M + \ell(K)$, one immediately gets $\bar{v} \in \operatorname{PMin}_{Bo}(M + \ell(K), K)$. Otherwise, there exist $v \in M$ and $k \in K \setminus \ell(K)$ such that $\bar{v} = v + k$. Then, for all $l \geq 1$, one gets $v_l = v + ((l-1)/l)(\bar{v} - v) \in M + K$, hence, like in Remark 3.7, it follows $v - \bar{v} \in \operatorname{cl} T_{M+K}(\bar{v}) \cap (-K)$, but, since $v - \bar{v} \in -K \setminus \ell(K)$, this contradicts the proper minimality of $\bar{v}$, consequently $\bar{v} \in M + \ell(K)$, therefore $\operatorname{PMin}_{Bo}(M + K, K) \subseteq \operatorname{PMin}_{Bo}(M + \ell(K), K)$. □

Another proper minimality notion introduced in order to extend Geoffrion's one for general convex cones originates from Benson's paper [11] and it can be extended for not necessarily pointed ordering cones as follows.

Definition 3.4. An element $\bar{v} \in M$ is said to be a *properly minimal element of M in the sense of Benson (regarding the partial ordering induced by K)* if $\operatorname{cl}\operatorname{cone}(M + K - \bar{v}) \cap (-K) \subseteq K$. The set of all properly minimal elements of M in the sense of Benson is denoted by $\operatorname{PMin}_{Be}(M, K)$.

Remark 3.8. Since $M - \bar{v} \subseteq \operatorname{cl}\operatorname{cone}(M + K - \bar{v})$, Remark 3.2(iii) yields that $\operatorname{PMin}_{Hu}(M, K) \subseteq \operatorname{Min}(M, K)$.

Proposition 3.4. *It holds* $\operatorname{PMin}_{Be}(M, K) \subseteq \operatorname{PMin}_{Be}(M + \ell(K), K) = \operatorname{PMin}_{Be}(M + K, K)$.

Proof. The convexity of K guarantees that $M + K = M + K + \ell(K) = (M + K) + K$. Therefore, for all $v \in V$ it holds $\operatorname{cone}(M + K - v) = \operatorname{cone}(M + K + \ell(K) - v) = \operatorname{cone}(M + K + K - v)$ and, taking into consideration that $M \subseteq M + \ell(K) \subseteq M + K$, it follows that $\operatorname{PMin}_{Be}(M, K) \subseteq \operatorname{PMin}_{Be}(M + \ell(K), K) \subseteq \operatorname{PMin}_{Be}(M + K, K)$.

Take $\bar{v} \in \mathrm{PMin}_{Be}(M + K, K)$. If $\bar{v} \in M + \ell(K)$, one immediately gets $\bar{v} \in \mathrm{PMin}_{Be}(M + \ell(K), K)$. Otherwise, there exist $v \in M$ and $k \in K \setminus \ell(K)$ such that $\bar{v} = v + k$. Then $v \in M + K$ and $v \leqq_K \bar{v}$, hence, via Remarks 3.8 and 3.1, $\bar{v} - v \in \ell(K)$. But $\bar{v} - v = k \notin \ell(K)$, therefore $\mathrm{PMin}_{Be}(M + K, K) \subseteq \mathrm{PMin}_{Be}(M + \ell(K), K)$. □

Remark 3.9. Because $T_{M+K}(v) \subseteq \mathrm{cl\,cone}(M + K - v)$ whenever $v \in M + K$, it follows that $\mathrm{PMin}_{Be}(M, K) \subseteq \mathrm{PMin}_{Bo}(M, K)$, which turns into an equality when $M + K$ is convex. Note also that for $\bar{v} \in M$ it holds $\mathrm{cone}(M + K - \bar{v}) \subseteq \mathrm{coneco}((M - \bar{v}) \cup K)$, the two sets coinciding if $M + K$ is convex. Thus we have in general that $\mathrm{PMin}_{Hu}(M, K) \subseteq \mathrm{PMin}_{Be}(M, K)$, while when $M + K$ is convex it follows that $\mathrm{PMin}_{Hu}(M, K) = \mathrm{PMin}_{Be}(M, K) = \mathrm{PMin}_{Bo}(M, K)$.

In the following we consider another proper minimality concept due to Borwein, originally given for not necessarily pointed ordering cones (cf. [17]), that is defined in a quite similar manner to the one of Benson.

Definition 3.5. An element $\bar{v} \in M$ is said to be a *properly minimal element of M in the global sense of Borwein (regarding the partial ordering induced by K)* if $\mathrm{cl}(\mathrm{cone}(M - \bar{v})) \cap (-K) \subseteq K$. The set of all properly minimal elements of M in the global sense of Borwein is denoted by $\mathrm{PMin}_{GB}(M, K)$.

Remark 3.10. Since $M - \bar{v} \subseteq \mathrm{cl\,cone}(M - \bar{v})$, Remark 3.2(iii) yields that $\mathrm{PMin}_{GB}(M, K) \subseteq \mathrm{Min}(M, K)$.

Remark 3.11. It holds $\mathrm{PMin}_{Be}(M, K) = \mathrm{PMin}_{GB}(M + K, K)$.

Proposition 3.5. *It holds* $\mathrm{PMin}_{GB}(M + K, K) \subseteq \mathrm{PMin}_{GB}(M + \ell(K), K)$.

Proof. When $\bar{v} \in \mathrm{PMin}_{GB}(M + K, K)$, assuming that $\bar{v} \notin M + \ell(K)$ implies the existence of a $v \in M$ and $k \in K \setminus \ell(K)$ such that $\bar{v} = v + k$. Then $v \in M + \ell(K)$ and $v \leqq \bar{v}$, hence via Remarks 3.10 and 3.1 one gets $\bar{v} - v \in \ell(K)$. But $\bar{v} - v = k \notin \ell(K)$, therefore $\bar{v} \in M + \ell(K)$.

Since $\mathrm{cone}(M + \ell(K) - \bar{v}) \subseteq \mathrm{cone}(M + K - \bar{v})$, from $\mathrm{cl\,cone}(M + K - \bar{v}) \cap (-K) \subseteq K$ one immediately gets $\mathrm{cl\,cone}(M + \ell(K) - \bar{v}) \cap (-K) = \{0\}$, consequently, $\bar{v} \in \mathrm{PMin}_{GB}(M + \ell(K), K)$. □

Remark 3.12. Taking into account the corresponding definitions, one immediately gets that $\mathrm{PMin}_{Be}(M, K) \subseteq \mathrm{PMin}_{GB}(M, K)$. Consequently, via Remark 3.11, one can complete Proposition 3.5 with the inclusion $\mathrm{PMin}_{GB}(M + K, K) \subseteq \mathrm{PMin}_{GB}(M, K)$. Of interest would be to determine whether between $\mathrm{PMin}_{GB}(M, K)$ and $\mathrm{PMin}_{GB}(M + \ell(K), K)$ can be established any inclusion in general. On the other hand, as noted in [48] for K pointed, no relation of inclusion between $\mathrm{PMin}_{Bo}(M, K)$ and $\mathrm{PMin}_{GB}(M, K)$ can be given in general. However, taking into account Remark 3.9, when $M + K$ is convex, then $\mathrm{PMin}_{Bo}(M, K) \subseteq \mathrm{PMin}_{GB}(M, K)$.

A formally different proper minimality approach is the following one that extends for not necessarily pointed ordering cones the notion introduced by Henig in [120] and Lampe in [153] by employing a nontrivial convex cone $K' \subseteq V$ that contains the given ordering cone K.

Definition 3.6. An element $\bar{v} \in M$ is said to be a *properly minimal element of M in the sense of Henig and Lampe (regarding the partial ordering induced by K)* if there exists a nontrivial convex cone $K' \subseteq V$ with $K \setminus \ell(K) \subseteq \operatorname{qri} K' \setminus \ell(K')$ such that $(M - \bar{v}) \cap (-K') \subseteq K'$. The set of all properly minimal elements of M in the sense of Henig and Lampe is denoted by $\operatorname{PMin}_{HL}(M, K)$.

Example 3.1. When $V = \mathbb{R}^2$ and $K = \mathbb{R} \times \{0\}$, then $\operatorname{qri} K = \operatorname{ri} K = K = -K = \ell(K)$. Therefore, in this situation, $\operatorname{qri} K \setminus \ell(K) \neq \operatorname{qri} K$.

Remark 3.13. Since for $\bar{v} \in M$ and K' as given in Definition 3.6 it holds $(M - \bar{v}) \cap (-K) \subseteq (M - \bar{v}) \cap (-K')$, Remark 3.2(iii) yields that $\operatorname{PMin}_{HL}(M, K) \subseteq \operatorname{Min}(M, K)$.

Remark 3.14. Note that $\bar{v} \in \operatorname{PMin}_{HL}(M, K)$ implies $\bar{v} \in \operatorname{Min}(M, K')$, while the existence of K' as asked in Definition 3.6 for which $\bar{v} \in \operatorname{Min}(M, K')$ yields $\bar{v} \in \operatorname{PMin}_{HL}(M, K)$.

Proposition 3.6. *It holds* $\operatorname{PMin}_{HL}(M, K) \subseteq \operatorname{PMin}_{HL}(M + K, K) = \operatorname{PMin}_{HL}(M + \ell(K), K)$.

Proof. The inclusion of $\operatorname{PMin}_{HL}(M, K)$ in each of the other two sets follows from Proposition 3.1 and Remark 3.14.

When $\bar{v} \in \operatorname{PMin}_{HL}(M + K, K)$, assuming that $\bar{v} \notin M + \ell(K)$ implies the existence of a $v \in M$ and $k \in K \setminus \ell(K)$ such that $\bar{v} = v + k$. Then $v \in M + \ell(K)$ and $v \leqq \bar{v}$, hence via Remarks 3.13 and 3.1 one gets $\bar{v} - v \in \ell(K)$. But $\bar{v} - v = k \notin \ell(K)$, therefore $\bar{v} \in M + \ell(K)$.

Moreover, there exists a nontrivial convex cone $K' \subseteq V$ such that $K \setminus \ell(K) \subseteq \operatorname{qri} K' \setminus \ell(K')$ and $(M + K - \bar{v}) \cap (-K') \subseteq \ell(K')$. Thus $(M + \ell(K) - \bar{v}) \cap (-K') \subseteq \ell(K')$, hence $\bar{v} \in \operatorname{PMin}_{HL}(M + \ell(K), K)$.

Vice versa, take $\bar{v} \in \operatorname{PMin}_{HL}(M + \ell(K), K)$. Then $\bar{v} \in M + \ell(K) \subseteq M + K$ and there exists a nontrivial convex cone $K' \subseteq V$ such that $K \setminus \ell(K) \subseteq \operatorname{qri} K' \setminus \ell(K')$ and $(M + \ell(K) - \bar{v}) \cap (-K') \subseteq \ell(K')$.

Assuming that $(M + K - \bar{v}) \cap (-K') \subseteq \ell(K')$ does not take place, there would exist $v \in M$, $k \in K$ and $k' \in K' \setminus \ell(K')$ such that $v + k - \bar{v} = -k'$. Then $v - \bar{v} = -(k + k') \in -(K + (K' \setminus \ell(K'))) \subseteq -((K' \setminus \ell(K')) + (K' \setminus \ell(K'))) = -K' \setminus \ell(K')$, but, on the other hand, $v - \bar{v} \in \ell(K')$ because $(M + \ell(K) - \bar{v}) \cap (-K') \subseteq \ell(K')$, therefore we reached a contradiction.

Consequently, $\bar{v} \in \operatorname{PMin}_{HL}(M + K, K)$ and the proof is complete. □

The following statement reveals the relation between the proper minimality notions in the sense of Benson and in the sense of Henig and Lampe.

Proposition 3.7. *It holds* $\operatorname{PMin}_{HL}(M, K) \subseteq \operatorname{PMin}_{Be}(M, K)$.

Proof. If $K = \{0\}$ the inclusion follows because of [48, Proposition 2.4.11]. Take further that $K \neq \{0\}$.

Let $\bar{v} \in \text{PMin}_{HL}(M, K)$. Then $\bar{v} \in M$ and there exists a nontrivial convex cone $K' \subseteq V$ such that $K \setminus \ell(K) \subseteq \text{qri}\, K' \setminus \ell(K')$ and $(M + K - \bar{v}) \cap (-K') \subseteq \ell(K')$.

Assume now that $\bar{v} \notin \text{PMin}_{Be}(M, K)$, i.e. that there exists a $k \in K \setminus \ell(K)$ such that $-k \in \text{cl}\,\text{cone}(M + K - \bar{v})$. Then $k \in \text{qri}\, K' \setminus \ell(K')$ and consequently there exist $v \in M$, $u \in K$ and $t > 0$ such that $k = t(v + u - \bar{v}) \in -\text{qri}\, K' \setminus \ell(K')$. Then $v - \bar{v} \in -(\text{qri}\, K' \setminus \ell(K')) - K \subseteq -\text{qri}\, K' \setminus \ell(K') - K' = -\text{qri}\, K' \setminus \ell(K')$. This yields that $v - \bar{v} \notin \ell(K')$, contradicting the fact that $(M - \bar{v}) \cap (-K') \subseteq \ell(K')$. □

The last proper minimality notion we introduce extends for not necessarily pointed ordering cones the classical one based on linear scalarization, that identifies minimal elements of M as solutions of scalar optimization problems.

Definition 3.7. An element $\bar{v} \in M$ is said to be a *properly minimal element of M in the sense of linear scalarization (regarding the partial ordering induced by K)* if there exists a $v^* \in \{x^* \in K^* : \langle x^*, x\rangle > 0\ \forall x \in K \setminus \ell(K)\}$ such that $\langle v^*, \bar{v}\rangle \leq \langle v^*, v\rangle$ for all $v \in M$. The set of properly minimal elements of M in the sense of linear scalarization is denoted by $\text{PMin}_{LS}(M, K)$.

The properly minimal elements of M in the sense of linear scalarization are also minimal, as the next result shows.

Remark 3.15. When $\bar{v} \in \text{PMin}_{LS}(M, K)$, there exists a $v^* \in \{x^* \in K^* : \langle x^*, x\rangle > 0\ \forall x \in K \setminus \ell(K)\}$ such that $\langle v^*, \bar{v}\rangle \leq \langle v^*, v\rangle$ for all $v \in M$. Assuming that $\bar{v} \notin \text{Min}(M, K)$, there must exist a $v \in M$ such that $v - \bar{v} \in -K \setminus \ell(K)$. Then $\langle v^*, \bar{v} - v\rangle > 0$, but this contradicts the previous inequality, consequently $\text{PMin}_{LS}(M, K) \subseteq \text{Min}(M, K)$.

Proposition 3.8. *It holds* $\text{PMin}_{LS}(M, K) \subseteq \text{PMin}_{LS}(M + K, K) = \text{PMin}_{LS}(M + \ell(K), K)$.

Proof. When $\bar{v} \in \text{PMin}_{LS}(M, K)$, there exists a $v^* \in \{x^* \in K^* : \langle x^*, x\rangle > 0\ \forall x \in K \setminus \ell(K)\}$ such that $\langle v^*, \bar{v}\rangle \leq \langle v^*, v\rangle$ for all $v \in M$. If $k \in K$ one has $\langle v^*, k\rangle \geq 0$, consequently $\langle v^*, \bar{v}\rangle \leq \langle v^*, v + k\rangle$ for all $v \in M$ and all $k \in K$, hence $\bar{v} \in \text{PMin}_{LS}(M + K, K)$.

If $\bar{v} \in \text{PMin}_{LS}(M + \ell(K), K)$, there exists a $v^* \in \{x^* \in K^* : \langle x^*, x\rangle > 0\ \forall x \in K \setminus \ell(K)\}$ such that $\langle v^*, \bar{v}\rangle \leq \langle v^*, v\rangle$ for all $v \in M + \ell(K)$. If $k \in K \setminus \ell(K)$ one has $\langle v^*, k\rangle > 0$, consequently $\langle v^*, \bar{v}\rangle \leq \langle v^*, v + q\rangle$ for all $v \in M$ and all $q \in K$, hence $\bar{v} \in \text{PMin}_{LS}(M + K, K)$.

When $\bar{v} \in \text{PMin}_{LS}(M + K, K)$, there exists a $v^* \in \{x^* \in K^* : \langle x^*, x\rangle > 0\ \forall x \in K \setminus \ell(K)\}$ such that $\langle v^*, \bar{v}\rangle \leq \langle v^*, v\rangle$ for all $v \in M + K$. As $\bar{v} \in M + K$ there exist $\bar{m} \in M$ and $\bar{k} \in K$ such that $\bar{v} = \bar{m} + \bar{k}$. Taking $v = \bar{m} \in M + K$ in the previous inequality one obtains $\langle v^*, \bar{k}\rangle \leq 0$, that cannot take place if $\bar{k} \notin \ell(K)$. Thus $\bar{v} \in M + \ell(K)$. Since $v^* \in \{x^* \in K^* : \langle x^*, x\rangle > 0\ \forall x \in K \setminus \ell(K)\}$ fulfills also $\langle v^*, \bar{v}\rangle \leq \langle v^*, v\rangle$ for all $v \in M + \ell(K)$, it follows that $\bar{v} \in \text{PMin}_{LS}(M + \ell(K), K)$, too. □

When K is pointed, the proper minimality in the sense of linear scalarization is the most restrictive among the considered proper minimality notions. Let us show that in the more general framework used here the properly minimal elements in the sense of linear scalarization are properly minimal elements in the senses of Hurwicz and Henig and Lampe, respectively, too.

Proposition 3.9. *It holds* $\mathrm{PMin}_{LS}(M, K) \subseteq \mathrm{PMin}_{Hu}(M, K)$.

Proof. If $K = \{0\}$ there is nothing to prove. Assume that $K \neq \{0\}$ and take $\bar{v} \in \mathrm{PMin}_{LS}(M, K)$. Then $\bar{v} \in M$ and there exists a $v^* \in \{x^* \in K^* : \langle x^*, x\rangle > 0\ \forall x \in K \setminus \ell(K)\}$ such that $\langle v^*, v\rangle \geq 0$ for all $v \in M - \bar{v}$. This yields that for all $v \in (M - \bar{v}) \cup K$ it holds $\langle v^*, v\rangle \geq 0$ and, consequently, $\langle v^*, v\rangle \geq 0$ for all $v \in \mathrm{cl}(\mathrm{coneco}(M - \bar{v}) \cup K)$.

Assuming that there exists a $k \in K \setminus \ell(K)$ such that $-k \in \mathrm{cl}(\mathrm{coneco}(M - \bar{v}) \cup K)$, we get $0 < \langle v^*, k\rangle \leq 0$, that is a contradiction, consequently $\bar{v} \in \mathrm{PMin}_{Hu}(M, K)$. □

Proposition 3.10. *It holds* $\mathrm{PMin}_{LS}(M, K) \subseteq \mathrm{PMin}_{HL}(M, K)$.

Proof. When $\bar{v} \in \mathrm{PMin}_{LS}(M, K)$ it holds $\bar{v} \in M$ and there exists a $v^* \in \{x^* \in K^* : \langle x^*, x\rangle > 0\ \forall x \in K \setminus \ell(K)\}$ such that $\langle v^*, v\rangle \geq 0$ for all $v \in M - \bar{v}$. Take $K' := \{v \in V : \langle v^*, v\rangle > 0\} \cup \{0\}$. Obviously, K' is a nontrivial pointed convex cone and $K \setminus \ell(K) \subseteq \mathrm{int}\, K' = \{v \in V : \langle v^*, v\rangle > 0\} = \mathrm{qri}\, K'$.

Assuming that there exists a $v \in M$ such that $v - \bar{v} \in -K' \setminus \ell(K') = -K' \setminus \{0\}$, it follows that $\langle v^*, \bar{v} - v\rangle > 0$. This contradicts the fact that $\langle v^*, \bar{v}\rangle \leq \langle v^*, v\rangle$ for all $v \in M$, consequently $\bar{v} \in \mathrm{PMin}_{HL}(M, K)$. □

Summarizing the results proven above, we obtain the following general inclusion scheme for the proper minimal sets introduced in this section.

Theorem 3.1. *It holds*

$$\mathrm{PMin}_{LS}(M, K) \subseteq \begin{matrix} \mathrm{PMin}_{Hu}(M, K) \\ \mathrm{PMin}_{HL}(M, K) \end{matrix} \subseteq \mathrm{PMin}_{Be}(M, K)$$

$$\subseteq \begin{matrix} \mathrm{PMin}_{GB}(M, K) \\ \mathrm{PMin}_{Bo}(M, K) \end{matrix} \subseteq \mathrm{Min}(M, K). \tag{3.2.1}$$

Moreover, if $M + K$ *is convex, then* (3.2.1) *turns into*

$$\mathrm{PMin}_{LS}(M, K) \subseteq \mathrm{PMin}_{HL}(M, K) \subseteq \mathrm{PMin}_{Hu}(M, K) = \mathrm{PMin}_{Be}(M, K)$$
$$= \mathrm{PMin}_{Bo}(M, K) \subseteq \mathrm{PMin}_{GB}(M, K) \subseteq \mathrm{Min}(M, K). \tag{3.2.2}$$

Remark 3.16. In Definition 3.11 we define a proper minimality notion with respect to general increasing scalarization functions, which contains as a special case the one in the sense of linear scalarization. It is not known yet whether this proper minimality notion can be brought into the inclusion schemes presented in Theorem 3.1 (see also Remark 4.27).

Remark 3.17. If K is pointed, because $\ell(K) = \{0\}$ the definitions and results provided in this section collapse, taking also $\operatorname{int} K'$ instead of $\operatorname{qri} K'$ in Definition 3.6, into their correspondents from [48, Subsection 2.4.3]. Note also that other results regarding inclusions that exist between the sets of different proper minimality notions attached to a set regarding the partial ordering induced by a pointed convex cone can be found in [113, 154].

3.3 Pointed Ordering Cones

Consider further that $\operatorname{qi} K^* \neq \emptyset$. Then $K^{*0} \neq \emptyset$ and, consequently, the cone K is pointed. The definition of a minimal element of a set becomes then the following one.

Definition 3.8. An element $\bar{v} \in M$ is said to be a *minimal element of M (regarding the partial ordering induced by K)* if there is no $v \in M$ satisfying $v \leq_K \bar{v}$. The set of all minimal elements of M is denoted by $\operatorname{Min}(M, K)$ and it is called the *minimal set of M (regarding the partial ordering induced by K)*.

The corresponding maximality notion is defined analogously. The elements of the set $\operatorname{Max}(M, K) := \operatorname{Min}(M, -K) = -\operatorname{Min}(-M, K)$ are called *maximal elements of M* (regarding the partial ordering induced by K). Analogously, one can consider corresponding maximality notions for all the other minimality concepts introduced within this section. As seen in the previous section, the minimality notion introduced above can be refined by defining the so-called properly minimal elements of a set with respect to the ordering cone.

3.3.1 Properly Minimal Elements

As mentioned in Remark 3.17, the definitions of properly minimal elements considered in Sect. 3.2 are simpler when K is pointed. For reader's convenience we formulate in the present framework the two of them with which we work within this subsection.

Definition 3.9. An element $\bar{v} \in M$ is said to be a *properly minimal element of M in the global sense of Borwein (regarding the partial ordering induced by K)* if $\operatorname{cl}(\operatorname{cone}(M - \bar{v})) \cap (-K) = \{0\}$. The set of all properly minimal elements of M in the global sense of Borwein is denoted by $\operatorname{PMin}_{GB}(M, K)$.

Definition 3.10. An element $\bar{v} \in M$ is said to be a *properly minimal element of M in the sense of linear scalarization (regarding the partial ordering induced by K)* if there exists a $v^* \in K^{*0}$ such that $\langle v^*, \bar{v}\rangle \leq \langle v^*, v\rangle$ for all $v \in M$. The set of properly minimal elements of M in the sense of linear scalarization (regarding the partial ordering induced by K) is denoted by $\operatorname{PMin}_{LS}(M, K)$.

Remark 3.18. Specializing Theorem 3.1 for the situation when the ordering cone K is pointed one rediscovers [48, Proposition 2.4.16]. The question when do the inclusions in (3.2.2) turn into equalities is not only of theoretical importance. For instance, having all but the last of them fulfilled as equalities would ensure characterizations through linear scalarization for all the other properly minimal elements of M. Several results in this direction were already obtained in the literature and were summarised in [48, Subsection 2.4.3 and Subsection 2.4.4]. Note also that results on characterizations by means of scalarizations of the properly minimal elements in the sense of Henig and Lampe can be found in [100, 175].

In the following we present other weak conditions which guarantee the coincidence of some of the minimality notions considered above. To this end, we begin with a separation statement.

Theorem 3.2. *Let S and T be closed convex cones in V fulfilling $T^* + S^{*0} = \operatorname{qi}(T^* + S^*)$. Then $S \cap T = \{0\}$ if and only if there exists a $v^* \in V^* \setminus \{0\}$ such that $\langle v^*, s\rangle < 0 \leq \langle v^*, t\rangle$ for all $s \in S \setminus \{0\}$ and all $t \in T$.*

Proof. "$\Leftarrow$" If $v^* \in V^* \setminus \{0\}$ fulfills $\langle v^*, s\rangle < 0 \leq \langle v^*, t\rangle$ for all $s \in S \setminus \{0\}$ and all $t \in T$ and we assume that there exists an $x \in S \cap T \setminus \{0\}$, then it follows that $\langle v^*, x\rangle < 0 \leq \langle v^*, x\rangle$, which cannot happen. The necessity is thus proven and one can easily observe that it is valid in the most general setting, i.e. when S and T are simple nonempty subsets of V.

"$\Rightarrow$" Assume that no $v^* \in V^* \setminus \{0\}$ satisfies $\langle v^*, s\rangle < 0 \leq \langle v^*, t\rangle$ for all $s \in S \setminus \{0\}$ and all $t \in T$. Consequently, $-S^{*0} \cap T^* = \emptyset$. Thus, by the hypothesis, $0 \notin \operatorname{qi}(T^* + S^*)$. As $0 \in T^* + S^*$, Lemma 1.2 yields that there exists an $x \in V \setminus \{0\}$ satisfying $\langle 0, x\rangle \leq \langle t^* + s^*, x\rangle$ for all $s^* \in S^*$ and all $t^* \in T^*$. Then $-\langle s^*, x\rangle \leq \langle t^*, x\rangle$ for all $s^* \in S^*$ and all $t^* \in T^*$, consequently, $x \in S^{**} \cap T^{**} = S \cap T$. As $x \neq 0$, the hypothesis $S \cap T = \{0\}$ is contradicted, thus there exists a $v^* \in V^* \setminus \{0\}$ such that $\langle v^*, s\rangle < 0 \leq \langle v^*, t\rangle$ for all $s \in S \setminus \{0\}$ and all $t \in T$. □

Remark 3.19. If A and B are convex subsets of V, recall that $\operatorname{int}(A + B) = B + \operatorname{int} A$ when $\operatorname{int} A \neq \emptyset$, and $\operatorname{core}(A + B) = B + \operatorname{core} A$ when $\operatorname{core} A \neq \emptyset$ (cf. [196, Theorem 2.1 and Theorem 2.2], see also [195]). Consequently, the assertion proven in Theorem 3.2 extends for cones whose duals have not necessarily a nonempty interior the separation statement for cones given in [16, Proposition 2] and [140, Theorem 3.22], that has as consequence, for instance, Lemma 1.3. A situation that stresses the applicability of our statement, while the other one fails is given below.

Example 3.2. Let $V = \ell^2(\mathbb{N})$, the real Hilbert space of the real sequences $(x_n)_{n\in\mathbb{N}}$ such that $\sum_{n=1}^{\infty} |x_n|^2 < +\infty$, equipped with the usual norm $\|\cdot\| : \ell^2 \to \mathbb{R}$, $\|x\| = \left(\sum_{n=1}^{\infty} |x_n|^2\right)^{1/2}$ for all $x = (x_n)_{n\in\mathbb{N}} \in \ell^2$. Take $T = \ell^2_+ = \{(x_n)_{n\in\mathbb{N}} \in \ell^2 : x_n \geq 0\ \forall n \in \mathbb{N}\}$, the positive cone of ℓ^2, and $S = -\ell^2_+$. Then $S \cap T = \{0\}$ and $T^* = -S^* = \ell^2_+$. It is known that $\operatorname{int} \ell^2_+ = \emptyset$, so [16, Proposition 2] and [140, Theorem 3.22] cannot be applied in this case, but $\operatorname{qi} \ell^2_+ = \{(x_n)_{n\in\mathbb{N}} \in \ell^2 : x_n > 0\ \forall n \in \mathbb{N}\}$. We also have $T^* + S^{*0} = \ell^2 = \operatorname{qi}(T^* + S^*)$. Theorem 3.2 yields then the existence of a $v^* \in \ell^2 \setminus \{0\}$ such that $\langle v^*, s\rangle < 0 \leq \langle v^*, t\rangle$ for all $s \in -\ell^2_+ \setminus \{0\}$ and all $t \in \ell^2_+$.

Using Theorem 3.2, one can show that under weak hypotheses all the classes of proper minimal elements introduced for M coincide. In order to give these results, let us introduce for the pair of sets (S, T), where $S, T \subseteq V$ such that S and $S + T$ are convex the following property

$$(QC) \qquad T + \operatorname{qi} S = \operatorname{qi}(S + T).$$

Remark 3.20. As mentioned in Remark 3.19, given the convex sets $A, B \subseteq V$, the pair (A, B) has automatically the property (QC) when $\operatorname{int} A$ or $\operatorname{core} A$ is nonempty. Moreover, in many situations where $\operatorname{qi} A \neq \emptyset$ unlike the interior or algebraic interior of A, it holds $B + \operatorname{qi} A = \operatorname{qi}(A + B)$, this equality being always satisfied when B contains a single element. Thus, the property (QC) seems quite natural. Note that it has been conjectured in [110] that when $A, B \subseteq V$ are convex sets with $\operatorname{qi} A \neq \emptyset$, it holds $B + \operatorname{qi} A = \operatorname{qi}(A + B)$, but in [223, Example 2] one can find a counterexample showing that this property is not true in general.

Theorem 3.3. *Let $\bar{v} \in M$ such that the pair $(K^*, (\operatorname{cone}(M - \bar{v}))^*)$ has the property (QC) and let M be convex and K be closed. Then $\bar{v} \in \operatorname{PMin}_{GB}(M, K)$ if and only if $\bar{v} \in \operatorname{PMin}_{LS}(M, K)$.*

Proof. The sufficiency follows via (3.2.1), even without assuming M convex and K closed, so we focus on the other implication. The convexity of M guarantees the same property for $\operatorname{cone}(M - \bar{v})$. As K is closed and the proper minimality in the global sense of Borwein of $\bar{v}$ means actually that $(\operatorname{cl}\operatorname{cone}(M - \bar{v})) \cap (-K) = \{0\}$, due to (QC) we can apply Theorem 3.2, obtaining the existence of a $\bar{v}^* \in V^* \setminus \{0\}$ for which

$$\langle v^*, -k\rangle < 0 \leq \langle v^*, v\rangle \ \forall k \in K \setminus \{0\} \ \forall v \in \operatorname{cl}\operatorname{cone}(M - \bar{v}). \tag{3.3.3}$$

Then the first inequality in (3.3.3) yields $v^* \in K^{*0}$, while from the second one can deduce that for all $m \in M$ one gets $0 \leq \langle v^*, m - \bar{v}\rangle$. Consequently, there exists a $v^* \in K^{*0}$ for which $\langle v^*, \bar{v}\rangle \leq \langle v^*, m\rangle$ for all $m \in M$, i.e. $\bar{v} \in \operatorname{PMin}_{LS}(M, K)$. □

Remark 3.21. One has $(\operatorname{cl}\operatorname{cone}(M - \bar{v}))^* = (\operatorname{cone}(M - \bar{v}))^* = \{v^* \in V^* : \langle v^*, t(v - \bar{v})\rangle \geq 0 \ \forall v \in M \ \forall t \geq 0\} = \{v^* \in V^* : \langle v^*, v\rangle \geq \langle v^*, \bar{v}\rangle \ \forall v \in M\}$.

Using Theorem 3.3 and the fact that under its hypotheses $M + K$ is convex, (3.2.2) yields the following consequence.

Corollary 3.1. *Let $\bar{v} \in M$ such that the pair $(K^*, (\operatorname{cone}(M - \bar{v}))^*)$ has the property (QC) and let M be convex and K be closed. Then $\bar{v} \in \operatorname{PMin}_{GB}(M, K)$ if and only if it is properly minimal to M in any other sense mentioned above.*

Remark 3.22. A result similar to the one displayed in Corollary 3.1 is available in [48, Theorem 2.4.27], where under the hypotheses $\operatorname{int} K^* \neq \emptyset$, K closed and $M + K$

convex it is shown that an element is properly minimal to M in the sense of Borwein if and only if it is properly minimal to M in any other sense except for the global one of Borwein. But that statement is not applicable, for instance, for the finance model with m investors trading securities and having identical expectations on the security payoffs which is modeled in [1] as a vector optimization problem whose objective function maps from a portfolio vector space to an ordered payoff vector space that is $L^p(\Omega, \Sigma, P)$, with $p \geq 1$, where (Ω, Σ, P) is an underlying probability space. Note that also in Theorem 3.3 we can weaken the assumption of convexity of M by taking only the cone $\operatorname{cone}(M - \bar{v})$ convex. This is enough to guarantee the proper minimality of a properly minimal element of M in the global sense of Borwein in all other senses, except for the one due to Borwein, which can be caught under the additional hypothesis $M + K$ convex (cf. [48, Proposition 2.4.16]).

Now let us give a condition that guarantees the proper minimality in the sense of linear scalarization of a minimal element of M.

Theorem 3.4. *Let* $\bar{v} \in \operatorname{Min}(M, K)$ *such that* $\operatorname{cone}(M - \bar{v})$ *is convex and closed and the pair* $(K^*, (\operatorname{cone}(M - \bar{v}))^*)$ *has the property (QC), and let K be closed. Then* $\bar{v} \in \operatorname{PMin}_{LS}(M, K)$.

Proof. One can write $\bar{v} \in \operatorname{Min}(M, K)$ equivalently as $(M - \bar{v}) \cap (-K) = \{0\}$. This yields $\operatorname{cone}(M - \bar{v}) \cap (-K) = \{0\}$. Applying Theorem 3.2, one gets the existence of a $\bar{v}^* \in V^* \setminus \{0\}$ for which

$$\langle v^*, -k \rangle < 0 \leq \langle v^*, v \rangle \ \forall k \in K \setminus \{0\} \ \forall v \in \operatorname{cone}(M - \bar{v}).$$

Then $v^* \in K^{*0}$ and for all $m \in M$ it holds $0 \leq \langle v^*, m - \bar{v} \rangle$, i.e. $\bar{v} \in \operatorname{PMin}_{LS}(M, K)$. □

A consequence of this statement follows. Note that different to Corollary 3.1, here it is not necessary to guarantee the convexity of $M + K$ in order to include the properly minimal elements of M in the sense of Borwein.

Corollary 3.2. *Let* $\bar{v} \in \operatorname{Min}(M, K)$ *such that* $\operatorname{cone}(M - \bar{v})$ *is convex and closed and the pair* $(K^*, (\operatorname{cone}(M - \bar{v}))^*)$ *has the property (QC), and let K be closed. Then* $\bar{v}$ *is properly minimal to M in every mentioned sense.*

Remark 3.23. The hypotheses of Theorem 3.4 and Corollary 3.2 are sufficient to guarantee that all the inclusions in (3.2.1) turn into equalities even without assuming the convexity of $M + K$, a condition than ensures (3.2.2). Note also that asking $\operatorname{cone}(M - \bar{v})$ to be closed together with the minimality of $\bar{v}$ in M guarantee that $\bar{v} \in \operatorname{PMin}_{GB}(M, K)$.

Remark 3.24. Like in the proof of Theorem 6.1, one can show, by employing Lemma 1.3, that if $V = \mathbb{R}^k$, K is closed and M is polyhedral, any minimal element of M is also properly minimal to M in the sense of linear scalarization. Different other separation statements may be employed for the same purpose, too. However, some of them may not deliver valuable hypotheses under which a minimal element

$v \in M$ of M is also properly minimal to M in the sense of linear scalarization. For instance, [48, Corollary 2.1.6] asks in order to separate $M - v$ from $-K$ that $v \notin \text{cl}(M + K)$, but when $v \in M$, then $v \in M + K$, too, so the mentioned condition cannot be fulfilled.

Next, we prove by means of scalar Lagrange duality another statement that guarantees the proper minimality in the sense of linear scalarization of a minimal element of M, without assuming the property (QC) fulfilled anymore. The technique used to prove the following statement was inspired by the similar investigations performed in [131] in the linear case.

Theorem 3.5. *Let M be convex and $\bar{v} \in \text{Min}(M, K)$ for which one of the following conditions*

$$(RCS_2)\Big|\ V \textit{ is a Fréchet space, } M \textit{ and } K \textit{ are closed, and } \bar{v} \in \text{sqri}(M + K),$$

$$(RCS_3)\Big|\ \dim(M + K - \bar{v}) < +\infty \textit{ and } \bar{v} \in \text{ri}(M + K),$$

and, respectively,

$$(RCS_4)\left|\begin{array}{l} M \textit{ and } K \textit{ are closed and there exists a } \lambda \in K^{*0} \textit{such that} \\ \bigcup\limits_{\eta \in K^* + \lambda} \big(\text{epi}\, \sigma_M + (\eta, \langle \eta, \bar{v}\rangle)\big) \textit{ is closed in the topology } \omega(V^*, V) \times \mathscr{R}, \end{array}\right.$$

is fulfilled. Then $\bar{v} \in \text{PMin}_{LS}(M, K)$.

Proof. Let $\lambda \in K^{*0}$ (the one that exists by (RCS_4) if this condition is satisfied). Consider the scalar optimization problem

$$(SP) \qquad \inf_{\substack{v \in M, \\ v - \bar{v} \in -K}} \langle \lambda, v - \bar{v}\rangle.$$

The satisfaction of any of the considered regularity conditions yields via Remark 2.10 that there is strong duality for (SP) and its Lagrange dual problem

$$(LD) \qquad \sup_{\mu \in K^*} \inf_{v \in M} \big[\langle \lambda, v - \bar{v}\rangle + \langle \mu, v - \bar{v}\rangle\big].$$

Consequently, there exists a $\bar{\mu} \in K^*$ such that

$$\inf_{\substack{v \in M, \\ v - \bar{v} \in -K}} \langle \lambda, v - \bar{v}\rangle = \inf_{v \in M} \langle \lambda + \bar{\mu}, v - \bar{v}\rangle.$$

Let $\bar{\eta} := \lambda + \bar{\mu} \in K^{*0}$.

As $\bar{v} \in \text{Min}(M, K)$, $\bar{v}$ is the only feasible point to the problem (SP) and, consequently, its optimal objective value is 0. Thus, $0 = \inf_{v \in M} \langle \bar{\eta}, v - \bar{v}\rangle$, which can be equivalently rewritten as $\langle \bar{\eta}, \bar{v}\rangle \leq \langle \bar{\eta}, v\rangle$ for all $v \in M$, i.e. $\bar{v} \in \text{PMin}_{LS}(M, K)$. □

A consequence of Theorem 3.5 follows.

Corollary 3.3. *Let M be convex and $\bar{v} \in \mathrm{Min}(M, K)$ for which one of the conditions (RCS_i), $i \in \{2, 3, 4\}$ is fulfilled. Then $\bar{v}$ is properly minimal to M in every mentioned sense.*

Remark 3.25. Taking a closer look at the proof of Theorem 3.5 one can notice that the considered regularity conditions are employed for guaranteeing the strong duality for (SP) and (LD). The classical Slater constraint qualification cannot be considered in Theorem 3.5 (and consequently neither in Corollary 3.3) even if $\operatorname{int} K \neq \emptyset$ since when $\bar{v} \in \mathrm{Min}(M, K)$ it follows that $\bar{v}$ is the only feasible element to the problem (SP) and consequently there exists no $v \in M$ for which $v - \bar{v} \in \operatorname{int} K$.

Moreover, the regularity condition (RCS_2) yields, by employing the definition of the strong quasi-relative interior, that $\operatorname{cone}(M + K - \bar{v})$ is closed. Using also the minimality of $\bar{v}$ in M, that yields $\bar{v} \in \mathrm{Min}(M + K, K)$, it follows that the only element $\operatorname{cone}(M + K - \bar{v})$ and $-K$ have in common is 0, consequently $\bar{v} \in \mathrm{PMin}_{GB}(M + K, K)$. By [48, Proposition 2.4.9] follows then $\bar{v} \in \mathrm{PMin}_{GB}(M, K)$.

Last but not least, it may be not so obvious how (RC_4^L) turns into (RCS_4) for (SP) in Theorem 3.5. The objective function of (SP) is lower semicontinuous because it is an affine function, while its constraint function is K-epi-closed since it is a linear function and K is closed. Considering for each $\mu \in K^*$ the function $\varphi_\mu : V \to \overline{\mathbb{R}}$ defined by $\varphi_\mu(v) = \langle \lambda, v - \bar{v} \rangle + \langle \mu, v - \bar{v} \rangle + \delta_M(v)$, one gets $\varphi_\mu(v) = \langle \lambda + \mu, v - \bar{v} \rangle + \delta_M(v)$. Then

$$\varphi_\mu(v^*) = \sup_{v \in M} \{ \langle v^*, v \rangle - \langle \lambda + \mu, v - \bar{v} \rangle \}$$
$$= \langle \lambda + \mu, \bar{v} \rangle + \sup_{v \in M} \langle v^* - \lambda - \mu, v \rangle = \langle \lambda + \mu, \bar{v} \rangle + \sigma_M(v^* - \lambda - \mu).$$

One has $(v^*, r) \in \operatorname{epi} \varphi^*$ if and only if $\sigma_M(v^* - \lambda - \mu) \leq r - \langle \lambda + \mu, \bar{v} \rangle$, which happens when $(v^*, r) \in \operatorname{epi} \sigma_M + (\lambda + \mu, \langle \lambda + \mu, \bar{v} \rangle)$.

After investigating some of the classical minimality concepts, let us introduce other proper minimality notions that can be characterized via scalarization, generalizing thus the proper minimality in the sense of linear scalarization. As the situation depicted in Example 3.3 shows, even in simple cases wrong choices of the linear scalarization function can lead to unconstrained scalar optimization problems which give no insights on the vector optimization problems they were derived from. Moreover, for various other purposes, from delivering optimality conditions for nonsmooth optimization problems (like in [81]) to investigating error bounds for convex inequality systems (cf. [22]), the linear scalarization did not bring any valuable results and other scalarization functions had to be employed.

Example 3.3 (cf. [15]). Let $V = \mathbb{R}^2$, $K = \mathbb{R}^2_+$, $M = \{(x_1^2 - x_2, x_2)^\top : (x_1, x_2) \in \mathbb{R}^2\}$. For all $\lambda = (\lambda_1, \lambda_2)^\top \in \operatorname{int} \mathbb{R}^2_+ = (\mathbb{R}^2_+)^{*0}$ with $\lambda_1 \neq \lambda_2$, one has $\inf\{\lambda^\top v : v \in M\} = -\infty$, therefore no properly minimal element in the sense of linear scalarization of the set M can be identified by using the corresponding scalarization functions. Only for $\bar{\lambda} = (\bar{\lambda}_1, \bar{\lambda}_2)^\top \in \operatorname{int} \mathbb{R}^2_+$ with $\bar{\lambda}_1 = \bar{\lambda}_2$ one gets that $\bar{\lambda}^\top \bar{v} \leq \bar{\lambda}^\top v$ for all $v \in M$, whenever $\bar{v} \in \{0\} \times \mathbb{R}$.

Motivated by this, let us consider an arbitrary nonempty set of scalarization functions defined on V and taking values in $\overline{\mathbb{R}}$ denoted by $\mathscr{M}$ and using them we introduce a more general proper minimality concept.

Definition 3.11. An element $\bar{v} \in M$ is said to be an *$\mathscr{M}$-properly minimal element of M* if there exists an $s \in \mathscr{M}$ such that $s(\bar{v}) \leq s(v)$ for all $v \in M$. The set of all $\mathscr{M}$-properly minimal elements of M is denoted by $\operatorname{PMin}_{\mathscr{M}}(M, K)$.

Remark 3.26. We will not deal with $\mathscr{M}$-properly minimal elements of M as introduced above. This very general definition will serve as basis for approaching vector optimization problems via duality with respect to $\mathscr{M}$-properly efficient solutions with the scalarization functions equipped with valuable properties, like convexity and K-monotonicity.

3.3.2 Weakly Minimal Elements

There are minimality notions weaker than the classical minimality, too, among which we recall the following one. In order to consider it we take within this subsection $\operatorname{qi} K \neq \emptyset$.

Definition 3.12. An element $\bar{v} \in M$ is said to be a *weakly minimal element of M (regarding the partial ordering induced by K)* if $(\bar{v} - \operatorname{qi} K) \cap M = \emptyset$. The set of all weakly minimal elements of M is denoted by $\operatorname{WMin}(M, K)$.

Remark 3.27. In the literature the weak minimality is usually considered for the cases $\operatorname{int} K \neq \emptyset$ or $\operatorname{core} K \neq \emptyset$, when the nonempty set coincides with $\operatorname{qi} K$. But there are vector optimization problems, as mentioned for instance in [1, 80, 100] where the image space is partially ordered by convex cones with empty interiors. Motivated by them, weakly minimal elements defined by means of the quasi interior were considered in works like [105, 106, 115, 198], being called *quasi-weakly minimal elements* or qi-*minimal elements*. However, in [110] and here we opted to name them simply weakly minimal elements because of two reasons. Firstly, in the literature there were already considered several types of *quasi-minimal elements* that do not coincide with the ones introduced in Definition 3.12 (see, for instance, [158]). On the other hand, if the pair (K, M) has the property *(QC)* (which automatically happens if $\operatorname{int} K \neq \emptyset$ or $\operatorname{core} K \neq \emptyset$) our results extend their counterparts known in the literature for the classical weakly minimal elements, as we shall see later in this chapter and also in Sect. 4.3. Note also that in the literature one can find generalized weakly minimal elements defined by means of the (quasi-)relative interior of the ordering cone (see, for instance, [6, 7, 115, 129, 219]) and we will deal with such elements in Sect. 3.3.3.

Analogously, $\bar{v} \in M$ is a *weakly maximal element* of M (regarding the partial ordering induced by K) if $(\bar{v} + \text{qi}\, K) \cap M = \emptyset$. We denote by $\text{WMax}(M, K)$ the set of all weakly maximal elements of the set M (regarding the partial ordering induced by K). One can prove that $\text{WMin}(M, -K) = -\text{WMin}(-M, K) = \text{WMax}(M, K)$.

Remark 3.28. The relation $(\bar{v} - \text{qi}\, K) \cap M = \emptyset$ in Definition 3.12 can be equivalently rewritten as $(M - \bar{v}) \cap (-\text{qi}\, K) = \emptyset$. Whenever the cone K is nontrivial we notice that if we consider as ordering cone in V the cone $\hat{K} = \text{qi}\, K \cup \{0\}$, then $\bar{v} \in \text{WMin}(M, K)$ if and only if $(\bar{v} - \hat{K}) \cap M = \{\bar{v}\}$, which actually means $\bar{v} \in \text{Min}(M, \hat{K})$.

Remark 3.29. If K is not dense in V, any minimal element of M is also weakly minimal to M since $(\bar{v} - K) \cap M = \{\bar{v}\}$ implies via Proposition 1.1(a) that $(\bar{v} - \text{qi}\, K) \cap M = \emptyset$. If $K = V$ then $\text{WMin}(M, K) = \emptyset$, as it happens also in the classical theory of weakly minimal elements where $\text{int}\, K \neq \emptyset$.

Proposition 3.11. *It holds*

$$\text{WMin}(M + K, K) \cap M \subseteq \text{WMin}(M, K) \subseteq \text{WMin}(M + K, K).$$

Proof. If $v \in \text{WMin}(M + K, K) \cap M$, $(v - \text{qi}\, K) \cap (M + K) = \emptyset$. As $(v - \text{qi}\, K) \cap M \subseteq (v - \text{qi}\, K) \cap (M + K)$ it follows $(v - \text{qi}\, K) \cap M = \emptyset$, too, therefore $v \in \text{WMin}(M, K)$.

Consider now an element $\bar{v} \in \text{WMin}(M, K)$ assumed not to be a weakly minimal element of the set $M + K$. Then there is an element $v \in (\bar{v} - \text{qi}\, K) \cap (M + K) \neq \emptyset$ and there is an $u \in M$ with $\bar{v} - v \in \text{qi}\, K$ and $v - u \in K$. Consequently, by using Proposition 1.1(c) we obtain that $\bar{v} - u \in \text{qi}\, K + K = \text{qi}\, K$, or alternatively $u \in (\bar{v} - \text{qi}\, K) \cap M$. Hence, $\bar{v}$ is not a weakly minimal element of the set M, and the conclusion follows by contradiction. □

Next we formulate some necessary and sufficient characterizations via linear scalarizations of the weakly minimal elements of the set M with respect to K.

Theorem 3.6. *If $M + K$ is convex, the pair (K, M) has the property (QC) and $\bar{v} \in \text{WMin}(M, K)$ then there exists a $v^* \in K^* \setminus \{0\}$ such that $\langle v^*, \bar{v} \rangle \leq \langle v^*, v \rangle$, for all $v \in M$.*

Proof. As $\bar{v} \in \text{WMin}(M, K)$, using also the property (QC), one gets $\bar{v} \notin M + \text{qi}\, K = \text{qi}(M + K)$. As $\bar{v} \in M + K$, we can apply Lemma 1.2, which guarantees the existence of a $v^* \in V^* \setminus \{0\}$ such that

$$\langle v^*, v + k \rangle \geq \langle v^*, \bar{v} \rangle \text{ for all } v \in M \text{ and } k \in K. \tag{3.3.4}$$

As K is a cone, (3.3.4) yields $v^* \in K^*$. Taking $k = 0$, (3.3.4) implies $\langle v^*, \bar{v} \rangle \leq \langle v^*, v \rangle$ for all $v \in M$. □

Proposition 3.12. *Let a function $f : V \to \overline{\mathbb{R}}$ which is strictly K-increasing on M. If there is an element $\bar{v} \in M$ fulfilling $f(\bar{v}) \leq f(v)$ for all $v \in M$, then $\bar{v} \in \mathrm{WMin}(M, K)$.*

Proof. If $\bar{v} \notin \mathrm{WMin}(M, K)$, then there exists a $v \in (\bar{v} - \mathrm{qi}\, K) \cap M$. This implies $f(v) < f(\bar{v})$, which contradicts the assumption. □

If the cone K is moreover closed, Proposition 3.12 and Example 1.2 yield the following statement.

Theorem 3.7. *If K is closed and there exist $v^* \in K^* \setminus \{0\}$ and $\bar{v} \in M$ such that for all $v \in M$ it holds $\langle v^*, \bar{v} \rangle \leq \langle v^*, v \rangle$, then $\bar{v} \in \mathrm{WMin}(M, K)$.*

From Theorems 3.6 and 3.7 we obtain the following equivalent characterization via linear scalarization of the weakly minimal elements of M with respect to K.

Theorem 3.8. *If $\bar{v} \in M$, $M + K$ is convex, the pair (K, M) has the property (QC) and K is closed, then $\bar{v} \in \mathrm{WMin}(M, K)$ if and only if there exists a $v^* \in K^* \setminus \{0\}$ satisfying $\langle v^*, \bar{v} \rangle \leq \langle v^*, v \rangle$ for all $v \in M$.*

Remark 3.30. In Theorem 3.8 we have extended the characterization via scalarization of the weakly minimal elements of a given set (regarding the partial ordering induced by a convex cone) for the case when only the quasi interior of the ordering cone is nonempty. In the literature such characterizations were previously known to hold only if the interior or the algebraic interior of the cone were nonempty, when the pair (K, M) has automatically the property (QC) when M is convex (cf. [195, 196], see also Remark 3.19), see for instance [48, Corollary 2.4.26]. Other attempts in order to extend these results were made by relaxing the convexity hypotheses, see for instance [91, Theorem 5.3].

3.3.3 Relatively Weakly Minimal Elements

There are minimality notions even weaker than the weak minimality that we treated in Sect. 3.3.2, like the following one, which requires only the nonemptiness of the quasi-relative interior of the ordering cone. Therefore, we take within this subsection $\mathrm{qri}\, K \neq \emptyset$.

Definition 3.13. An element $\bar{v} \in M$ is said to be a *relatively minimal element of M (regarding the partial ordering induced by K)* if $(\bar{v} - \mathrm{qri}\, K) \cap M = \emptyset$. The set of all relatively minimal elements of M is denoted by $\mathrm{RMin}(M, K)$.

Remark 3.31. As mentioned in Remark 3.27, one can find in the literature vector optimization problems where the image space is partially ordered by convex cones with empty interiors and in such situations generalizations of the classical weakly minimal elements of the corresponding image sets can be employed. Generalized weakly minimal elements defined by means of the quasi-relative interior of the ordering cone as we have done in Definition 3.13 can be found, for instance, in works like [6, 7, 115, 129, 219], where they are called *quasi relative minimal* or *weakly minimal*, respectively. However, as the quasi-relative interior collapses into the relative interior in finitely dimensional spaces and in order to avoid not necessary complications, we opted for the name given in Definition 3.13.

Analogously, $\bar{v} \in M$ is a *relatively maximal element* of M (regarding the partial ordering induced by K) if $(\bar{v} + \operatorname{qri} K) \cap M = \emptyset$. We denote by $\operatorname{RMax}(M, K)$ the set of all relatively maximal elements of the set M (regarding the partial ordering induced by K). One can prove that $\operatorname{RMin}(M, -K) = -\operatorname{RMin}(-M, K) = \operatorname{RMax}(M, K)$.

Remark 3.32. The relation $(\bar{v} - \operatorname{qri} K) \cap M = \emptyset$ in Definition 3.13 can be equivalently rewritten as $(M - \bar{v}) \cap (-\operatorname{qri} K) = \emptyset$. If K is nontrivial, considering also the cone $\tilde{K} = \operatorname{qri} K \cup \{0\}$ one has $\bar{v} \in \operatorname{RMin}(M, K)$ if and only if $\bar{v} \in \operatorname{Min}(M, \tilde{K})$.

Employing Proposition 1.1(b), one can easily prove the following statement.

Proposition 3.13. *If* $\operatorname{cl} K$ *is pointed, then* $\operatorname{Min}(M, K) \subseteq \operatorname{RMin}(M, K)$, *while when* $K = V$ *it holds* $\operatorname{RMin}(M, K) = \emptyset$.

We give now a statement similar to Proposition 3.11, but for the relatively minimal elements of M and $M + K$.

Proposition 3.14. *It holds* $\operatorname{RMin}(M + K, K) \cap M \subseteq \operatorname{RMin}(M, K) \subseteq \operatorname{RMin}(M + K, K)$.

Proof. If $\bar{v} \in \operatorname{RMin}(M + K, K) \cap M$, then $(\bar{v} - \operatorname{qri} K) \cap (M + K) = \emptyset$. As $(\bar{v} - \operatorname{qri} K) \cap M \subseteq (\bar{v} - \operatorname{qri} K) \cap (M + K)$, it follows that $(\bar{v} - \operatorname{qri} K) \cap M = \emptyset$, too, therefore $\bar{v} \in \operatorname{RMin}(M, K)$.

If $\bar{v} \in \operatorname{RMin}(M, K) \setminus \operatorname{RMin}(M + K, K)$, then there exist $v \in (\bar{v} - \operatorname{qri} K) \cap (M + K) \neq \emptyset$ and $u \in M$ such that $v - u \in K$. Then $\bar{v} - v \in \operatorname{qri} K$, thus Proposition 1.1(c) yields $\bar{v} - u \in \operatorname{qri} K + K = \operatorname{qri} K$, consequently $u \in (\bar{v} - \operatorname{qri} K) \cap M$. Hence $\bar{v} \notin \operatorname{RMin}(M, K)$ and the conclusion follows by contradiction. □

Analogously to the property (QC) considered in Sects. 3.3.1 and 3.3.2 for the case $\operatorname{qi} K \neq \emptyset$, in order to characterize via scalarization the relatively minimal elements of M with respect to K we introduce for the pair of sets (S, T) where $S, T \subseteq V$ such that S and $S + T$ are convex the following property

$$(QR) \qquad T + \operatorname{qri} S = \operatorname{qri}(S + T).$$

Remark 3.20 applies if $\operatorname{qi} K \neq \emptyset$. Regarding the property (QR), one can easily find pairs of sets that have it and some that do not, as seen below. Note also that Proposition 1.2(b) yields the fulfillment of (QR) for any pair $(S, \{v\})$, where $v \in V$.

Example 3.4. If $V = \mathbb{R}^2$, the pair consisting of $K = \{0\} \times \mathbb{R}_+$ and $T = (0, 1) \times \{0\}$ has the property (QR), while the same K and $C = [0, 1] \times \{0\}$ do not. Note that $\operatorname{qi} K = \operatorname{int} K = \emptyset$ in this case.

Next we formulate some necessary and sufficient characterizations via linear scalarization of the relatively minimal elements of M with respect to K.

Theorem 3.9. *If $M + K$ is convex, the pair (K, M) has the property (QR) and $\bar{v} \in \mathrm{RMin}(M, K)$, then there exists a $v^* \in K^* \setminus \{0\}$ such that $\langle v^*, \bar{v}\rangle \leq \langle v^*, v\rangle$ for all $v \in M$.*

Proof. As $\bar{v} \in \mathrm{RMin}(M, K)$ one gets that $\bar{v} \notin v + \mathrm{qri}\, K$ for all $v \in M$. Thus, $\bar{v} \notin M + \mathrm{qri}\, K$, consequently, via (QR), $\bar{v} \notin \mathrm{qri}(M + K)$. As $\bar{v} \in M + K$, Lemma 1.1 grants the existence of a $v^* \in V^* \setminus \{0\}$ such that

$$\langle v^*, \bar{v}\rangle \leq \langle v^*, v + k\rangle \ \forall v \in M \ \forall k \in K,$$

that, because K is a cone, yields that $v^* \in K^* \setminus \{0\}$ and, moreover, for $k = 0$, $\langle v^*, \bar{v}\rangle \leq \langle v^*, v\rangle$ for all $u \in M$. □

In case $V = \mathbb{R}^n$ the hypotheses of Theorem 3.9 can be simplified as follows.

Theorem 3.10. *If $M, K \subseteq \mathbb{R}^n$ with K a nontrivial convex cone and $M + K$ convex and $\bar{v} \in \mathrm{RMin}(M, K)$, then there exists a $v^* \in K^* \setminus \{0\}$ such that $\langle v^*, \bar{v}\rangle \leq \langle v^*, v\rangle$, for all $v \in M$.*

Proof. As $\bar{v} \in \mathrm{RMin}(M, K)$, Proposition 3.14 yields $\bar{v} \in \mathrm{RMin}(M + K, K)$, i.e. $(\bar{v} - \mathrm{ri}\, K) \cap (M + K) = \emptyset$. Then, $\mathrm{ri}(\bar{v} - K) \cap \mathrm{ri}(M + K) = \emptyset$ and the classical separation theorem due to Rockafellar (cf. [178, Theorem 11.3]) yields the existence of a $v^* \in V^* \setminus \{0\}$ such that

$$\langle v^*, \bar{v} - p\rangle \leq \langle v^*, v + k\rangle \ \forall v \in M \ \forall k, p \in K.$$

As K is a cone, this inequality implies $v^* \in K^* \setminus \{0\}$ and, when $p = k = 0$, also $\langle v^*, \bar{v}\rangle \leq \langle v^*, v\rangle$ for all $v \in M$. □

Theorem 3.11. *Let a function $f : V \to \overline{\mathbb{R}}$ that is relatively strictly K-increasing on M. If there is an element $\bar{v} \in M$ fulfilling $f(\bar{v}) \leq f(v)$ for all $v \in M$, then $\bar{v} \in \mathrm{RMin}(M, K)$.*

Proof. If $\bar{v} \notin \mathrm{RMin}(M, K)$, then there exists a $v \in (\bar{v} - \mathrm{qri}\, K) \cap M$. This implies $f(v) < f(\bar{v})$, which contradicts the assumption. □

Using Theorem 3.11 and Example 1.2 one can prove the next statement.

Theorem 3.12. *If $K^0 \neq \emptyset$ and there exist $v^* \in K^* \setminus \{0\}$ and $\bar{v} \in M$ such that for all $v \in M$ it holds $\langle v^*, \bar{v}\rangle \leq \langle v^*, v\rangle$, then $\bar{v} \in \mathrm{RMin}(M, K)$.*

Combining Theorems 3.9 and 3.12 we obtain an equivalent characterization via linear scalarization for the relatively minimal elements of M with respect to K.

Theorem 3.13. *Let $\bar{v} \in M$, $K^0 \neq \emptyset$, $M + K$ be convex and assume that the pair (K, M) has the property (QR). Then $\bar{v} \in \mathrm{RMin}(M, K)$ if and only if there exists a $v^* \in K^* \setminus \{0\}$ satisfying $\langle v^*, \bar{v}\rangle \leq \langle v^*, v\rangle$ for all $v \in M$.*

Remark 3.33. If K is closed and $\text{qi}\, K \neq \emptyset$, the investigations from this subsection collapse into the corresponding ones from Sect. 3.3.2.

Remark 3.34. One can define elements similar to the relatively minimal ones by means of other generalized relative interiors of the ordering cone, like the strong quasi-relative interior. But, as noted in Example 3.4, not even all the pairs consisting in a convex cone and a convex subset of $\mathbb{R}^n$ have the property (QR) (for the relative interior in this case), therefore it should be imposed for the ordering cone in each case and the corresponding results would be only special cases of the ones proven in this subsection. However, it makes sense to consider relatively minimal type elements defined by means of other generalized relative interiors of the ordering cone for other purposes, as done for instance in [6, 7].

Chapter 4
Vector Duality via Scalarization for Vector Optimization Problems

4.1 Historical Overview and Motivation

As seen in Chap. 3, one can consider different minimality notions for sets and these can be employed in different situations in order to serve various purposes, among which one can find the theory of vector optimization. Solving a vector optimization problem amounts of determining its feasible elements where the value of its objective function satisfies the desired minimality property within the image set of the feasible set through the objective function, the so-called *image set* of the vector optimization problem. These elements are usually called *efficient solutions*.

One can find in the literature different types of efficient solutions, called after the minimality notions the values taken there by the objective functions satisfy in the image sets of the vector optimization problems in discussion. The best known and most used of these efficiency notions is the classical *(Pareto-)efficiency*, but the corresponding *efficiency set*, i.e. set of efficient solutions, is sometimes not so easy to determine. In order to avoid such a situation, various other types of efficient solutions were proposed in the literature, following the existing minimality concepts, or, in some cases, generating them, as several minimality notions were actually derived from efficiency notions proposed for vector optimization problems. Some of the efficiency notions are weaker than the classical one, as it is the case for the *weak efficiency* and its generalizations or for the *ε-efficiency* or *approximate efficiency*, while most of them are more restrictive than it, like the *ideal efficiency*, *strong efficiency* or *proper efficiency*.

In the literature one can find several weak efficiency notions, that are defined with respect to a partial ordering induced by some (generalized) interior of the initial ordering cone, and more proper efficiency ones, derived from the corresponding proper minimality concepts, the most important of which were considered in Chap. 3. With respect to the considered efficient solutions one can attach to the investigated vector optimization problem one or more vector dual problems. In the literature one can find several ways to do this, for instance by means of

S.-M. Grad, *Vector Optimization and Monotone Operators via Convex Duality*, Vector Optimization, DOI 10.1007/978-3-319-08900-3_4

set-valued techniques, by using geometric properties or sometimes quite strange constructions or, last but not least, via conjugate functions. Besides the large number of papers dedicated to or touching the subject, one can find sections or chapters dedicated vector duality in many respected books dedicated to vector optimization such as [140, 155, 193], and the most complete overview on vector duality methods and techniques at the moment is the monograph [48]. There one can find quite complete investigations concerning vector duality of conjugate type via scalarization, of Wolfe and Mond-Weir type and of set-valued type. For vector duality investigations regarding constrained vector optimization problems we refer also to [54, 55, 101, 138, 140, 170, 200], while in contributions like [26, 58, 59, 102] duality statements for unconstrained optimization problems are given.

The arguably most usual way to approach a vector optimization problem is by scalarizing it, i.e. by attaching to it a scalar optimization problem or a family of such problems, whose solutions are hoped to deliver the desired efficient solutions of the original vector optimization problem or at least valuable insights regarding them. Among the scalarization methods the linear one is by far the most known and widely used, but in different circumstances (as, for instance, in the ones presented in [22, 81]) it may deliver no valuable results regarding the vector optimization problems and, on the other hand, an unfortunate choice of its scalarization parameters can lead to unbounded scalar optimization problems (see, for instance [15,92]). That is why several other functions with similar properties, i.e. strongly or strictly cone-monotone increasing, were employed for the same purpose in works like [31, 37, 92, 93, 97, 98, 102, 139, 140, 166], giving birth to new proper and weak minimality notions, from which corresponding proper and weak efficiency notions were derived.

These facts motivated us to introduce in Sect. 3.3.1 a general scheme for defining properly minimal elements with respect to different scalarization functions, that will be used in this chapter for introducing different types of properly efficient solutions with respect to which vector dual problems are assigned to the original vector optimization problems.

We propose in Sect. 4.2 a scheme for defining properly efficient elements of a general vector optimization problem via a very general set of scalarization functions and with respect to them several vector dual problems are attached to the primal one, by employing the idea of construction considered in [58, 59, 140], following our research from [31, 37, 48]. Then we specialize in Sect. 4.3 the scalarization function to be one of the scalarization functions considered in the literature and, depending on its cone-monotonicity properties, the corresponding vector duals and vector duality statements are particularized, too. Finally, in Sect. 4.4 we specialize the primal problem to be constrained and unconstrained, respectively, and for each of them we obtain from the general case new vector duals with respect to properly efficient solutions in the sense of different scalarizations. The corresponding necessary and sufficient optimality conditions are delivered in each case, too.

4.2 Vector Duality via a General Scalarization for General Vector Optimization Problems

After introducing in Chap. 3 several minimality notions for sets, we consider within this section and the following ones the counterparts of some of them for vector optimization problems, generically called *efficient solutions*. Then we assign vector dual problems to these with respect to some types of properly efficient and weakly efficient solutions.

Consider again a Hausdorff locally convex space V partially ordered by the nontrivial pointed convex cone $K \subseteq V$. Let X and Y be two Hausdorff locally convex vector spaces. The vector optimization problem we shall first work with is the general vector-minimization problem

$$(PVG) \qquad \underset{x \in X}{\operatorname{Min}} F(x),$$

where $F : X \to V^\bullet$ is a proper vector function. The solution concepts we consider for (PVG) follow from the ones introduced for sets in Chap. 3.

Definition 4.1. An element $\bar{x} \in \operatorname{dom} F$ is said to be an *efficient solution* to the vector optimization problem (PVG) if $F(\bar{x}) \in \operatorname{Min}(F(\operatorname{dom} F), K)$. The set of all efficient solutions to (PVG) is called the *efficiency set of* (PVG), being denoted by $\mathscr{E}(PVG)$.

Definition 4.2. When $\operatorname{qi} K \neq \emptyset$, an element $\bar{x} \in \operatorname{dom} F$ is said to be a *weakly efficient solution* to the vector optimization problem (PVG) if $F(\bar{x}) \in \operatorname{WMin}(F(\operatorname{dom} F), K)$. The set of all weakly efficient solutions to (PVG) is called the *weak efficiency set of* (PVG), being denoted by $\mathscr{W}\mathscr{E}(PVG)$.

In order to introduce also properly efficient solutions to (PVG), consider a set of *scalarization functions*

$$\mathscr{S} \subseteq \Big\{ s : V^\bullet \to \overline{\mathbb{R}} : F(\operatorname{dom} F) + K \subseteq \operatorname{dom} s \text{ and } s \text{ is proper, convex and}$$
$$\text{strongly } K\text{-increasing on } F(\operatorname{dom} F) + K \text{ and } s(\infty_K) = +\infty \Big\}.$$

Definition 4.3. An element $\bar{x} \in X$ is said to be an *$\mathscr{S}$-properly efficient solution* to the vector optimization problem (PVG) if $F(\bar{x}) \in \operatorname{PMin}_{\mathscr{S}}(F(\operatorname{dom} F), K)$. The set of all $\mathscr{S}$-properly efficient solutions to (PVG) is said to be the *$\mathscr{S}$-proper efficiency set of* (PVG), being denoted by $\mathscr{P}\mathscr{E}_{\mathscr{S}}(PVG)$.

Remark 4.1. Every $\mathscr{S}$-properly efficient solution to (PVG) belongs to $\operatorname{dom} F$ and it is also an efficient solution to the same vector optimization problem. If $\operatorname{qi} K \neq \emptyset$ and $\operatorname{cl} K \neq V$, each efficient solution to (PVG) is a weakly efficient one, too.

In order to deal with (PVG) via duality, consider now the proper vector perturbation function $\Phi : X \times Y \to V^\bullet$ which fulfills $\Phi(x, 0) = F(x)$ for all $x \in X$. Like in the scalar case, Y is the perturbation space and its elements the perturbation variables. Then $0 \in \Pr_Y(\operatorname{dom} \Phi)$. The primal vector optimization problem introduced above can be reformulated as

$$(PVG) \qquad \operatorname*{Min}_{x \in X} \Phi(x, 0).$$

When $s \in \mathscr{S}$, the scalarized optimization problem attached to (PVG) is

$$(PG^s) \qquad \inf_{x \in X} s \circ \Phi(x, 0),$$

to which one can assign the following conjugate dual problems

$$(DG_1^s) \qquad \sup_{\substack{v^* \in K^*, \\ y^* \in Y^*}} \big\{ - s^*(v^*) - (v^*\Phi)^*(0, y^*) \big\},$$

and, respectively,

$$(DG_2^s) \qquad \sup_{y^* \in Y^*} \big\{ - (s \circ \Phi)^*(0, y^*) \big\}.$$

Using them, we attach to (PVG) the following dual vector problems with respect to $\mathscr{S}$-properly efficient solutions

$$(DVG_1^{\mathscr{S}}) \qquad \operatorname*{Max}_{(s, v^*, y^*, v) \in \mathscr{B}_1^{G\mathscr{S}}} h_1^{G\mathscr{S}}(s, v^*, y^*, v),$$

where

$$\mathscr{B}_1^{G\mathscr{S}} = \Big\{ (s, v^*, y^*, v) \in \mathscr{S} \times K^* \times Y^* \times V : s(v) \leq -s^*(v^*) - (v^*\Phi)^*(0, y^*) \Big\}$$

and

$$h_1^{G\mathscr{S}}(s, v^*, y^*, v) = v,$$

and, respectively,

$$(DVG_2^{\mathscr{S}}) \qquad \operatorname*{Max}_{(s, y^*, v) \in \mathscr{B}_2^{G\mathscr{S}}} h_2^{G\mathscr{S}}(s, y^*, v),$$

where

$$\mathscr{B}_2^{G\mathscr{S}} = \Big\{ (s, y^*, v) \in \mathscr{S} \times Y^* \times V : s(v) \leq -(s \circ \Phi)^*(0, y^*) \Big\}$$

and

$$h_2^{G\mathscr{S}}(s, y^*, v) = v.$$

Remark 4.2. When s is also lower semicontinuous and Φ is K-convex and K-lower semicontinuous, one can replace $(s \circ \Phi)^*(0, y^*)$ in the definition of $\mathscr{B}_2^{G\mathscr{S}}$ with $\overline{\inf_{v^* \in K^*}\{-s^*(v^*) - (v^*\Phi)^*(\cdot,\cdot)\}}(0, y^*)$ (cf. [49, Theorem 3.1]). Similar observations can be formulated later for the other vector duals of this type that appear in our presentation.

Without resorting to the vector perturbation function Φ one can also attach to (PVG) another vector dual, inspired by a vector dual considered for constrained vector problems in [54, 55] and [48, Subsection 4.3.2], namely

$$(DVG_3^{\mathscr{S}}) \qquad \operatorname*{Max}_{(s,v)\in\mathscr{B}_3^{G\mathscr{S}}} h_3^{G\mathscr{S}}(s, v),$$

where

$$\mathscr{B}_3^{G\mathscr{S}} = \left\{(s, v) \in \mathscr{S} \times V : s(v) \leq \inf_{x\in X} s(F(x))\right\}$$

and

$$h_3^{G\mathscr{S}}(s, v) = v.$$

Remark 4.3. It is a simple verification to show that in general it holds $(s \circ \Phi)^* \leq \inf_{v^* \in K^*}[s^*(v^*) + (v^*\Phi)^*(\cdot)]$, thus whenever $(s, v^*, y^*, v) \in \mathscr{B}_1^{G\mathscr{S}}$ we have $s(v) \leq -(s\circ\Phi)^*(0, y^*)$, which yields $(s, y^*, v) \in \mathscr{B}_2^{G\mathscr{S}}$. Consequently, $h_1^{G\mathscr{S}}(\mathscr{B}_1^{G\mathscr{S}}) \subseteq h_2^{G\mathscr{S}}(\mathscr{B}_2^{G\mathscr{S}})$. Sufficient conditions for having equality in this inclusion can be found in [48, Theorem 3.5.2]; we mention here only one, namely that, provided that Φ is proper and K-convex, for each $s \in \mathscr{S}$ there exists a point $\bar{x} \in \operatorname{dom} F$ such that s is continuous at $F(\bar{x})$. But the scalarization functions most used in the literature (see [31, 37] or Sect. 4.3) are also continuous at least over the sets on which they are strongly or strictly K-increasing and also the regularity condition considered earlier for the strong duality statement regarding $(DVG_1^{\mathscr{S}})$ and (PVG) covers the hypotheses mentioned above. Since in this case the first two vector duals introduced above have the same images of their feasible sets through their objective vector functions, it is not necessary to particularize both of them when dealing with concrete scalarization functions from the literature in Sect. 4.3. However, we treat them in the general case since the scalarization functions need not be continuous. In Example 4.1 one can find a possible scalarization function of this kind, while in Example 4.2 we deliver another one, in a situation where $(DVG_1^{\mathscr{S}})$ and $(DVG_2^{\mathscr{S}})$ do not coincide.

Example 4.1. Let $X = \mathbb{R}$, $V = \mathbb{R}^2$, $K = \mathbb{R}^2_+$,

$$\mathscr{S} = \left\{s : (\mathbb{R}^2)^\bullet \to \overline{\mathbb{R}} : s(x,y) = x^2 + y^2 + \delta_{\mathbb{R}^2_+}(x,y)\right\}$$

and $F : \mathbb{R} \to (\mathbb{R}^2)^\bullet$, $F(x) = (x,0)^\top$ if $x \in (0,1)$ and $F(x) = \infty_{\mathbb{R}^2_+}$ otherwise. One can easily see that $F(\operatorname{dom} F) = (0,1) \times \{0\}$ and $\operatorname{dom} s = \mathbb{R}^2_+$, thus the condition $F(\operatorname{dom} F) + \mathbb{R}^2_+ \subseteq \operatorname{dom} s$ is satisfied. Moreover, the scalarization function s is proper, convex and strongly $\mathbb{R}^2_+$-increasing on $\mathbb{R}^2_+$, but it is not continuous on $F(\operatorname{dom} F)$.

Example 4.2. Let $X = \mathbb{R}$, $Y = \mathbb{R}$, $V = \mathbb{R}^2$, $K = \{(0,0)^\top\}$,

$$\mathscr{S} = \left\{ s : (\mathbb{R}^2)^\bullet \to \overline{\mathbb{R}},\ s(x,y) = \begin{cases} x \ln x - x + \frac{y^2}{2}, & \text{if } x > 0, y \leq 0,\\ \frac{y^2}{2}, & \text{if } x = 0, y \leq 0,\\ +\infty, & \text{otherwise} \end{cases} \right\}$$

and

$$\Phi : \mathbb{R}^2 \to (\mathbb{R}^2)^\bullet,\ \Phi(x,y) = \begin{cases} \begin{pmatrix} x \\ x \end{pmatrix}, & \text{if } x = y = 0 \text{ or } (x \neq 0 \text{ and } y \neq 0),\\ \infty_{\{(0,0)^\top\}}, & \text{otherwise.} \end{cases}$$

Then the scalarization function s is proper, convex and strongly K-increasing on its domain and $(\Phi(\cdot,0))(\operatorname{dom}\Phi(\cdot,0)) + K = \{(0,0)\} \subseteq [0,+\infty)\times(-\infty,0] = \operatorname{dom} s$. Regarding the conjugates that appear in the formulation of $(DVG_1^{\mathscr{S}})$ and $(DVG_2^{\mathscr{S}})$, we have

$$s^*(v_1^*, v_2^*) = \begin{cases} e^{v_1^*} + \frac{(v_2^*)^2}{2}, & \text{if } v_1^* \in \mathbb{R}, v_2^* \leq 0,\\ e^{v_1^*}, & \text{if } v_1^* \in \mathbb{R}, v_2^* > 0, \end{cases}$$

$$\left((v_1^*, v_2^*)^\top \Phi\right)^*(0, y^*) = \begin{cases} 0, & \text{if } v_1^* + v_2^* = y^* = 0,\\ +\infty, & \text{otherwise,} \end{cases}$$

and $(s \circ \Phi)^*(0,y^*) = 0$ for all $y^* \in \mathbb{R}$. It is straightforward to see that $s(0,0) = 0 = -(s \circ \Phi)^*(0,y^*)$ for all $y^* \in \mathbb{R}$, thus $(0,0) \in h_2^{G\mathscr{S}}(\mathscr{B}_2^{G\mathscr{S}})$. On the other hand, $s^*(v_1^*, v_2^*) > 0$ for all $v_1^*, v_2^* \in \mathbb{R}$, thus $-s^*(v_1^*, v_2^*) - \left((v_1^*, v_2^*)^\top \Phi\right)^*(0,y^*) < 0$ whenever $v_1^*, v_2^*, y^* \in \mathbb{R}$. As $s(0,0) = 0$, it is obvious that $(0,0) \notin h_1^{G\mathscr{S}}(\mathscr{B}_1^{G\mathscr{S}})$. Consequently, $(DVG_1^{\mathscr{S}})$ and $(DVG_2^{\mathscr{S}})$ do not coincide in this situation.

Remark 4.4. Note also that we have $\inf_{x\in X} s(F(x)) \geq -(s \circ \Phi)^*(0,y^*)$ for all $(s, y^*, v) \in \mathscr{B}_2^{G\mathscr{S}}$, thus $h_2^{G\mathscr{S}}(\mathscr{B}_2^{G\mathscr{S}}) \subseteq h_3^{G\mathscr{S}}(\mathscr{B}_3^{G\mathscr{S}})$. To show that the opposite inclusion does not always hold, we consider the situation presented in Example 4.3. However, as it will be seen further in Theorem 4.14, for $(DVG_3^{\mathscr{S}})$ strong duality holds whenever (*PVG*) has an $\mathcal{S}$-properly efficient element, so we will not insist much on this vector dual.

Example 4.3. Let $X = \mathbb{R}^2$, $Y = \mathbb{R}$, $V = \mathbb{R}^2$, $K = \mathbb{R}^2_+$,

$$U = \left\{(x,y)^\top \in \mathbb{R}^2 : 0 \leq x \leq 2, \begin{array}{l} 3 \leq y \leq 4, \text{ if } x = 0, \\ 1 \leq y \leq 4, \text{ if } x \in (0,2] \end{array} \right\},$$

$$F : \mathbb{R}^2 \to (\mathbb{R}^2)^\bullet,\ F(x,y) = \begin{cases} \begin{pmatrix} y \\ y \end{pmatrix}, \text{ if } (x,y)^\top \in U, x \leq 0, \\ \infty_{\mathbb{R}^2_+}, \text{ otherwise,} \end{cases}$$

$$\Phi : \mathbb{R}^2 \times \mathbb{R} \to (\mathbb{R}^2)^\bullet,\ \Phi(x,y,z) = \begin{cases} \begin{pmatrix} y \\ y \end{pmatrix}, \text{ if } (x,y)^\top \in U, x - z \leq 0, \\ \infty_{\mathbb{R}^2_+}, \text{ otherwise,} \end{cases}$$

and (see also Sect. 4.3)

$$\mathcal{S} = \left\{s : (\mathbb{R}^2)^\bullet \to \overline{\mathbb{R}},\ s(x,y) = ax + by\ :\ (a,b) \in \operatorname{int} \mathbb{R}^2_+,\ s(\infty_{\mathbb{R}^2_+}) = +\infty\right\}.$$

Note first that $F(x,y) = (y,y)^\top$ if $x = 0$ and $3 \leq y \leq 4$, while otherwise $F(x,y) = \infty_{\mathbb{R}^2_+}$. Whenever $s \in \mathcal{S}$ there exist $(v_1^*, v_2^*)^\top \in \operatorname{int} \mathbb{R}^2_+$ such that $s \circ F = (v_1^*, v_2^*)^\top F$. We have $(v_1^*, v_2^*)^\top F(x,y) \geq 3(v_1^* + v_2^*)$ for all $(x,y)^\top \in \mathbb{R}^2$. Consequently, $(3,3)^\top \in h_3^{G\mathcal{S}}(\mathcal{B}_3^{G\mathcal{S}})$. Assuming that $(3,3)^\top \in h_2^{G\mathcal{S}}(\mathcal{B}_2^{G\mathcal{S}})$, it follows that there exist $(v_1^*,\ v_2^*)^\top \in \operatorname{int} \mathbb{R}^2_+$ and $y^* \in \mathbb{R}$ such that $3(v_1^* + v_2^*) \leq -((v_1^*, v_2^*)^\top \Phi)^*(0, y^*)$, i.e. $((v_1^*, v_2^*)^\top\ \Phi)^*(0, y^*) \leq -3(v_1^* + v_2^*)$. For all $(v_1^*, v_2^*)^\top \in \operatorname{int} \mathbb{R}^2_+$ and all $y^* \in \mathbb{R}$ we have

$$\left((v_1^*, v_2^*)^\top \Phi\right)^*(0, y^*) = \sup_{\substack{(x,y)^\top \in U, \\ x - z \leq 0, z \in \mathbb{R}}} \{y^* z - y(v_1^* + v_2^*)\}$$

$$= \sup_{(x,y)^\top \in U} \left\{-y(v_1^* + v_2^*) + \sup_{z \geq x} y^* z\right\} = -(v_1^* + v_2^*) + \delta_{(-\infty,0]}(y^*) > -3(v_1^* + v_2^*).$$

Therefore, our assumption is false, i.e. $(3,3)^\top \notin h_2^{G\mathcal{S}}(\mathcal{B}_2^{G\mathcal{S}})$. Consequently, the dual problems $(DVG_2^{\mathcal{S}})$ and $(DVG_3^{\mathcal{S}})$ do not coincide in this situation.

For the dual vector-maximization problems introduced above we consider efficient solutions, defined below for $(DVG_1^{\mathcal{S}})$ and analogously for the others.

Definition 4.4. An element $(\bar{s}, \bar{v}^*, \bar{y}^*, \bar{v}) \in \mathcal{B}_1^{G\mathcal{S}}$ is said to be an *efficient solution* to the vector optimization problem $(DVG_1^{\mathcal{S}})$ if $(\bar{s}, \bar{v}^*, \bar{y}^*, \bar{v}) \in \operatorname{dom} h_1^{G\mathcal{S}}$ and $h_1^{G\mathcal{S}}(\bar{s}, \bar{v}^*, \bar{y}^*, \bar{v}) \in \operatorname{Max}(h_1^{G\mathcal{S}}(\operatorname{dom} h_1^{G\mathcal{S}}), K)$. The set of all efficient solutions to $(DVG_1^{\mathcal{S}})$ is called the *efficiency set of* $(DVG_1^{\mathcal{S}})$, being denoted by $\mathcal{E}(DVG_1^{\mathcal{S}})$.

Remark 4.5. Replacing the inequalities from the feasible sets of the vector duals to (*PVG*) by the corresponding equalities we obtain other vector duals to (*PVG*) which have smaller feasible sets. All the investigations done within this section can be considered for those vector duals, too. However, we will not consider further these vector dual problems since one can easily show that $(s, v^*, y^*, v) \in \mathscr{E}(DVG_1^{\mathscr{S}})$ yields $s(v) = -s^*(v^*) - (v^*\Phi)^*(0, y^*)$, $(s, y^*, v) \in \mathscr{E}(DVG_2^{\mathscr{S}})$ implies $s(v) = -(s \circ \Phi)^*(0, y^*)$, while when $(s, v) \in \mathscr{E}(DVG_3^{\mathscr{S}})$, one gets $s(v) = \inf_{x \in X} s(F(x))$. Of course, this observation can be extended for all the special instances of these three vector duals considered later in this chapter.

Let us now show that for the just introduced vector dual problems to (*PVG*) there is weak duality.

Theorem 4.1. *There are no $x \in X$ and $(s, v) \in \mathscr{B}_3^{G\mathscr{S}}$ such that $F(x) \leq_K h_3^{G\mathscr{S}}(s, v)$.*

Proof. Assume to the contrary that there exist $x \in X$ and $(s, v) \in \mathscr{B}_3^{G\mathscr{S}}$ fulfilling $F(x) \leq_K h_3^{G\mathscr{S}}(s, v)$. Then $x \in \operatorname{dom} F$ and it follows $s(F(x)) < s(v)$ since $s \in \mathscr{S}$. But from the way the feasible set of the vector dual is defined, we get $s(v) \leq \inf_{z \in X} s(F(z))$ and combining these two inequalities we reach a contradiction. □

The weak duality statements for the other two vector duals can be obtained as consequences of Theorem 4.1, having in mind the inclusions from Remarks 4.3 and 4.4.

Theorem 4.2. *There are no $x \in X$ and $(s, v^*, y^*, v) \in \mathscr{B}_1^{G\mathscr{S}}$ such that $F(x) \leq_K h_1^{G\mathscr{S}}(s, v^*, y^*, v)$.*

Theorem 4.3. *There are no $x \in X$ and $(s, y^*, v) \in \mathscr{B}_2^{G\mathscr{S}}$ such that $F(x) \leq_K h_2^{G\mathscr{S}}(s, y^*, v)$.*

Next we turn our attention to strong duality for the vector duals introduced in this paper. Due to the way it is constructed, for $(DVG_3^{\mathscr{S}})$ strong duality follows at once, without any additional assumption.

Theorem 4.4. *If $\bar{x} \in \mathscr{PE}_{\mathscr{S}}(PVG)$, there exists an $\bar{s} \in \mathscr{S}$ such that $(\bar{s}, F(\bar{x})) \in \mathscr{E}(DVG_3^{\mathscr{S}})$ and $F(\bar{x}) = h_3^{G\mathscr{S}}(\bar{s}, F(\bar{x}))$.*

Proof. As $\bar{x} \in X$ is an $\mathscr{S}$-properly efficient solution to (*PVG*), $F(\bar{x}) \in V$ and there exists a function $\bar{s} \in \mathscr{S}$ such that $\bar{s}(F(\bar{x})) \leq \bar{s}(F(x))$ for all $x \in X$. Thus $\bar{s}(F(\bar{x})) \leq \inf_{x \in X} \bar{s}(F(x))$. Consequently, $(\bar{s}, F(\bar{x})) \in \mathscr{B}_3^{G\mathscr{S}}$ and $F(\bar{x}) = h_3^{G\mathscr{S}}(\bar{s}, \bar{F}(\bar{x}))$. The efficiency of $(\bar{s}, F(\bar{x}))$ to $(DVG_3^{\mathscr{S}})$ follows immediately via Theorem 4.1. □

To obtain strong duality for the other two vector duals we assigned to (*PVG*) we need some additional hypotheses. Thus, we take the function Φ to be K-convex and we impose the fulfillment of a suitable regularity condition. One can introduce different regularity conditions inspired from (RC_i^G), $i \in \{1, 2, 3, 4\}$ (see for instance [21, 45, 48]), but, in order to avoid unnecessary complications given the complexity of the considered problem, we consider here only a classical one involving continuity, namely

$$(RCV^{\mathscr{S}}) \left| \begin{array}{l} \forall s \in \mathscr{S} \exists x' \in X \text{ such that } (x',0) \in \operatorname{dom} \Phi, \Phi(x',\cdot) \text{ is} \\ \text{continuous at } 0 \text{ and s is continuous at } \Phi(x',0), \end{array} \right.$$

that can be weakened if $\operatorname{int} K \neq \emptyset$, as we will see later. This regularity condition guarantees, as can be seen in the proof of the next statement, on the one hand that there is strong duality for the scalarized problem attached to (*PVG*) and, on the other hand, that the conjugate functions of s and Φ can be separated. The strong duality statements for the mentioned two vector duals to (*PVG*) follow.

Theorem 4.5. *If Φ is a K-convex vector function, the regularity condition $(RCV^{\mathscr{S}})$ is fulfilled and $\bar{x} \in \mathscr{PE}_{\mathscr{S}}(PVG)$, there exist $\bar{s} \in \mathscr{S}$, $\bar{v}^* \in K^*$ and $\bar{y}^* \in Y^*$ such that $(\bar{s}, \bar{v}^*, \bar{y}^*, F(\bar{x})) \in \mathscr{E}(DVG_1^{\mathscr{S}})$, $(\bar{s}, \bar{y}^*, F(\bar{x})) \in \mathscr{E}(DVG_2^{\mathscr{S}})$ and $F(\bar{x}) = h_1^{G\mathscr{S}}(\bar{s}, \bar{v}^*, \bar{y}^*, F(\bar{x})) = h_2^{G\mathscr{S}}(\bar{s}, \bar{y}^*, F(\bar{x}))$.*

Proof. As $\bar{x} \in X$ is an $\mathscr{S}$-properly efficient solution to (*PVG*), $F(\bar{x}) \in V$ and there exists a scalarization function $\bar{s} \in \mathscr{S}$ such that $\bar{s}(F(\bar{x})) \leq \bar{s}(F(x))$ for all $x \in X$. Thus $\bar{s}(F(\bar{x})) = \min_{x \in X} \bar{s}(F(x))$.

Remark 2.5 yields then the existence of a $\bar{y}^* \in Y^*$ such that $\sup_{y^* \in Y^*}\{-(\bar{s} \circ \Phi)^*(0, y^*)\}$ is attained at $\bar{y}^*$ and $\bar{s}(F(\bar{x})) = -(\bar{s} \circ \Phi)^*(0, \bar{y}^*)$. Thus $(\bar{s}, \bar{y}^*, F(\bar{x})) \in \mathscr{B}_2^{G\mathscr{S}}$ and $F(\bar{x}) = h_2^{G\mathscr{S}}(\bar{s}, \bar{y}^*, F(\bar{x}))$.

On the other hand, the hypotheses yield (see Remark 4.3) also the existence of a $\bar{v}^* \in K^*$ such that $(\bar{s} \circ \Phi)^*(0, \bar{y}^*) = \bar{s}^*(\bar{v}^*) + (\bar{v}^* \Phi)^*(0, \bar{y}^*)$, thus $h_1^{G\mathscr{S}}(\bar{s}, \bar{v}^*, \bar{y}^*, F(\bar{x})) = F(\bar{x})$, too, and $(\bar{s}, \bar{v}^*, \bar{y}^*, F(\bar{x})) \in \mathscr{B}_1^{G\mathscr{S}}$. The efficiency of $(\bar{s}, \bar{v}^*, \bar{y}^*, F(\bar{x})) \in \mathscr{B}_1^{G\mathscr{S}}$ to $(DVG_1^{\mathscr{S}})$ follows immediately by Theorem 4.2, while the efficiency of $(\bar{s}, \bar{y}^*, F(\bar{x})) \in \mathscr{B}_2^{G\mathscr{S}}$ to $(DVG_2^{\mathscr{S}})$ is a consequence of Theorem 4.3. □

Moreover, one can formulate necessary and sufficient optimality conditions for (*PVG*) and its vector dual problems introduced above. We begin with $(DVG_3^{\mathscr{S}})$, since the statement regarding it does not require the fulfillment of a regularity condition.

Theorem 4.6. *(a) If $\bar{x} \in \mathscr{PE}_{\mathscr{S}}(PVG)$, there exists a pair $(\bar{s}, \bar{v}) \in \mathscr{E}(DVG_3^{\mathscr{S}})$ such that*

(i) $F(\bar{x}) = \bar{v}$;
(ii) $\bar{s}(\bar{v}) = \bar{s}(F(\bar{x})) = \min_{x \in X} \bar{s}(F(x))$.

(b) Assume that $\bar{x} \in X$ and $(\bar{s}, \bar{v}) \in \mathscr{S} \times V$ fulfill the relations (i)–(ii). Then $\bar{x} \in \mathscr{PE}_{\mathscr{S}}(PVG)$ and $(\bar{s}, \bar{v}) \in \mathscr{E}(DVG_3^{\mathscr{S}})$.

Proof. (a) The existence of a pair $(\bar{s}, \bar{v}) \in \mathscr{E}(DVG_3^{\mathscr{S}})$ and (i) follow directly from Theorem 4.4. The first equality in (ii) follows directly from (i), while the second one is a direct consequence of the fact that $\bar{x} \in X$ is an $\mathscr{S}$-properly efficient solution to (*PVG*) and $\bar{s}$ is the corresponding scalarization function.

(b) The second equality in (*ii*) yields the $\mathscr{S}$-properly efficiency of $\bar{x}$ to (*PVG*), while the first one implies $(\bar{s}, \bar{v}) \in \mathscr{B}_3^{G_\mathscr{S}}$. Having these, (*i*) and Theorem 4.1 guarantee the efficiency of $(\bar{s}, \bar{v}) \in \mathscr{B}_3^{G_\mathscr{S}}$ to $(DVG_3^{\mathscr{S}})$. □

Remark 4.6. The optimality conditions (*i*)–(*ii*) in Theorem 4.6 can be equivalently written as $F(\bar{x}) = \bar{v}$ and $0 \in \partial(\bar{s} \circ F)(\bar{x})$.

Theorem 4.7. *(a) When Φ is a K-convex vector function, the regularity condition $(RCV^{\mathscr{S}})$ is fulfilled and $\bar{x} \in \mathscr{PE}_{\mathscr{S}}(PVG)$, there exists $(\bar{s}, \bar{v}^*, \bar{y}^*, \bar{v}) \in \mathscr{E}(DVG_1^{\mathscr{S}})$ such that*

(i) $F(\bar{x}) = \bar{v}$;
(ii) $\bar{s}(\bar{v}) + \bar{s}^*(\bar{v}^*) = \langle \bar{v}^*, \bar{v} \rangle$;
(iii) $(\bar{v}^* F)(\bar{x}) + (\bar{v}^* \Phi)^*(0, \bar{y}^*) = 0$.

(b) Assume that $\bar{x} \in X$ and $(\bar{s}, \bar{v}^*, \bar{y}^*, \bar{v}) \in \mathscr{S} \times V^* \times Y^* \times V$ fulfill the relations (i)–(iii). Then $\bar{x} \in \mathscr{PE}_{\mathscr{S}}(PVG)$ and $(\bar{s}, \bar{v}^*, \bar{y}^*, \bar{v}) \in \mathscr{E}(DVG_1^{\mathscr{S}})$.

Proof. (a) The existence of an element $(\bar{s}, \bar{v}^*, \bar{y}^*, \bar{v}) \in \mathscr{E}(DVG_1^{\mathscr{S}})$ and (*i*) are guaranteed by Theorem 4.5. Employing Remark 4.5, one gets $\bar{s}(\bar{v}) = -\bar{s}^*(\bar{v}^*) - (\bar{v}^* \Phi)^*(0, \bar{y}^*)$. But the Young-Fenchel inequality yields $\bar{s}(\bar{v}) + \bar{s}^*(\bar{v}^*) \geq \langle \bar{v}^*, \bar{v} \rangle = (\bar{v}^* F)(\bar{x}) = (\bar{v}^* \Phi)(\bar{x}, 0)$ and $(\bar{v}^* \Phi)(\bar{x}, 0) + (\bar{v}^* \Phi)^*(0, \bar{y}^*) \geq 0$, so (*ii*) and (*iii*) are fulfilled.

(b) From (*ii*) it follows that $\bar{v}^* \in \operatorname{dom} \bar{s}^*$ and by Remark 1.6 it follows $\bar{v}^* \in K^*$. Summing up (*ii*) and (*iii*) one gets $\bar{s}(\bar{v}) + \bar{s}^*(\bar{v}^*) + (\bar{v}^* F)(\bar{x}) + (\bar{v}^* \Phi)^*(0, \bar{y}^*) = \langle \bar{v}^*, \bar{v} \rangle$, which, employing (*i*), turns into $-\bar{s}(F(\bar{x})) = \bar{s}^*(\bar{v}^*) + (\bar{v}^* \Phi)^*(0, \bar{y}^*)$, which yields $(\bar{s}, \bar{v}^*, \bar{y}^*, \bar{v}) \in \mathscr{B}_1^{G_\mathscr{S}}$. Using that $(\bar{s} \circ \Phi)^* \leq \bar{s}^*(v^*) + (v^* \Phi)^*(\cdot)$ for all $v^* \in K^*$, the previous equality yields $\bar{s}(F(\bar{x})) \leq -(\bar{s} \circ \Phi)^*(0, \bar{y}^*)$. But $-(\bar{s} \circ \Phi)^*(0, \bar{y}^*) \leq (\bar{s} \circ \Phi)(x, 0)$ for all $x \in X$, therefore $\bar{s}(F(\bar{x})) \leq (\bar{s} \circ \Phi)(x, 0)$ for all $x \in X$, i.e. $\bar{x} \in \mathscr{PE}_{\mathscr{S}}(PVG)$. The efficiency of $(\bar{s}, \bar{v}^*, \bar{y}^*, \bar{v}) \in \mathscr{B}_1^{G_\mathscr{S}}$ to $(DVG_1^{\mathscr{S}})$ follows immediately by (*i*) and Theorem 4.2. □

Remark 4.7. The optimality conditions (*i*)–(*iii*) in Theorem 4.7 can be equivalently written as $F(\bar{x}) = \bar{v}$, $\bar{v}^* \in \partial \bar{s}(F(\bar{x}))$ and $(0, \bar{y}^*) \in \partial(\bar{v}^* \Phi)(\bar{x}, 0)$.

Analogously, one can prove the corresponding optimality conditions statement for (*PVG*) and $(DVG_2^{\mathscr{S}})$.

Theorem 4.8. *(a) When Φ is a K-convex vector function, the regularity condition $(RCV^{\mathscr{S}})$ is fulfilled and $\bar{x} \in \mathscr{PE}_{\mathscr{S}}(PVG)$, there exists $(\bar{s}, \bar{y}^*, \bar{v}) \in \mathscr{E}(DVG_2^{\mathscr{S}})$ such that*

(i) $F(\bar{x}) = \bar{v}$;
(ii) $\bar{s}(\bar{v}) + (\bar{s} \circ \Phi)^*(0, \bar{y}^*) = 0$.

(b) *Assume that* $\bar{x} \in X$ *and* $(\bar{s}, \bar{y}^*, \bar{v}) \in \mathscr{S} \times Y^* \times V$ *fulfill the relations* (i)–(ii). *Then* $\bar{x} \in \mathscr{PE}_{\mathscr{S}}(PVG)$ *and* $(\bar{s}, \bar{y}^*, \bar{v}) \in \mathscr{E}(DVG_2^{\mathscr{S}})$.

Remark 4.8. As mentioned in the proof of Theorem 4.5, the hypotheses of Theorem 4.8(a) yield the existence of a $\bar{v}^* \in K^*$ such that $(\bar{s} \circ \Phi)^*(0, \bar{y}^*) = \bar{s}^*(\bar{v}^*) + (\bar{v}^*\Phi)^*(0, \bar{y}^*)$, i.e. one obtains actually the optimality conditions (i)–(iii) from Theorem 4.7. However, we choose to give Theorem 4.8 as done above because the optimality conditions formulated there are not only necessary but also sufficient and, on the other hand, the assertion (a) can be provided under weaker hypotheses that do not necessarily guarantee the existence of the mentioned $\bar{v}^* \in K^*$. Similar observations can be made regarding the optimality conditions of the other vector duals of this type considered within this chapter, too.

Remark 4.9. The optimality conditions (i)–(ii) in Theorem 4.8 can be equivalently written as $F(\bar{x}) = \bar{v}$ and $(0, \bar{y}^*) \in \partial(\bar{s} \circ \Phi)(\bar{x}, 0)$.

Often, when $\operatorname{qi} K \neq \emptyset$, the scalarization functions considered in the literature are not strongly K-increasing, but strictly K-increasing. Following ideas from [31, 37, 48], one can notice that such scalarization functions can be brought into the vector duality framework we treat here by employing the nontrivial pointed convex cone $\hat{K} = \operatorname{qi} K \cup \{0\}$, already mentioned in Remark 3.28. It can also be verified that every function which is strictly K-increasing on $F(\operatorname{dom} F) + K$ is also strongly $\hat{K}$-increasing on $F(\operatorname{dom} F) + K$. In the remaining part of the section let $\operatorname{qi} K \neq \emptyset$. In order to capture the mentioned strictly K-increasing scalarization functions within the duality framework dealt with in this section, consider another set of scalarization functions, namely

$$\mathscr{T} \subseteq \Big\{ s : V^\bullet \to \overline{\mathbb{R}} : F(\operatorname{dom} F) + K \subseteq \operatorname{dom} s \text{ and } s \text{ is proper, convex}$$
$$\text{and strictly } K\text{-increasing on } F(\operatorname{dom} F) + K,\ s(\infty_K) = +\infty \Big\}.$$

Definition 4.5. If $\operatorname{qi} K \neq \emptyset$, we say that an element $\bar{x} \in X$ is a *$\mathscr{T}$-properly efficient solution* to (PVG) if $F(\bar{x}) \in \operatorname{PMin}_{\mathscr{T}}(F(\operatorname{dom} F), K)$. The set of all $\mathscr{T}$-properly efficient solutions to (PVG) is said to be the *$\mathscr{T}$-proper efficiency set of* (PVG), being denoted by $\mathscr{PE}_{\mathscr{T}}(PVG)$.

Remark 4.10. The elements introduced in Definition 4.5 were considered so far in the literature (see [31,37,48]) only when $\operatorname{int} K \neq \emptyset$. But, as we have seen above (for instance in Remark 3.19), the framework can be extended to the case $\operatorname{qi} K \neq \emptyset$. One may argue that it would be more adequate to call the $\mathscr{T}$-properly efficient solutions to (PVG) actually $\mathscr{T}$-weakly efficient solutions, since they are defined only in case $\operatorname{qi} K \neq \emptyset$, like the classical weakly efficient solutions, and for the vector duals to (PVG) introduced with respect to them we consider weakly efficient solutions. However, we opted to stay consequent to the name used in Definition 3.11, from which Definition 4.5 originates.

With respect to the $\mathscr{T}$-properly efficient solutions of the primal problem (PVG) one can define three vector duals that are obtained analogously to $(DVG_i^{\mathscr{S}})$, $i = \{1,2,3\}$, namely

$$(DVG_1^{\mathscr{T}}) \qquad \underset{(s,v^*,y^*,v)\in\mathscr{B}_1^{G\mathscr{T}}}{\text{WMax}} \; h_1^{G\mathscr{T}}(s,v^*,y^*,v),$$

where

$$\mathscr{B}_1^{G\mathscr{T}} = \Big\{(s,v^*,y^*,v) \in \mathscr{T} \times K^* \times Y^* \times V : s(v) \leq -s^*(v^*) - (v^*\Phi)^*(0,y^*)\Big\}$$

and

$$h_1^{G\mathscr{T}}(s,v^*,y^*,v) = v,$$

$$(DVG_2^{\mathscr{T}}) \qquad \underset{(s,y^*,v)\in\mathscr{B}_2^{G\mathscr{T}}}{\text{WMax}} \; h_2^{G\mathscr{T}}(s,y^*,v),$$

where

$$\mathscr{B}_2^{G\mathscr{T}} = \Big\{(s,y^*,v) \in \mathscr{T} \times Y^* \times V : s(v) \leq -(s\circ\Phi)^*(0,y^*)\Big\}$$

and

$$h_2^{G\mathscr{T}}(s,y^*,v) = v,$$

and, respectively,

$$(DVG_3^{\mathscr{T}}) \qquad \underset{(s,v)\in\mathscr{B}_3^{G\mathscr{T}}}{\text{WMax}} \; h_3^{G\mathscr{T}}(s,v),$$

where

$$\mathscr{B}_3^{G\mathscr{T}} = \Big\{(s,v) \in \mathscr{T} \times V : s(v) \leq \inf_{x\in X} s(F(x))\Big\}$$

and

$$h_3^{G\mathscr{T}}(s,v) = v.$$

An observation similar to Remark 4.5 can be given for the vector duals to (PVG) with respect to $\mathscr{T}$-properly efficient solutions of the primal problem (PVG), too. Moreover, analogously to Remarks 4.3 and 4.4, one can easily show that

$$h_1^{G\mathscr{T}}(\mathscr{B}_1^{G\mathscr{T}}) \subseteq h_2^{G\mathscr{T}}(\mathscr{B}_2^{G\mathscr{T}}) \subseteq h_3^{G\mathscr{T}}(\mathscr{B}_3^{G\mathscr{T}}). \tag{4.2.1}$$

To these problems we consider weakly efficient solutions, directly defined only for $(DVG_1^{\mathscr{T}})$, since for the other two vector duals they can be given analogously.

Definition 4.6. An element $(\bar{s}, \bar{v}^*, \bar{y}^*, \bar{v}) \in \mathscr{B}_1^{G\mathscr{T}}$ is said to be a *weakly efficient solution* to the vector optimization problem $(DVG_1^{\mathscr{T}})$ if $(\bar{s}, \bar{v}^*, \bar{y}^*, \bar{v}) \in \operatorname{dom} h_1^{G\mathscr{T}}$ and $h_1^{G\mathscr{T}}(\bar{s}, \bar{v}^*, \bar{y}^*, \bar{v}) \in \operatorname{WMax}(h_1^{G\mathscr{T}}(\operatorname{dom} h_1^{G\mathscr{T}}), K)$. The set of all weakly efficient solutions to $(DVG_1^{\mathscr{T}})$ is called the *weak efficiency set of* $(DVG_1^{\mathscr{T}})$, being denoted by $\mathscr{W}\mathscr{E}(DVG_1^{\mathscr{T}})$.

The weak and strong duality as well as the optimality conditions statements concerning (PVG) and these vector duals can be obtained as direct consequences of their counterparts for (PVG) and its vector duals with respect to $\mathscr{S}$-properly efficient solutions, by employing the cone $\hat{K}$.

Theorem 4.9. *There are no $x \in X$ and $(s, v) \in \mathscr{B}_3^{G\mathscr{T}}$ such that $F(x) <_K h_3^{G\mathscr{T}}(s, v)$.*

Theorem 4.10. *There are no $x \in X$ and $(s, v^*, y^*, v) \in \mathscr{B}_1^{G\mathscr{T}}$ such that $F(x) <_K h_1^{G\mathscr{T}}(s, v^*, y^*, v)$.*

Theorem 4.11. *There are no $x \in X$ and $(s, y^*, v) \in \mathscr{B}_2^{G\mathscr{T}}$ such that $F(x) <_K h_2^{G\mathscr{T}}(s, y^*, v)$.*

Theorem 4.12. *If $\bar{x} \in \mathscr{P}\mathscr{E}_{\mathscr{T}}(PVG)$, there exists an $\bar{s} \in \mathscr{T}$ such that $(\bar{s}, F(\bar{x})) \in \mathscr{W}\mathscr{E}(DVG_3^{\mathscr{T}})$ and $F(\bar{x}) = h_3^{G\mathscr{T}}(\bar{s}, F(\bar{x}))$.*

Theorem 4.13. *If Φ is a K-convex vector function, the regularity condition $(RCV^{\mathscr{S}})$ is fulfilled and $\bar{x} \in \mathscr{P}\mathscr{E}_{\mathscr{T}}(PVG)$, there exist $\bar{s} \in \mathscr{T}$, $\bar{v}^* \in K^*$ and $\bar{y}^* \in Y^*$ such that $(\bar{s}, \bar{v}^*, \bar{y}^*, F(\bar{x})) \in \mathscr{W}\mathscr{E}(DVG_1^{\mathscr{T}})$, $(\bar{s}, \bar{y}^*, F(\bar{x})) \in \mathscr{W}\mathscr{E}(DVG_2^{\mathscr{T}})$ and $F(\bar{x}) = h_1^{G\mathscr{T}}(\bar{s}, \bar{v}^*, \bar{y}^*, F(\bar{x})) = h_2^{G\mathscr{T}}(\bar{s}, \bar{y}^*, F(\bar{x}))$.*

Theorem 4.14. *(a) If $\bar{x} \in \mathscr{P}\mathscr{E}_{\mathscr{T}}(PVG)$, there exists $(\bar{s}, \bar{v}) \in \mathscr{W}\mathscr{E}(DVG_3^{\mathscr{T}})$ such that*

(i) $F(\bar{x}) = \bar{v}$;
(ii) $\bar{s}(\bar{v}) = \bar{s}(F(\bar{x})) = \min_{x \in X} \bar{s}(F(x))$.

(b) Assume that $\bar{x} \in X$ and $(\bar{s}, \bar{v}) \in \mathscr{T} \times V$ fulfill the relations (i)–(ii). Then $\bar{x} \in \mathscr{P}\mathscr{E}_{\mathscr{T}}(PVG)$ and $(\bar{s}, \bar{v}) \in \mathscr{W}\mathscr{E}(DVG_3^{\mathscr{T}})$.

Theorem 4.15. *(a) If Φ is a K-convex vector function, the regularity condition $(RCV^{\mathscr{S}})$ is fulfilled and $\bar{x} \in \mathscr{P}\mathscr{E}_{\mathscr{T}}(PVG)$, there exists $(\bar{s}, \bar{v}^*, \bar{y}^*, \bar{v}) \in \mathscr{W}\mathscr{E}(DVG_1^{\mathscr{T}})$ such that*

(i) $F(\bar{x}) = \bar{v}$;
(ii) $\bar{s}(\bar{v}) + \bar{s}^*(\bar{v}^*) = \langle \bar{v}^*, \bar{v} \rangle$;
(iii) $(\bar{v}^* F)(\bar{x}) + (\bar{v}^* \Phi)^*(0, \bar{y}^*) = 0$.

(b) Assume that $\bar{x} \in X$ *and* $(\bar{s}, \bar{v}^*, \bar{y}^*, \bar{v}) \in \mathscr{T} \times V^* \times Y^* \times V$ *fulfill the relations (i)–(iii). Then* $\bar{x} \in \mathscr{PE}_{\mathscr{T}}(PVG)$ *and* $(\bar{s}, \bar{v}^*, \bar{y}^*, \bar{v}) \in \mathscr{WE}(DVG_1^{\mathscr{T}})$.

Theorem 4.16. *(a) If* Φ *is a* K*-convex vector function, the regularity condition* $(RCV^{\mathscr{S}})$ *is fulfilled and* $\bar{x} \in \mathscr{PE}_{\mathscr{T}}(PVG)$*, there exists* $(\bar{s}, \bar{y}^*, \bar{v}) \in \mathscr{WE}(DVG_2^{\mathscr{T}})$ *such that*

(i) $F(\bar{x}) = \bar{v}$;
(ii) $\bar{s}(\bar{v}) + (\bar{s} \circ \Phi)^*(0, \bar{y}^*) = 0.$

(b) Assume that $\bar{x} \in X$ *and* $(\bar{s}, \bar{y}^*, \bar{v}) \in \mathscr{T} \times Y^* \times V$ *fulfill the relations (i)–(ii). Then* $\bar{x} \in \mathscr{PE}_{\mathscr{T}}(PVG)$ *and* $(\bar{s}, \bar{y}^*, \bar{v}) \in \mathscr{WE}(DVG_2^{\mathscr{T}})$.

Remark 4.11. The optimality conditions provided in Theorems 4.14–4.16 can be equivalently written in the same way as done in Remarks 4.6, 4.7 and 4.9, respectively.

Remark 4.12. If int $K \neq \emptyset$, the regularity condition $(RCV^{\mathscr{S}})$ can be weakened in the hypotheses of Theorems 4.13, 4.15 and 4.16 to

$$(RCV_0^{\mathscr{S}}) \big| \exists x' \in X \text{ such that } (x', 0) \in \operatorname{dom} \Phi \text{ and } \Phi(x', \cdot) \text{ is continuous at } 0,$$

because the continuity assumptions for s are no longer necessary under the mentioned hypothesis, as it can be seen below. The optimization problem $\inf_{x \in X} \bar{s}(F(x))$ is actually nothing else than

$$\inf_{\substack{x \in X, y \in Y, \\ \Phi(x,0) - y \in -K}} \bar{s}(y).$$

The Lagrange dual of the latter is

$$\sup_{v^* \in K^*} \inf_{\substack{x \in X, \\ y \in Y}} \left[\bar{s}(y) + \langle v^*, \Phi(x, 0) - y \rangle \right],$$

and it can be rewritten as $\sup_{v^* \in K^*} \big\{ -\bar{s}^*(v^*) - ((v^*\Phi)(\cdot, 0))^*(0) \big\}$. The regularity condition $(RCV_0^{\mathscr{S}})$ yields the existence of an $x' \in \operatorname{dom} F$ such that $\Phi(x', 0) + \operatorname{int} K \subseteq \operatorname{dom} \bar{s}$ and also a $y' \in \operatorname{dom} \bar{s}$ such that $\Phi(x', 0) - y' \in -\operatorname{int} K$. Using now [48, Theorem 3.2.9], one obtains that for the primal-dual pair of scalar optimization problems introduced above there is strong duality, thus there exists a $\bar{v}^* \in K^*$ such that $\bar{s}(F(\bar{x})) = -\bar{s}^*(\bar{v}^*) - ((\bar{v}^*\Phi)(\cdot, 0))^*(0)$. Applying Remark 2.5, $(RCV_0^{\mathscr{S}})$ also yields the existence of a $\bar{y}^* \in Y^*$ such that $((\bar{v}^*\Phi)(\cdot, 0))^*(0) = (\bar{v}^*\Phi)^*(0, \bar{y}^*)$. The same weaker regularity condition can be used when the considered scalarization functions are continuous, too.

4.3 Vector Duality via Different Scalarizations for General Vector Optimization Problems

In this section we consider several concrete scalarization functions, obtaining vector duals and corresponding duality statements by particularizing the set $\mathcal{S}$ or $\mathcal{T}$, respectively. More exactly, we present the linear scalarization, the maximum(-linear) scalarization, the set scalarization, the (semi)norm scalarization, the oriented distance scalarization and the quadratic scalarization. Since all these scalarizations are made with continuous functions, taking into consideration Remark 4.3 and Theorem 4.4, we will not deal in this section with all the vector duals we considered to (*PVG*), working only with ($DVG_1^{\mathcal{S}}$) and ($DVG_1^{\mathcal{T}}$), respectively.

Before proceeding, let us mention that different other scalarizations were considered in the literature, from which we recall some here. From the scalarizations involving strongly K-increasing functions we mention the one using continuous sublinear functions from [183] and the one containing penalty functions from [213]. Regarding the scalarizations involving strictly K-increasing functions, we have, for instance, the one with continuous sublinear functions from [157] and the bottleneck scalarization considered in [119]. Other scalarizations can be found for instance in [165].

4.3.1 *Linear Scalarization*

The *linear scalarization* is the simplest and, consequently, most often used scalarization method in the literature and it operates with strongly or strictly K-increasing linear continuous functions. From the huge amount of works where it appears we mention here only [48, 140, 166]. We first deal with the case of the strongly K-increasing linear functions. Take the set of scalarization functions

$$\mathcal{S}_l = \left\{ s_{v^*} : V^\bullet \to \overline{\mathbb{R}} \,:\, v^* \in K^{*0}, s_{v^*}(v) = \langle v^*, v \rangle \ \forall v \in V^\bullet \right\}.$$

Each $s_{v^*} \in \mathcal{S}_l$ is a linear continuous strongly K-increasing function with $\operatorname{dom} s_{v^*} = V$.

An element $\bar{x} \in X$ is said to be an *$\mathcal{S}_l$-properly efficient solution to* (*PVG*) if there exists a $v^* \in K^{*0}$ such that $\langle v^*, F(\bar{x}) \rangle \leq \langle v^*, F(x) \rangle$ for all $x \in X$. Note that the $\mathcal{S}_l$-properly efficient solutions to (*PVG*) are actually the classical properly efficient solutions to it in the sense of linear scalarization from the literature, that can be introduced via Definition 3.10 and consequently we denote their set by $\mathcal{PE}_{LS}(PVG)$. Noticing that for all $k^* \in K^*$ one has $s_{v^*}^*(k^*) = \delta_{\{v^*\}}(k^*)$, the dual vector problem ($DVG_1^{\mathcal{S}}$) becomes

$$(DVG_1^{\mathscr{S}_l}) \qquad \operatorname*{Max}_{(v^*,y^*,v)\in\mathscr{B}_1^{G\mathscr{S}_l}} h_1^{G\mathscr{S}_l}(v^*,y^*,v),$$

where

$$\mathscr{B}_1^{G\mathscr{S}_l} = \left\{(v^*,y^*,v) \in K^{*0} \times Y^* \times V : \langle v^*, v\rangle \leq -(v^*\Phi)^*(0,y^*)\right\}$$

and

$$h_1^{G\mathscr{S}_l}(v^*,y^*,v) = v.$$

Note that this is actually the vector dual to (PVG) considered for instance in [48, Section 4.3] and [101]. The weak and strong duality statements for (PVG) and $(DVG_1^{\mathscr{S}_l})$ follow from Theorems 4.2 and 4.5, with the observation that due to the continuity of the scalarization function the regularity condition we consider is $(RCV_0^{\mathscr{S}})$.

Theorem 4.17. *(a) There are no $x \in X$ and $(v^*,y^*,v) \in \mathscr{B}_1^{G\mathscr{S}_l}$ such that $F(x) \leq_K h_1^{G\mathscr{S}_l}(v^*,y^*,v)$.*
(b) If Φ is a K-convex vector function, the regularity condition $(RCV_0^{\mathscr{S}})$ is fulfilled and $\bar{x} \in \mathscr{PE}_{LS}(PVG)$, there exist $\bar{v}^* \in K^{*0}$ and $\bar{y}^* \in Y^*$ such that $(\bar{v}^*,\bar{y}^*,F(\bar{x})) \in \mathscr{E}(DVG_1^{\mathscr{S}_l})$ and $F(\bar{x}) = h_1^{G\mathscr{S}_l}(\bar{v}^*,\bar{y}^*,F(\bar{x}))$.

Moreover, from Theorem 4.7 one can obtain the following necessary and sufficient optimality conditions regarding (PVG) and $(DVG_1^{\mathscr{S}_l})$.

Theorem 4.18. *(a) If Φ is a K-convex vector function, the regularity condition $(RCV_0^{\mathscr{S}})$ is fulfilled and $\bar{x} \in \mathscr{PE}_{LS}(PVG)$, there exists $(\bar{v}^*,\bar{y}^*,\bar{v}) \in \mathscr{E}(DVG_1^{\mathscr{S}_l})$ such that*

(i) $F(\bar{x}) = \bar{v}$;
(ii) $(\bar{v}^*F)(\bar{x}) + (\bar{v}^*\Phi)^*(0,\bar{y}^*) = 0$.

(b) Assume that $\bar{x} \in X$ and $(\bar{v}^*,\bar{y}^*,\bar{v}) \in \mathscr{B}_1^{G\mathscr{S}_l}$ fulfill the relations (i)–(ii). Then $\bar{x} \in \mathscr{PE}_{LS}(PVG)$ and $(\bar{v}^*,\bar{y}^*,\bar{v}) \in \mathscr{E}(DVG_1^{\mathscr{S}_l})$.

On the other hand, when $\operatorname{qi} K \neq \emptyset$ and K is closed and the pair $(K, F(\operatorname{dom} F))$ has the property (QC), one can take as set of scalarization functions also

$$\mathscr{T}_l = \left\{s_{v^*} : V^\bullet \to \overline{\mathbb{R}} : v^* \in K^* \setminus \{0\}, s_{v^*}(v) = \langle v^*, v\rangle \ \forall v \in V^\bullet\right\}.$$

Each $s_{v^*} \in \mathscr{T}_l$ is a linear continuous strictly K-increasing function with $\operatorname{dom} s_{v^*} = V$. An element $\bar{x} \in X$ is said to be a *$\mathscr{T}_l$-properly efficient solution to (PVG)* if there exists a $v^* \in K^* \setminus \{0\}$ such that $\langle v^*, F(\bar{x})\rangle \leq \langle v^*, F(x)\rangle$ for all $x \in X$, and we denote this by $\bar{x} \in \mathscr{PE}_{\mathscr{T}_l}(PVG)$. Note that when $F(X) + K$

is convex, that happens for instance when F is a K-convex function, one has $\mathscr{PE}_{\mathscr{T}_l}(PVG) = \mathscr{WE}(PVG)$. The dual vector problem $(DVG_1^{\mathscr{T}})$ becomes

$$(DVG_1^{\mathscr{T}_l}) \qquad \operatorname*{WMax}_{(v^*,y^*,v)\in\mathscr{B}_1^{G\mathscr{T}_l}} h_1^{G\mathscr{T}_l}(v^*,y^*,v),$$

where

$$\mathscr{B}_1^{G\mathscr{T}_l} = \Big\{(v^*,y^*,v) \in (K^*\setminus\{0\}) \times Y^* \times V : \langle v^*, v\rangle \leq -(v^*\Phi)^*(0,y^*)\Big\}$$

and

$$h_1^{G\mathscr{T}_l}(v^*,y^*,v) = v.$$

Note that this is actually the vector dual to (PVG) considered in [110] and, for the case $\operatorname{int} K \neq \emptyset$, for instance in [48, 140]. The weak and strong duality statements for (PVG) and $(DVG_1^{\mathscr{T}_l})$ follow from Theorems 4.10 and 4.13, again with the observation that due to the continuity of the scalarization function the regularity condition we consider is $(RCV_0^{\mathscr{S}})$.

Theorem 4.19. *(a) There are no $x \in X$ and $(v^*,y^*,v) \in \mathscr{B}_1^{G\mathscr{T}_l}$ such that $F(x) <_K h_1^{G\mathscr{T}_l}(v^*,y^*,v)$.*
(b) If Φ is a K-convex vector function, the regularity condition $(RCV_0^{\mathscr{S}})$ is fulfilled and $\bar{x} \in \mathscr{WE}(PVG)$, there exist $\bar{v}^* \in K^* \setminus \{0\}$ and $\bar{y}^* \in Y^*$ such that $(\bar{v}^*,\bar{y}^*,F(\bar{x})) \in \mathscr{WE}(DVG_1^{\mathscr{T}_l})$ and $F(\bar{x}) = h_1^{G\mathscr{T}_l}(\bar{v}^*,\bar{y}^*,F(\bar{x}))$.

The corresponding necessary and sufficient optimality conditions follow from Theorem 4.15.

Theorem 4.20. *(a) If Φ is a K-convex vector function, the regularity condition $(RCV_0^{\mathscr{S}})$ is fulfilled and $\bar{x} \in \mathscr{WE}(PVG)$, there exists $(\bar{v}^*,\bar{y}^*,\bar{v}) \in \mathscr{WE}(DVG_1^{\mathscr{T}_l})$ such that*

(i) $F(\bar{x}) = \bar{v}$;
(ii) $(\bar{v}^*F)(\bar{x}) + (\bar{v}^*\Phi)^*(0,\bar{y}^*) = 0$.

(b) Assume that $\bar{x} \in X$ and $(\bar{v}^*,\bar{y}^*,\bar{v}) \in \mathscr{B}_1^{G\mathscr{T}_l}$ fulfill the relations (i)–(ii). Then $\bar{x} \in \mathscr{PE}_{\mathscr{T}_l}(PVG)$ and $(\bar{v}^*,\bar{y}^*,\bar{v}) \in \mathscr{WE}(DVG_1^{\mathscr{T}_l})$.

Remark 4.13. The optimality conditions (i)–(ii) in both Theorems 4.18 and 4.20 can be equivalently written as $F(\bar{x}) = \bar{v}$ and $(0,\bar{y}^*) \in \partial(\bar{v}^*\Phi)(\bar{x},0)$.

Remark 4.14. In case $\operatorname{int} K \neq \emptyset$ or $\operatorname{core} K \neq \emptyset$ the duality statements with respect to the $\mathscr{T}_l$-properly efficient solutions to (PVG) remain valid even if the cone K is not necessarily closed.

4.3.2 Maximum(-Linear) Scalarization

In case V is a finitely dimensional space one of the scalarizations one can meet especially in the applications of vector optimization is the so-called *Tchebyshev (or, maximum) scalarization* (cf. [88, 140, 144, 164, 197]). We deal here with a more general scalarization function defined by combining a weighted maximum scalarization function with a linear function, as used for instance in [165, 168]. Let $V = \mathbb{R}^k$ and $K = \mathbb{R}^k_+$. In this case let the components of the multiobjective function F be the proper functions $F_i : X \to \overline{\mathbb{R}}$, $i = 1, \ldots, k$, such that $\cap_{i=1}^k \operatorname{dom} F_i \neq \emptyset$, and $F : X \to \mathbb{R}^k \cup \{\infty_{\mathbb{R}^k_+}\}$ is defined by

$$F(x) = \begin{cases} (F_1(x), \ldots, F_k(x))^\top, & \text{if } x \in \bigcap_{i=1}^k \operatorname{dom} F_i, \\ \infty_{\mathbb{R}^k_+}, & \text{otherwise.} \end{cases}$$

Let also be $\eta \geq 0$. For $w = (w_1, \ldots, w_k)^\top \in \operatorname{int} \mathbb{R}^k_+$ and $a = (a_1, \ldots, a_k)^\top \in \mathbb{R}^k$ we consider the scalarization function $s_{w,a} : (\mathbb{R}^k)^\bullet \to \overline{\mathbb{R}}$, defined by

$$s_{w,a}(y) = \begin{cases} \max\limits_{j=1,\ldots,k} \{w_j(y_j - a_j)\} + \eta \sum\limits_{j=1}^k w_j y_j, & \text{if } y = (y_1, \ldots, y_k)^\top \in \mathbb{R}^k, \\ +\infty, & \text{otherwise.} \end{cases}$$

For all $w \in \operatorname{int} \mathbb{R}^k_+$ and $a \in \mathbb{R}^k$, $s_{w,a}$ is convex and strictly $\mathbb{R}^k_+$-increasing and fulfills $F\left(\cap_{i=1}^k \operatorname{dom} F_i\right) + \mathbb{R}^k_+ \subseteq \mathbb{R}^k = \operatorname{dom} s$. We introduce the following set of scalarization functions

$$\mathscr{T}_{ml} = \left\{ s_{w,a} : \mathbb{R}^k \cup \{\infty_{\mathbb{R}^k_+}\} \to \overline{\mathbb{R}} : (w, a) \in \operatorname{int} \mathbb{R}^k_+ \times \mathbb{R}^k \right\}.$$

Then an element $\bar{x} \in X$ is said to be a *$\mathscr{T}_{ml}$-properly efficient solution to* (PVG) if there exist $w \in \operatorname{int} \mathbb{R}^k_+$ and $a \in \mathbb{R}^k$ such that $\max\{w_j(F_j(\bar{x}) - a_j) : j = 1, \ldots, k\} + \eta \sum_{j=1}^k w_j F_j(\bar{x}) \leq \max\{w_j(F_j(x) - a_j) : j = 1, \ldots, k\} + \eta \sum_{j=1}^k w_j F_j(x)$ for all $x \in X$, and we denote this by $\bar{x} \in \mathscr{PE}_{\mathscr{T}_{ml}}(PVG)$.

Let be $w = (w_1, \ldots, w_k)^\top \in \operatorname{int} \mathbb{R}^k_+$ and $a = (a_1, \ldots, a_k)^\top \in \mathbb{R}^k$ fixed. Since the conjugate function of $s_{w,a} \in \mathscr{T}_{ml}$ is, for $k^* \in \mathbb{R}^k$,

$$s^*_{w,a}(k^*) = \begin{cases} (k^* - \eta w)^\top a, & \text{if } \eta w \leqq k^* \text{ and } \sum\limits_{j=1}^k \frac{k^*_j}{w_j} = k\eta + 1, \\ +\infty, & \text{otherwise,} \end{cases}$$

the dual vector problem to (PVG) with respect to $\mathscr{T}_{ml}$-properly efficient solutions is

$$(DVG_1^{\mathcal{T}_{ml}}) \qquad \underset{(w,a,v^*,y^*,v)\in\mathcal{B}_1^{G\mathcal{T}_{ml}}}{\text{WMax}} \; h_1^{G\mathcal{T}_{ml}}(w,a,v^*,y^*,v),$$

where

$$\mathcal{B}_1^{G\mathcal{T}_{ml}} = \Big\{(w,a,v^*,y^*,v) \subset \operatorname{int}\mathbb{R}^k_+ \times \mathbb{R}^k \times \mathbb{R}^k_+ \times Y^* \times \mathbb{R}^k : \eta w \leqq v^*, \sum_{j=1}^k \frac{v_j^*}{w_j} = k\eta+1,$$

$$\max_{j=1,\ldots,k}\{w_j(v_j - a_j)\} + \eta\sum_{j=1}^k w_j v_j \leq -(v^* - \eta w)^\top a - (v^*\Phi)^*(0,y^*)\Big\}$$

and

$$h_1^{G\mathcal{T}_{ml}}(w,a,v^*,y^*,v) = v.$$

Remark 4.15. If $w \in \operatorname{int}\mathbb{R}^k_+$ and $a \in \mathbb{R}^k$, one can note that $s^*_{w,a}(0) = +\infty$, consequently $\operatorname{dom} s^*_{w,a} \subseteq \mathbb{R}^k_+ \setminus \{0\}$.

The weak and strong duality statements for (PVG) and $(DVG_1^{\mathcal{T}_{ml}})$ follow from Theorems 4.10 and 4.13, again with the observation that due to the continuity of the scalarization function the regularity condition we consider is $(RCV_0^{\mathcal{S}})$.

Theorem 4.21. *(a) There are no $x \in X$ and $(w,a,v^*,y^*,v) \in \mathcal{B}_1^{G\mathcal{T}_{ml}}$ such that $F(x) <_K h_1^{G\mathcal{T}_{ml}}(w,a,v^*,y^*,v)$.*
(b) If Φ is a K-convex vector function, the regularity condition $(RCV_0^{\mathcal{S}})$ is fulfilled and $\bar{x} \in \mathcal{PE}_{\mathcal{T}_{ml}}(PVG)$, there exist $\bar{v}^* \in K^* \setminus \{0\}$, $\bar{w} \in \operatorname{int}\mathbb{R}^k_+$, $\bar{a} \in \mathbb{R}^k$ and $\bar{y}^* \in Y^*$ such that $(\bar{w},\bar{a},\bar{v}^*,\bar{y}^*,F(\bar{x})) \in \mathcal{WE}(DVG_1^{\mathcal{T}_{ml}})$ and $F(\bar{x}) = h_1^{G\mathcal{T}_{ml}}(\bar{w},\bar{a},\bar{v}^*,\bar{y}^*,F(\bar{x}))$.

The corresponding necessary and sufficient optimality conditions follow from Theorem 4.15.

Theorem 4.22. *(a) If Φ is a K-convex vector function, the regularity condition $(RCV_0^{\mathcal{S}})$ is fulfilled and $\bar{x} \in \mathcal{PE}_{\mathcal{T}_{ml}}(PVG)$, there exists $(\bar{w},\bar{a},\bar{v}^*,\bar{y}^*,\bar{v}) \in \mathcal{WE}(DVG_1^{\mathcal{T}_{ml}})$ such that*

(i) $F(\bar{x}) = \bar{v}$;
(ii) $\max_{j=1,\ldots,k}\{\bar{w}_j(\bar{v}_j - \bar{a}_j)\} + \eta\bar{w}^\top\bar{v} + (\bar{v}^* - \eta\bar{w})^\top\bar{a} + \delta_{\mathbb{R}^k_+}(\bar{v}^* - \eta\bar{w}) = \bar{v}^{*\top}\bar{v}$;
(iii) $\sum_{j=1}^k \frac{\bar{v}_j^*}{\bar{w}_j} = k\eta + 1$;
(iv) $(\bar{v}^*F)(\bar{x}) + (\bar{v}^*\Phi)^*(0,\bar{y}^*) = 0$.

(b) Assume that $\bar{x} \in X$ and $(\bar{w},\bar{a},\bar{v}^*,\bar{y}^*,\bar{v}) \in \mathcal{B}_1^{G\mathcal{T}_{ml}}$ fulfill the relations (i)–(iv). Then $\bar{x} \in \mathcal{PE}_{\mathcal{T}_{ml}}(PVG)$ and $(\bar{w},\bar{a},\bar{v}^*,\bar{y}^*,\bar{v}) \in \mathcal{WE}(DVG_1^{\mathcal{T}_{ml}})$.

Remark 4.16. The optimality conditions (i)–(iv) in Theorem 4.22 can be equivalently written as $F(\bar{x}) = \bar{v}$, $\bar{v}^* \in \partial s_{\bar{w},\bar{a}}(F(\bar{x}))$ and $(0, \bar{y}^*) \in \partial(\bar{v}^*\Phi)(\bar{x}, 0)$. Considering the set $L(w,a,y) = \{l \in \{1,\ldots,k\} \cap \mathbb{N} : w_l(y_l - a_l) = \max_{1\le j\le k} w_j(y_j - a_j)\}$ for $(w,a,y) \in \operatorname{int}\mathbb{R}^k_+ \times \mathbb{R}^k \times \mathbb{R}^k$, one can show that

$$\partial s_{w,a}(y) = \Big\{k^* = (k_1^*,\ldots,k_k^*)^\top \in \mathbb{R}^k : k_j^* = \eta w_j \ \forall j \in \{1,\ldots,k\} \setminus L(w,a,y),$$
$$\sum_{j\in L(w,a,y)} \frac{k_j^*}{w_j} = 1 + k\eta\Big\},$$

consequently the optimality condition $\bar{v}^* \in \partial s_{\bar{w},\bar{a}}(F(\bar{x}))$ can be equivalently written as $\bar{v}^* = (\bar{v}_1^*,\ldots,\bar{v}_k^*)^\top$ with $\bar{v}_j^* = \eta w_j$ when $j \in \{1,\ldots,k\} \setminus L(\bar{w},\bar{a},F(\bar{x}))$ and $\sum_{j\in L(\bar{w},\bar{a},F(\bar{x}))}(\bar{v}_j^*/\bar{w}_j) = 1 + k\eta$.

In case $\eta = 0$ the maximum-linear scalarization becomes the *weighted Tchebyshev scalarization*, that can also be seen as a special case (see also [203]) of the set scalarization that will be presented in Sect. 4.3.3. In the more particular case $w_j = 1$ and $a_j = 0$ for all $j = 1,\ldots,k$, the set of scalarization functions has only one element, namely the extended maximum function

$$\mathcal{T}_m = \left\{ s_m : \mathbb{R}^k \cup \{\infty_{\mathbb{R}^k_+}\} \to \overline{\mathbb{R}} : s_m(y) = \begin{cases} \max\limits_{j=1,\ldots,k} y_j, & \text{if } y \in \mathbb{R}^k, \\ +\infty, & \text{otherwise} \end{cases} \right\},$$

with $\operatorname{dom} s_m = \mathbb{R}^k$. Consequently an element $\bar{x} \in X$ is said to be a *$\mathcal{T}_m$-properly efficient solution to (PVG)* if $\max_{j=1,\ldots,k} F_j(\bar{x}) \le \max_{j=1,\ldots,k} F_j(x)$ for all $x \in X$, and we denote this by $\bar{x} \in \mathcal{PE}_{\mathcal{T}_m}(PVG)$. The dual vector problem to (PVG) with respect $\mathcal{T}_m$-properly efficient solutions is

$$(DVG_1^{\mathcal{T}_m}) \qquad \operatorname*{WMax}_{(v^*,y^*,v)\in\mathcal{B}_1^{G\mathcal{T}_m}} h_1^{G\mathcal{T}_m}(v^*,y^*,v),$$

where

$$\mathcal{B}_1^{G\mathcal{T}_m} = \left\{(v^*,y^*,v) \in \mathbb{R}^k_+ \times Y^* \times \mathbb{R}^k : \sum_{j=1}^k v_j^* = 1,\ \max_{j=1,\ldots,k}\{v_j\} \le -(v^*\Phi)^*(0,y^*)\right\}$$

and

$$h_1^{G\mathcal{T}_m}(v^*,y^*,v) = v.$$

The weak and strong duality statements for (PVG) and $(DVG_1^{\mathcal{T}_m})$ are particular instances of Theorem 4.21, while the necessary and sufficient optimality conditions can be derived via Theorem 4.22.

4.3.3 Set Scalarization

As *set scalarizations* we understand the scalarization approaches for which the scalarization functions are defined by means of some given sets. We consider here a quite general scalarization function inspired by the one introduced in [99] under the hypothesis $\operatorname{int} K \neq \emptyset$ and used in various formulations for dealing with different problems in vector optimization in works like [37,90,92,93,114,129,143,152,182, 183,185,188,191,192,194,203,204,214]. In this subsection $\operatorname{qi} K$ is taken nonempty.

Consider a fixed nonempty convex set $E \subseteq V$ which satisfies $\operatorname{cl} E + \operatorname{qi} K \subseteq \operatorname{core} E$. This condition is quite naturally fulfilled in different circumstances, for instance in case $E = K$ (when we replace the quasi interior of K with its algebraic interior) treated later as it induces the scalarization with conical sets or in mathematical economics where a variant of it is called the *free disposal assumption* (see [79,92,93]). For each $\mu \in \operatorname{qi} K$ we define the scalarization function $s_\mu : V^\bullet \to \overline{\mathbb{R}}$ by

$$s_\mu(v) = \inf\{t \in \mathbb{R} : v \in t\mu - \operatorname{cl} E\}.$$

Notice that $s_\mu(\infty_K) = +\infty$ and $\operatorname{dom} s_\mu = V$. According to [99, 203], in case $\operatorname{int} K \neq \emptyset$, for each $\mu \in \operatorname{int} K$ the function s_μ is convex, continuous and strictly K-increasing. These properties remain valid in the framework we use, too.

Proposition 4.1. *Whenever $\mu \in \operatorname{qi} K$, the function s_μ is strictly K-increasing.*

Proof. Let $p, q \in V$ such that $p <_K q$ and $\bar{r} = s_\mu(q)$. Then $p \in q - \operatorname{qi} K \subseteq r\mu - \operatorname{cl} E - \operatorname{qi} K \subseteq r\mu - \operatorname{core} E$ for all $r > \bar{r}$. Let us show now that $p \in r\mu - \operatorname{core} E$ for all $r > \bar{r}$ if and only if $s_\mu(p) < \bar{r}$, that would yield the strict K-monotonicity of the function s_μ.

If $s_\mu(p) < \bar{r}$, let $\bar{t} = s_\mu(p)$. Then for all $t \in (\bar{t}, \bar{r})$ one has $p \in t\mu - \operatorname{cl} E = \bar{r}\mu - (\bar{r} - t)\mu - \operatorname{cl} E \subseteq \bar{r}\mu - \operatorname{core} E$ since $(\bar{r} - t)\mu \in \operatorname{qi} K$, consequently $p \in r\mu - \operatorname{core} E$ whenever $r > \bar{r}$.

Viceversa, for any $r > \bar{r}$, $p \in r\mu - \operatorname{core} E$ yields the existence of a $w \in \operatorname{core} E$ such that $p = r\mu - w$. Then for a convenient choice of $r > \bar{r}$ there exists an $\alpha \in (0, \bar{r})$ such that $w - (r + \alpha - \bar{r})\mu \in E$ and let $t = \bar{r} - \alpha > 0$. But $p = (r - \bar{r} + \alpha)\mu + t\mu - w \in t\mu - E \subseteq t\mu - \operatorname{cl} E$, consequently $s_\mu(p) \le t < \bar{r}$. □

The set of scalarization functions we consider in this case is then

$$\mathcal{T}_s = \left\{s_\mu : V^\bullet \to \overline{\mathbb{R}} : \mu \in \operatorname{qi} K\right\}.$$

Then an element $\bar{x} \in X$ is said to be a *$\mathcal{T}_s$-properly efficient solution to* (*PVG*) if there exists a $\mu \in \operatorname{qi} K$ such that $s_\mu(F(\bar{x})) \le s_\mu(F(x))$ for all $x \in X$, and we denote this by $\bar{x} \in \mathcal{PE}_{\mathcal{T}_s}(PVG)$. Since for $\mu \in \operatorname{qi} K$ the conjugate function of s_μ is (cf. [37,48])

$$s_\mu^*: V^* \to \overline{\mathbb{R}},\ s_\mu^*(k^*) = \begin{cases} \sigma_{-\operatorname{cl} E}(k^*), & \text{if } \langle k^*, \mu\rangle = 1,\\ +\infty, & \text{otherwise,}\end{cases}$$

the dual vector problem attached to (PVG) via the set scalarization turns out to be

$$(DVG_1^{\mathcal{T}_s}) \qquad \operatorname*{WMax}_{(\mu,v^*,y^*,v)\in\mathcal{B}_1^{G\mathcal{T}_s}} h_1^{G\mathcal{T}_s}(\mu, v^*, y^*, v),$$

where

$$\mathcal{B}_1^{G\mathcal{T}_s} = \Big\{(\mu, v^*, y^*, v) \in \operatorname{qi} K \times K^* \times Y^* \times V : \langle v^*, \mu\rangle = 1,$$
$$\inf\big\{t \in \mathbb{R} : v \in t\mu - \operatorname{cl} E\big\} \leq -\sigma_{-\operatorname{cl} E}(v^*) - (v^*\Phi)^*(0, y^*)\Big\}$$

and

$$h_1^{G\mathcal{T}_s}(\mu, v^*, y^*, v) = v.$$

Remark 4.17. If $\mu \in \operatorname{qi} K$, one can note that $s_\mu^*(0) = +\infty$, consequently $\operatorname{dom} s_\mu^* \subseteq K^* \setminus \{0\}$.

The weak and strong duality statements for (PVG) and $(DVG_1^{\mathcal{T}_l})$ follow from Theorems 4.10 and 4.13.

Theorem 4.23. *(a) There are no $x \in X$ and $(\mu, v^*, y^*, v) \in \mathcal{B}_1^{G\mathcal{T}_s}$ such that $F(x) <_K h_1^{G\mathcal{T}_s}(\mu, v^*, y^*, v)$.*
(b) If Φ is a K-convex vector function, the regularity condition $(RCV_0^{\mathcal{S}})$ is fulfilled and $\bar{x} \in \mathcal{PE}_{\mathcal{T}_s}(PVG)$, there exist $\bar{\mu} \in \operatorname{qi} K$, $\bar{v}^* \in K^* \setminus \{0\}$ and $\bar{y}^* \in Y^*$ such that $(\bar{\mu}, \bar{v}^*, \bar{y}^*, F(\bar{x})) \in \mathcal{WE}(DVG_1^{\mathcal{T}_s})$ and $F(\bar{x}) = h_1^{G\mathcal{T}_s}(\bar{\mu}, \bar{v}^*, \bar{y}^*, F(\bar{x}))$.

The corresponding necessary and sufficient optimality conditions follow from Theorem 4.15.

Theorem 4.24. *(a) When Φ is a K-convex vector function, the regularity condition $(RCV_0^{\mathcal{S}})$ is fulfilled and $\bar{x} \in \mathcal{PE}_{\mathcal{T}_s}(PVG)$, there exists $(\bar{\mu}, \bar{v}^*, \bar{y}^*, \bar{v}) \in \mathcal{WE}(DVG_1^{\mathcal{T}_s})$ such that*

(i) $F(\bar{x}) = \bar{v}$;
(ii) $\inf\big\{t \in \mathbb{R} : \bar{v} \in t\bar{\mu} - \operatorname{cl} E\big\} + \sigma_{-\operatorname{cl} E}(\bar{v}^*) = \langle \bar{v}^*, \bar{v}\rangle$;
(iii) $\langle \bar{v}^*, \bar{\mu}\rangle = 1$;
(iv) $(\bar{v}^*F)(\bar{x}) + (\bar{v}^*\Phi)^*(0, \bar{y}^*) = 0$.

(b) Assume that $\bar{x} \in X$ and $(\bar{\mu}, \bar{v}^*, \bar{y}^*, \bar{v}) \in \mathcal{B}_1^{G\mathcal{T}_s}$ fulfill the relations (i)–(iv). Then $\bar{x} \in \mathcal{PE}_{\mathcal{T}_s}(PVG)$ and $(\bar{\mu}, \bar{v}^*, \bar{y}^*, \bar{v}) \in \mathcal{WE}(DVG_1^{\mathcal{T}_s})$.

Remark 4.18. The optimality conditions (i)–(iv) in Theorem 4.24 can be equivalently written, taking into consideration the formula of $\partial s_{\bar{\mu}}$ given in [81], as

$F(\bar{x}) = \bar{v}$, $\bar{v}^* \in \{x^* \in K^* : \langle x^*, \bar{\mu}\rangle = 1, s_{\bar{\mu}}(F(\bar{x})) = \langle x^*, F(\bar{x})\rangle\}$ and $(0, \bar{y}^*) \in \partial(\bar{v}^*\Phi)(\bar{x}, 0)$.

Remark 4.19. In the literature there are some interesting special cases of the set scalarization, from which we mention here the *scalarization with conical sets*, mentioned in papers like [183, 191], closely related to the so-called *Pascoletti-Serafini scalarization* (cf. [129]), the *scalarization with sets generated by norms* for which we refer to [143, 194, 214], having as a subcase the situation when oblique norms are employed (see [185, 194]), and, finally, the *scalarization with polyhedral sets* in finitely dimensional spaces treated in [204]. As mentioned in [89], a function very similar to the one we employed in the set scalarization is used in production theory where it is called *shortage function*. Note also that in [203] a deeper analysis of an approach for embedding other classical scalarization functions into the set scalarization concept can be found and that the set scalarization with its special instances was employed into vector duality in [31, 37, 48].

Let us present now two of the mentioned special cases of the set scalarization. We begin with the so-called *set scalarization with conical sets*, where we assume that core $K \neq \emptyset$ and E is taken to coincide with K. Since the latter is a convex cone, the condition $\operatorname{cl} E + \operatorname{core} K \subseteq \operatorname{core} K$ is automatically fulfilled, actually as an equality. For all $\nu \in \operatorname{core} K$ we define the scalarization function $s_\nu : V^\bullet \to \overline{\mathbb{R}}$ by

$$s_\nu(v) = \inf\{t \in \mathbb{R} : v \in t\nu - \operatorname{cl} K\}.$$

From the definition it follows that $s_\nu(\infty_K) = +\infty$ and $\operatorname{dom} s_\nu = V$. The set of scalarization functions is then

$$\mathcal{T}_{sc} = \left\{s_\nu : V^\bullet \cup \{\infty_K\} \to \overline{\mathbb{R}} : \nu \in \operatorname{core} K\right\}.$$

Then an element $\bar{x} \in X$ is said to be a *$\mathcal{T}_{sc}$-properly efficient solution to (PVG)* if there exists a $\nu \in \operatorname{core} K$ such that $s_\nu(F(\bar{x})) \leq s_\nu(F(x))$ for all $x \in X$, and we denote this by $\bar{x} \in \mathcal{PE}_{\mathcal{T}_{sc}}(PVG)$. Since $\sigma_{-\operatorname{cl} K} = \delta_{K^*}$, for all $\nu \in \operatorname{core} K$ the conjugate function of s_ν at $k^* \in V^*$ is

$$s_\nu^*(k^*) = \begin{cases} 0, & \text{if } k^* \in K^*,\ \langle k^*, \nu\rangle = 1, \\ +\infty, & \text{otherwise,} \end{cases}$$

leading to the following dual vector problem to (*PVG*)

$$(DVG_1^{\mathcal{T}_{sc}}) \qquad \operatorname*{WMax}_{(\nu, v^*, y^*, v) \in \mathcal{B}_1^{G\mathcal{T}_{sc}}} h_1^{G\mathcal{T}_{sc}}(\nu, v^*, y^*, v),$$

where

$$\mathscr{B}_1^{G\mathscr{T}_{sc}} = \Big\{(\nu, v^*, y^*, v) \in \operatorname{core} K \times K^* \times Y^* \times V : \langle v^*, \nu\rangle = 1,$$
$$\inf\big\{t \in \mathbb{R} : v \in t\nu - \operatorname{cl} K\big\} \leq -(v^*\Phi)^*(0, y^*)\Big\}$$

and

$$h_1^{G\mathscr{T}_{sc}}(\nu, v^*, y^*, v) = v.$$

The weak and strong duality statements for (PVG) and $(DVG_1^{\mathscr{T}_{sc}})$ are particular instances of Theorem 4.23, while the necessary and sufficient optimality conditions can be derived via Theorem 4.24.

A second special case of the set scalarization we present here is the so-called *scalarization with sets generated by norms* in finitely dimensional spaces. To this end we have to introduce several notions and present some results, following [185, 194]. Let $V = \mathbb{R}^k$ and $K \subseteq \mathbb{R}^k$ a convex cone with nonempty interior.

Definition 4.7. A norm $\gamma : \mathbb{R}^k \to \mathbb{R}$ is called *block norm* if its unit ball B_γ is polyhedral.

Definition 4.8. A norm $\gamma : \mathbb{R}^k \to \mathbb{R}$ is called *absolute* if whenever $\bar{y} \in \mathbb{R}^k$ one has $\gamma(y) = \gamma(\bar{y})$ for all $y \in \big\{z = (z_1, \dots, z_k)^\top \in \mathbb{R}^k : |z_j| = |\bar{y}_j|\ \forall j = 1, \dots, k\big\}$.

Definition 4.9. A block norm $\gamma : \mathbb{R}^k \to \mathbb{R}$ is called *oblique* if it is absolute and satisfies $\big(y - \mathbb{R}^k_+\big) \cap \mathbb{R}^k_+ \cap \operatorname{bd} B_\gamma = \{y\}$ for all $y \in \mathbb{R}^k_+ \cap \operatorname{bd} B_\gamma$.

Example 4.4. The Euclidean norm $\|\cdot\|_2$ in $\mathbb{R}^k$ is absolute, but not block, thus not oblique.

According to [185, 194], for a block norm γ there are some $r \in \mathbb{N}$, $a_i \in \mathbb{R}^k \setminus \{0\}$ and $\eta_i \in \mathbb{R}$, $i = 1, \dots, r$, such that the unit ball generated by γ is

$$B_\gamma = \big\{y \in \mathbb{R}^k : a_i^\top y \leq \eta_i, i = 1, \dots, r\big\}.$$

In order to introduce the scalarization function considered in this case, one needs also the sets

$$I_\gamma = \Big\{i \in \{1, \dots, r\} : \big\{y \in \mathbb{R}^k : a_i^\top y = \eta_i\big\} \cap B_\gamma \cap \operatorname{int} \mathbb{R}^k_+ \neq \emptyset\Big\}$$

and

$$E_\gamma = \big\{y \in \mathbb{R}^k : a_i^\top y \leq \eta_i\ \forall i \in I_\gamma\big\}.$$

If γ is an absolute norm on $\mathbb{R}^k$, $l \in \operatorname{int} \mathbb{R}^k_+$ and $w \in \mathbb{R}^k$, consider the scalarization function

$$\zeta_{\gamma,l,w} : (\mathbb{R}^k)^\bullet \to \overline{\mathbb{R}},\ \zeta_{\gamma,l,w}(y) = \inf\big\{t \in \mathbb{R} : y \in tl + E_\gamma + w\big\},$$

which fulfills $\zeta_{\gamma,l,w}(\infty_K) = +\infty$ and $\operatorname{dom} \zeta_{\gamma,l,w} = \mathbb{R}^k$. According to [185, 194], it is convex and strictly K-increasing when $\operatorname{bd} E_\gamma - (K \setminus \{0\}) \subseteq \operatorname{int} E_\gamma$. Moreover, when γ is an absolute block norm, the function $\zeta_{\gamma,l,w}$ is strictly $\mathbb{R}^k_+$-increasing for any $l \in \operatorname{int} \mathbb{R}^k_+$ and $w \in \mathbb{R}^k$, while when γ is an oblique norm, $\zeta_{\gamma,l,w}$ is strongly $\mathbb{R}^k_+$-increasing whenever $l \in \operatorname{int} \mathbb{R}^k_+$ and $w \in \mathbb{R}^k$.

Denote by $\mathscr{O}$ the set of the absolute norms $\gamma : \mathbb{R}^k \to \mathbb{R}$ for which $\operatorname{bd} E_\gamma - \operatorname{int} K \subseteq \operatorname{int} E_\gamma$ and consider the following set

$$\mathscr{T}_{sn} = \Big\{\zeta_{\gamma,l,w} : (\mathbb{R}^k)^\bullet \to \overline{\mathbb{R}} : \gamma \in \mathscr{O}, l \in \operatorname{int} \mathbb{R}^k_+, w \in \mathbb{R}^k,$$
$$\zeta_{\gamma,l,w}(y) = \inf\big\{t \in \mathbb{R} : y \in tl + E_\gamma + w\big\} \ \forall y \in \mathbb{R}^k\Big\}.$$

Then an element $\bar{x} \in X$ is said to be a *$\mathscr{T}_{sn}$-properly efficient solution to* (*PVG*) if there are an absolute norm $\gamma \in \mathscr{O}$, and some $l \in \operatorname{int} \mathbb{R}^k_+$ and $w \in \mathbb{R}^k$ such that $\zeta_{\gamma,l,w}(F(\bar{x})) \leq \zeta_{\gamma,l,w}(F(x))$ for all $x \in X$, and we denote this by $\bar{x} \in \mathscr{PE}_{\mathscr{T}_{sn}}(PVG)$.

Remark 4.20. Restricting moreover the set $\mathscr{T}_{sn}$ to contain only functions $\zeta_{\gamma,l,w}$ where γ is an absolute block norm or an oblique norm, one can get other scalarizations which could be treated separately, too. Actually, in the latter situation one would actually be able to consider corresponding $\mathscr{S}$-properly efficient solutions to the primal problem (*PVG*), due to the $\mathbb{R}^k_+$-strong monotonicity of the scalarization function (cf. [185,194]), thus it would not necessarily be a special case of the general set scalarization.

For some $(\gamma, l, w) \in \mathscr{O} \times \operatorname{int} \mathbb{R}^k_+ \times \mathbb{R}^k$, the conjugate of the corresponding scalarization function at $k^* \in \mathbb{R}^k$ is

$$\begin{aligned}
\zeta^*_{\gamma,l,w}(k^*) &= \sup_{y \in \mathbb{R}^k} \Big\{{k^*}^\top y - \inf\big\{t \in \mathbb{R} : y \in tl + E_\gamma + w\big\}\Big\} \\
&= \sup_{y \in \mathbb{R}^k} \Big\{{k^*}^\top y + \sup\big\{-t \in \mathbb{R} : y \in tl + E_\gamma + w\big\}\Big\}.
\end{aligned}$$

Denoting $w = y - tl - w$, one gets

$$\begin{aligned}
\zeta^*_{\gamma,l,w}(k^*) &= \sup_{t \in \mathbb{R}} \Big\{-t + \sup_{w \in E_\gamma} \Big\{{k^*}^\top (w + tl + w)\Big\}\Big\} \\
&= \sup_{t \in \mathbb{R}} \Big\{-t + t{k^*}^\top l + \sup_{w \in E_\gamma} {k^*}^\top w\Big\} + {k^*}^\top w \\
&= \sup_{t \in \mathbb{R}} \Big\{t\Big({k^*}^\top l - 1\Big)\Big\} + \sigma_{E_\gamma}(k^*) + {k^*}^\top w \\
&= \begin{cases} \sigma_{E_\gamma}(k^*) + {k^*}^\top w, & \text{if } {k^*}^\top l = 1, \\ +\infty, & \text{otherwise.} \end{cases}
\end{aligned}$$

Remark 4.21. According to [127, Example V.3.4.4], σ_{E_γ} is the lower semicontinuous hull of the function

$$d \mapsto \begin{cases} \inf\left\{ \sum_{i\in I_\gamma} t_i\eta_i : \sum_{i\in I_\gamma} t_i a_i = d,\ t_i \geq 0, i \in I_\gamma \right\} & \text{if } d \in \operatorname{cone}\{a_i : i \in I_\gamma\}, \\ +\infty, & \text{otherwise.} \end{cases}$$

But, since it would be quite complicated to work with this function, we choose to use further its shorter form σ_{E_γ}.

Now one can assign to *(PVG)* the following vector dual problem with respect to $\mathscr{T}_{sn}$-properly efficient solutions

$$(DVG_1^{\mathscr{T}_{sn}}) \qquad \operatorname*{WMax}_{(w,\gamma,l,v^*,y^*,v)\in\mathscr{B}_1^{G\mathscr{T}_{sn}}} h_1^{G\mathscr{T}_{sn}}(w,\gamma,l,v^*,y^*,v),$$

where

$$\mathscr{B}_1^{G\mathscr{T}_{sn}} = \Big\{(w,\gamma,l,v^*,y^*,v) \in \mathbb{R}^k \times \mathscr{O} \times \operatorname{int}\mathbb{R}^k_+ \times K^* \times Y^* \times \mathbb{R}^k : v^{*\top} l = 1,$$
$$\inf\big\{t \in \mathbb{R} : v \in tl + E_\gamma + w\big\} \leq -\sigma_{E_\gamma}(v^*) - v^{*\top}w - (v^*\Phi)^*(0,y^*)\Big\}$$

and

$$h_1^{G\mathscr{T}_{sn}}(w,\gamma,l,v^*,y^*,v) = v.$$

The weak and strong duality statements for *(PVG)* and $(DVG_1^{\mathscr{T}_{sn}})$ are particular instances of Theorem 4.23, while the necessary and sufficient optimality conditions can be derived via Theorem 4.24.

4.3.4 *(Semi-)Norm Scalarization*

The *(semi)norm scalarization* has its roots in the fact that in some circumstances some (semi)norms on V turn out to be strongly K-increasing functions, as noted in different works from which we recall here only [140, 143, 146, 185, 213]. The scalarization functions we investigate in the following are based on strongly K-increasing gauges. This kind of scalarization functions has been used in [202] for location problems and in [69] for goal programming, but also papers like [78, 139, 166, 218] can be mentioned here since they contain different scalarizations involving (semi)norms.

Assume that there exists a $b \in V$ such that $\Phi(\operatorname{dom}\Phi) \subseteq b + K$. We consider a convex set $E \subseteq V$ such that $0 \in \operatorname{core} E$ and its *gauge (Minkowski function)* $\gamma_E : V \to \mathbb{R}$, defined by $\gamma_E(x) = \inf\{t \geq 0 : x \in tE\}$, is strongly K-increasing on K. Since $0 \in \operatorname{int} E$, $\gamma_E(v) \in \mathbb{R}$ for all $v \in V$. For every $a \in b - K$ define the scalarization function $s_a : V^\bullet \to \overline{\mathbb{R}}$ by

$$s_a(v) = \begin{cases} \gamma_E(v - a), & \text{if } v \in b + K, \\ +\infty, & \text{otherwise}, \end{cases}$$

therefore $\operatorname{dom} s_a = V$. All these functions are convex, sublinear, strongly K-increasing on $b + K$ and, if it additionally holds $0 \in \operatorname{int} E$, also continuous on V. Note also that $F(\operatorname{dom} F) \subseteq b + K$.

The family of scalarization functions we chose here is

$$\mathscr{S}_g = \{s_a : V^\bullet \to \overline{\mathbb{R}} : a \in b - K\},$$

and an element $\bar{x} \in X$ is said to be an *$\mathscr{S}_g$-properly efficient solution to (PVG)* if there exists an $a \in b - K$ such that $s_a(F(\bar{x})) \leq s_a(F(x))$ for all $x \in X$, situation denoted by $\bar{x} \in \mathscr{PE}_{\mathscr{S}_g}(PVG)$. Since for $k^* \in V^*$ one has

$$(\gamma_E)^*(k^*) = \begin{cases} 0, & \text{if } \sigma_E(k^*) \leq 1, \\ +\infty, & \text{otherwise}, \end{cases}$$

and $(\delta_{b+K})^*(k^*) = \langle k^*, b\rangle + (\delta_K)^*(k^*)$, it follows that for $a \in b - K$ and $k^* \in V^*$ one has

$$(s_a)^*(k^*) = \min_{\substack{w^* \in -K^*, \\ \sigma_E(k^* - w^*) \leq 1}} \big[\langle k^* - w^*, a\rangle + \langle w^*, b\rangle\big] = \langle k^*, a\rangle + \min_{\substack{w^* \in -K^*, \\ \sigma_E(k^* - w^*) \leq 1}} \langle w^*, b - a\rangle.$$

Remark 4.22. When $0 \in \operatorname{int} E$, using [48, Theorem 3.5.3(a)] one can show that the continuity of γ_E yields a simpler formula for the conjugate of s_a, for $a \in b - K$, namely $(s_a)^*(k^*) = \langle k^*, a\rangle + \min_{w^* \in k^* - K^*}[\sigma_{b+K}(w^*) - \langle w^*, a\rangle]$.

Then the dual vector problem to (PVG) with respect to $\mathscr{S}_g$-properly efficient solutions is

$$(DVG_1^{\mathscr{S}_g}) \qquad \operatorname*{Max}_{(a,v^*,y^*,w^*,v) \in \mathscr{B}_1^{G\mathscr{S}_g}} h_1^{G\mathscr{S}_g}(a, v^*, y^*, w^*, v),$$

where

$$\mathscr{B}_1^{G\mathscr{S}_g} = \Big\{(a, v^*, y^*, w^*, v) \in (b - K) \times K^* \times Y^* \times K^* \times (b + K) : \sigma_E(v^* + w^*) \leq 1, \gamma_E(v - a) \leq \langle w^*, b - a\rangle - \langle v^*, a\rangle - (v^*\Phi)^*(0, y^*)\Big\}$$

and

$$h_1^{G\mathscr{S}_g}(a, v^*, y^*, w^*, v) = v.$$

The weak and strong duality statements for (PVG) and $(DVG_1^{\mathscr{S}_g})$ follow from Theorems 4.2 and 4.5, with the observation that when $0 \in \operatorname{int} E$ because of the continuity of the scalarization function one can consider the regularity condition $(RCV_0^{\mathscr{S}})$.

Theorem 4.25. *(a) There are no $x \in X$ and $(a, v^*, y^*, w^*, v) \in \mathscr{B}_1^{G\mathscr{S}_g}$ such that $F(x) \leq_K h_1^{G\mathscr{S}_g}(a, v^*, y^*, w^*, v)$.*

(b) If Φ is a K-convex vector function, the regularity condition $(RCV^{\mathscr{S}})$ is fulfilled and $\bar{x} \in \mathscr{P}\mathscr{E}_{\mathscr{S}_g}(PVG)$, there exist $\bar{a} \in b - K$, $\bar{v}^*, \bar{w}^* \in K^*$ and $\bar{y}^* \in Y^*$ such that $(\bar{a}, \bar{v}^*, \bar{y}^*, \bar{w}^*, F(\bar{x})) \in \mathscr{E}(DVG_1^{\mathscr{S}_g})$ and $F(\bar{x}) = h_1^{G\mathscr{S}_g}(\bar{a}, \bar{v}^*, \bar{y}^*, \bar{w}^*, F(\bar{x}))$.

Moreover, from Theorem 4.7 one can obtain the following necessary and sufficient optimality conditions regarding (PVG) and $(DVG_1^{\mathscr{S}_g})$.

Theorem 4.26. *(a) When Φ is a K-convex vector function, the regularity condition $(RCV^{\mathscr{S}})$ is fulfilled and $\bar{x} \in \mathscr{P}\mathscr{E}_{\mathscr{S}_g}(PVG)$, there exists $(\bar{a}, \bar{v}^*, \bar{y}^*, \bar{w}^*, \bar{v}) \in \mathscr{E}(DVG_1^{\mathscr{S}_g})$ such that*

(i) $F(\bar{x}) = \bar{v}$;
(ii) $\gamma_E(\bar{v} - \bar{a}) + \langle \bar{w}^*, b - \bar{a} \rangle = \langle \bar{v}^*, \bar{v} - \bar{a} \rangle$;
(iii) $\sigma_E(\bar{v}^* + \bar{w}^*) \leq 1$;
(iv) $(\bar{v}^* F)(\bar{x}) + (\bar{v}^* \Phi)^*(0, \bar{y}^*) = 0$.

(b) Assume that $\bar{x} \in X$ and $(\bar{a}, \bar{v}^*, \bar{y}^*, \bar{w}^*, \bar{v}) \in \mathscr{B}_1^{G\mathscr{S}_g}$ fulfill the relations (i)–(iv). Then $\bar{x} \in \mathscr{P}\mathscr{E}_{\mathscr{S}_g}(PVG)$ and $(\bar{a}, \bar{v}^*, \bar{y}^*, \bar{w}^*, \bar{v}) \in \mathscr{E}(DVG_1^{\mathscr{S}_g})$.

Remark 4.23. The optimality conditions (i)–(iv) in Theorem 4.26 can be equivalently written as $F(\bar{x}) = \bar{v}$, $\bar{v}^* \in \partial s_{\bar{a}}(F(\bar{x}))$ and $(0, \bar{y}^*) \in \partial(\bar{v}^*\Phi)(\bar{x}, 0)$. Noting that $\partial \gamma_E = F_{E^0}$ (see, for instance, [127, 151]) and, when γ_E is continuous, $\partial s_{\bar{a}} = \partial \delta_{b+K} + \cup_{z^* \in \partial \gamma_E(\cdot - \bar{a})} \partial \langle z^*, \cdot - \bar{a} \rangle = N_{b+K} + F_{E^0}(\cdot - \bar{a})$, the optimality condition $\bar{v}^* \in \partial s_{\bar{a}}(F(\bar{x}))$ can be equivalently written when $0 \in \operatorname{int} E$ as $\bar{v}^* \in N_{b+K}(F(\bar{x})) + F_{E^0}(F(\bar{x}) - \bar{a})$.

Remark 4.24. Note that σ_E defines the so-called *dual gauge* to γ_E. The duality approach described in this subsection can be considered in case V is a normed space and γ_E is a norm with the unit ball E, too, when σ_E gives actually the corresponding *dual norm*. If V is a Hilbert space, then the norm of V is strongly K-increasing on K if and only if $K \subseteq K^*$ (cf. [140]). This is the case if, for instance, $V = \mathbb{R}^k$ and K is the nonnegative orthant in $\mathbb{R}^k$. Not only the Euclidean norm is strongly $\mathbb{R}^k_+$-increasing on $\mathbb{R}^k_+$, but also the oblique norms are strongly $\mathbb{R}^k_+$-increasing on $\mathbb{R}^k_+$ (cf. [185, 194]). Other conditions which ensure that a norm is strongly K-increasing on a given set have been investigated in [139, 140, 213].

4.3.5 Oriented Distance Scalarization

Another scalarization function employed in the literature (see, for instance, [22, 80, 167, 216]) in order to deal with vector optimization problems is based on the so-called *oriented distance* function, introduced by Hiriart-Urruty in [124, 125]. It has not been yet considered for conjugate vector duality in the literature, because of the difficulty to compute its conjugate. As the technique used in [74] to finally provide a formula for the latter is specific to finitely dimensional spaces, let $X = \mathbb{R}^n$ and $V = \mathbb{R}^k$, the latter partially ordered by the nontrivial pointed closed convex cone $K \subseteq \mathbb{R}^k$ with a nonempty interior. On $\mathbb{R}^k$ we work with the Euclidean norm $\|\cdot\|_2$ and its associated distance function.

Definition 4.10. Given a metric space (Z, d) and a nonemptyset $U \subsetneq Z$, the function $\Delta_U : Z \to \mathbb{R}$ given by $\Delta_U(z) = d_U(z) - d_{Z\setminus U}(z)$, $z \in Z$, is said to be the *oriented distance* corresponding to the set U.

Different properties of the function Δ_U can be found in [74, 124, 125, 167, 216], for instance that it is Lipschitz continuous, and we recall here the ones needed for our investigation. When U is a convex set, Δ_U is a convex function, while when U is a closed convex cone with a nonempty interior, Δ_{-U} is strictly U-increasing.

We consider the scalarization function

$$s_d : (\mathbb{R}^k)^\bullet \to \overline{\mathbb{R}},\ s_d(y) = \begin{cases} \Delta_{-K}(y), & \text{if } y \in \mathbb{R}^k, \\ +\infty, & \text{otherwise,} \end{cases}$$

which is proper, convex and strictly K-increasing, with $\operatorname{dom} s_d = \mathbb{R}^k$. The set of the scalarization functions is given in this case by

$$\mathscr{T}_d = \left\{ s_d : (\mathbb{R}^k)^\bullet \to \overline{\mathbb{R}} \right\}.$$

Then an element $\bar{x} \in X$ is said to be a *$\mathscr{T}_d$-properly efficient solution to (PVG)* if $\Delta_{-K}(F(\bar{x})) \leq \Delta_{-K}(F(x))$ for all $x \in X$, and we denote this by $\bar{x} \in \mathscr{PE}_{\mathscr{T}_d}(PVG)$.

One has $s_d^* = \Delta_{-K}^*$, while the conjugate function of Δ_{-K} at $k^* \in \mathbb{R}^k$ is (cf. [74])

$$\Delta_{-K}^*(k^*) = \inf\left\{ \sum_{j=1}^{l} \delta_{K^*}(x_j^*) : 1 \leq l \leq n+2,\ x_j^* \in \mathbb{R}^k, j = 1, \ldots, l, \right.$$
$$\left. k^* = \sum_{j=1}^{l} x_j^*,\ \sum_{j=1}^{l} \|x_j^*\|_2 = 1 \right\}.$$

When $v \in \mathbb{R}^k$, $v^* \in K^*$ and $y^* \in Y^*$, the inequality $s_d(v) \leq -s_d^*(v^*) - (v^*\Phi)^*(0, y^*)$ that appears in the constraints of the vector dual to (PVG) obtained as a special case of ($DVG_1^{\mathscr{T}}$) when working with the oriented distance scalarization means $\Delta_{-K}(v) \leq \sup\left\{ \sum_{j=1}^{l} -\delta_{K^*}(x_j^*) : 1 \leq l \leq n+2, x_j^* \in \mathbb{R}^k, j = 1, \ldots, l, \right.$

$k^* = \sum_{j=1}^{l} x_j^*, \sum_{j=1}^{l} \|x_j^*\|_2 = 1\} - (v^*\Phi)^*(0, y^*)$. Note that the set that appears in the right-hand side of this inequality contains only two elements, 0 and $-\infty$. Consequently, when there exist $l \in [1, n+2] \cap \mathbb{N}$ and $x_j^* \in K^*$, $j = 1, \dots, l$, such that $v^* = \sum_{j=1}^{l} x_j^*$ and $\sum_{j=1}^{l} \|x_j^*\|_2 = 1$, its supremum is 0 and it is actually attained. The vector dual attached to (PVG) with respect to $\mathcal{T}_d$-properly efficient solutions is then

$$(DVG_1^{\mathcal{T}_d}) \qquad \underset{(v^*,l,x^*,y^*,v) \in \mathcal{B}_1^{G\mathcal{T}_d}}{\text{WMax}} h_1^{G\mathcal{T}_d}(v^*, l, x^*, y^*, v),$$

where

$$\mathcal{B}_1^{G\mathcal{T}_d} = \Big\{ (v^*, l, x^*, y^*, v) \in K^* \times \mathbb{N} \times (K^*)^l \times Y^* \times \mathbb{R}^k : \Delta_{-K}(v) \leq -(v^*\Phi)^*(0, y^*),$$

$$x^* = (x_1^*, \dots, x_l^*), \sum_{j=1}^{l} \|x_j^*\|_2 = 1, 1 \leq l \leq n+2, v^* = \sum_{j=1}^{l} x_j^* \Big\}$$

and

$$h_1^{G\mathcal{T}_d}(v^*, l, x^*, y^*, v) = v.$$

The weak and strong duality statements for (PVG) and $(DVG_1^{\mathcal{T}_d})$ follow from Theorems 4.10 and 4.13, while the corresponding necessary and sufficient optimality conditions can be derived from Theorem 4.15.

Theorem 4.27. *(a) There are no $x \in X$ and $(v^*, l, x^*, y^*, v) \in \mathcal{B}_1^{G\mathcal{T}_d}$ such that $F(x) <_K h_1^{G\mathcal{T}_d}(v^*, l, x^*, y^*, v)$.*

(b) If Φ is a K-convex vector function, the regularity condition $(RCV_0^{\mathcal{S}})$ is fulfilled and $\bar{x} \in \mathcal{PE}_{\mathcal{T}_d}(PVG)$, there exist $\bar{v}^* \in K^* \setminus \{0\}$, $\bar{l} \in \mathbb{N}$, $\bar{x}^* \in (K^*)^l$ and $\bar{y}^* \in Y^*$ such that $(\bar{v}^*, \bar{l}, \bar{x}^*, \bar{y}^*, F(\bar{x})) \in \mathcal{WE}(DVG_1^{\mathcal{T}_d})$ and $F(\bar{x}) = h_1^{G\mathcal{T}_d}(\bar{v}^*, \bar{l}, \bar{x}^*, \bar{y}^*, F(\bar{x}))$.

Theorem 4.28. *(a) When Φ is a K-convex vector function, the regularity condition $(RCV_0^{\mathcal{S}})$ is fulfilled and $\bar{x} \in \mathcal{PE}_{\mathcal{T}_d}(PVG)$, there exists $(\bar{v}^*, \bar{l}, \bar{x}^*, \bar{y}^*, \bar{v}) \in \mathcal{WE}(DVG_1^{\mathcal{T}_d})$ such that*

(i) $F(\bar{x}) = \bar{v}$;

(ii) $\Delta_{-K}(\bar{v}) = \bar{v}^{*\top}\bar{v}$;

(iii) $\bar{v}^* = \sum_{j=1}^{\bar{l}} \bar{x}_j^*$;

(iv) $\sum_{j=1}^{\bar{l}} \|\bar{x}_j^*\|_2 = 1$;

(v) $\bar{v}^{*\top}F(\bar{x}) + (\bar{v}^*\Phi)^*(0, \bar{y}^*) = 0$.

(b) *Assume that* $\bar{x} \in X$ *and* $(\bar{v}^*, \bar{l}, \bar{x}^*, \bar{y}^*, \bar{v}) \in \mathscr{B}_1^{G\mathscr{T}_d}$ *fulfill the relations* (i)–(v). *Then* $\bar{x} \in \mathscr{PE}_{\mathscr{T}_d}(PVG)$ *and* $(\bar{v}^*, \bar{l}, \bar{x}^*, \bar{y}^*, \bar{v}) \in \mathscr{WE}(DVG_1^{\mathscr{T}_d})$.

Remark 4.25. The optimality conditions (i)–(v) in Theorem 4.28 can be equivalently written as $F(\bar{x}) = \bar{v}$, $\bar{v}^* \in \partial\Delta_{-K}(\bar{v})$ and $(0, \bar{y}^*) \in \partial(\bar{v}^*\Phi)(\bar{x}, 0)$, where (cf. [74])

$$\partial\Delta_{-K}(\bar{v}) = \begin{cases} \overline{\text{co}}\{x^* \in \mathbb{R}^n : x^* \in -N_K(\bar{v}),\ \|x^*\|_2 = 1\}, & \text{if } \bar{v} \in -\operatorname{bd} K, \\ \overline{\text{co}}\left\{\frac{1}{\|p-\bar{v}\|_2}(p - \bar{v}) : p \in \Pi_{-K}(\bar{v})\right\}, & \text{if } \bar{v} \in -\operatorname{int} K, \\ \left\{\frac{1}{\|\bar{v}-P_{-K}(\bar{v})\|_2}(\bar{v} - P_{-K}(\bar{v}))\right\}, & \text{if } \bar{v} \notin -K, \end{cases}$$

and $\Pi_{-K}(\bar{v}) = \{y \in \mathbb{R}^n : \|y - \bar{v}\|_2 = \min_{z \in -\operatorname{bd} K} \|z - \bar{v}\|_2\}$. Note that in [80] one can find a simpler formula for $\partial\Delta_{-K}$ that moreover holds in Banach spaces, but under the assumption $\operatorname{int} K = \emptyset$. But, on the other hand, at the moment it is known that Δ_{-K} is strictly K-increasing only when $\operatorname{int} K \neq \emptyset$, therefore we cannot use the mentioned subdifferential formula in this framework.

4.3.6 Quadratic Scalarization

The last scalarization we consider in our investigation is based on a quadratic function and was considered for instance in [37, 87]. We work again in finitely dimensional spaces, taking $X = \mathbb{R}^n$ and $V = \mathbb{R}^k$, the latter partially ordered by the nontrivial pointed closed convex cone $K \subseteq \mathbb{R}^k$. Let $Q \in \mathscr{S}_+^k$ and $D \subseteq \mathbb{R}^k$ a *relatively open* set, i.e. $D = \operatorname{ri} D$. Denote by L the subspace parallel to $\operatorname{aff} D$. If $\operatorname{int}(K^* + L^\perp) \neq \emptyset$ and $QD \subseteq K^* + L^\perp$, where $L^\perp$ is the orthogonal subspace to L, then the function

$$s_q : (\mathbb{R}^k)^\bullet \to \overline{\mathbb{R}},\ s_q(y) = \begin{cases} y^\top Q y, & \text{if } y \in D, \\ +\infty, & \text{otherwise,} \end{cases}$$

is strongly K-increasing on D. Assume further that $F(\operatorname{dom} F) + K \subseteq D = \operatorname{dom} s_q$. The set of scalarization functions will consist in this case of a single element, namely

$$\mathscr{S}_q = \left\{s_q : (\mathbb{R}^k)^\bullet \to \overline{\mathbb{R}}\right\}.$$

An element $\bar{x} \in X$ is said to be an *$\mathscr{S}_q$-properly efficient solution to* (*PVG*) if $F(\bar{x})^\top QF(\bar{x}) \leq F(x)^\top QF(x)$ for all $x \in X$, and we denote this by $\bar{x} \in \mathscr{PE}_{\mathscr{S}_q}(PVG)$.

In order to formulate the vector dual problem to (*PVG*) that arises in this case we need the conjugate function of s_q. So far in the literature this conjugate was computed only for D being a subspace, so we assume further this hypothesis, too.

According to [128, p.331], the conjugate of the scalarization function is

$$s_q^* : \mathbb{R}^k \to \overline{\mathbb{R}},\ s_q^*(v^*) = \begin{cases} \frac{1}{4} v^{*\top}(P_D \circ Q \circ P_D)^\dagger v^*, & \text{if } v^* \in \operatorname{Im} Q + D^\perp, \\ +\infty, & \text{otherwise,} \end{cases}$$

$\operatorname{Im} Q$ is the image of Q seen as a symmetric positive semidefinite mapping on $\mathbb{R}^k$ and $Q^\dagger$ is the *Moore-Penrose pseudo-inverse* of Q.

Then the dual vector problem to (*PVG*) with respect to $\mathscr{S}_q$-properly efficient solution is

$$(DVG_1^{\mathscr{S}_q}) \qquad \operatorname*{Max}_{(v^*, y^*, v) \in \mathscr{B}_1^{G\mathscr{S}_q}} h_1^{G\mathscr{S}_q}(v^*, y^*, v),$$

where

$$\mathscr{B}_1^{G\mathscr{S}_q} = \Big\{ (v^*, y^*, v) \in K^* \times Y^* \times D : v^* \in \operatorname{Im} Q + D^\perp, \\ v^\top Q v \leq \tfrac{1}{4} v^{*\top}(P_D \circ Q \circ P_D)^\dagger v^* - (v^* \Phi)^*(0, y^*) \Big\}$$

and

$$h_1^{G\mathscr{S}_q}(v^*, y^*, v) = v.$$

The weak and strong duality statements for (*PVG*) and $(DVG_1^{\mathscr{S}_q})$ follow from Theorems 4.2 and 4.5.

Theorem 4.29. *(a) There are no $x \in X$ and $(v^*, y^*, v) \in \mathscr{B}_1^{G\mathscr{S}_q}$ such that $F(x) \leq_K h_1^{G\mathscr{S}_q}(v^*, y^*, v)$.*
(b) If Φ is a K-convex vector function, the regularity condition $(RCV_0^{\mathscr{S}})$ is fulfilled and $\bar{x} \in \mathscr{PE}_{\mathscr{S}_q}(PVG)$, there exist $\bar{v}^* \in K^*$ and $\bar{y}^* \in Y^*$ such that $(\bar{v}^*, \bar{y}^*, F(\bar{x})) \in \mathscr{E}(DVG_1^{\mathscr{S}_q})$ and $F(\bar{x}) = h_1^{G\mathscr{S}_q}(\bar{v}^*, \bar{y}^*, F(\bar{x}))$.

Moreover, from Theorem 4.7 one can obtain the following necessary and sufficient optimality conditions regarding (*PVG*) and $(DVG_1^{\mathscr{S}_q})$.

Theorem 4.30. *(a) If Φ is a K-convex vector function, the regularity condition $(RCV_0^{\mathscr{S}})$ is fulfilled and $\bar{x} \in \mathscr{PE}_{\mathscr{S}_q}(PVG)$, there exists $(\bar{v}^*, \bar{y}^*, \bar{v}) \in \mathscr{E}(DVG_1^{\mathscr{S}_q})$ such that*

(i) $F(\bar{x}) = \bar{v}$;
(ii) $\bar{v}^\top Q \bar{V} + \frac{1}{4}\bar{v}^{*\top}(P_D \circ Q \circ P_D)^\dagger \bar{v}^* = \bar{v}^{*\top}\bar{v}$;
(iii) $\bar{v}^{*\top} F(\bar{x}) + (\bar{v}^* \Phi)^*(0, \bar{y}^*) = 0$.

(b) Assume that $\bar{x} \in X$ and $(\bar{v}^*, \bar{y}^*, \bar{v}) \in \mathscr{B}_1^{G\mathscr{S}_q}$ fulfill the relations (i)–(iii). Then $\bar{x} \in \mathscr{PE}_{\mathscr{S}_q}(PVG)$ and $(\bar{v}^*, \bar{y}^*, \bar{v}) \in \mathscr{E}(DVG_1^{\mathscr{S}_q})$.

Remark 4.26. The optimality conditions (i)–(iii) in Theorem 4.30 can be equivalently written as $F(\bar{x}) = \bar{v}$, $\bar{v}^* \in \partial s_q(F(\bar{x}))$ and $(0, \bar{y}^*) \in \partial(\bar{v}^*\Phi)(\bar{x}, 0)$. When $\operatorname{int} D \neq \emptyset$ and consequently D is an open set, it holds $\partial s_q(y) = \{\nabla s_q(y)\} = \{2Qy\}$ for any $y \in D$ and $\partial s_q(y) = \emptyset$ when $y \notin D$. However, when $\operatorname{int} D = \emptyset$, it is not so easy to determine a simple formula for $\partial s_q(y)$ when $y \in D$.

Remark 4.27. We mentioned before that the scalarization functions considered in the literature serve different purposes, as noted, for instance, in [22, 81]. We have shown in this section how some of them can be employed for delivering vector dual problems to the original primal vector optimization problem and, under certain hypotheses, optimality conditions for it. One can also try to compare the resulting vector dual problems to (PVG). If $s \in \mathscr{T}$ (when $s \in \mathscr{S}$ the discussion is analogous), the Young-Fenchel inequality yields $s^*(v^*) + s(v) \geq \langle v^*, v\rangle$ for all $v \in V$ and all $v^* \in V^*$. Moreover, it is known that $\operatorname{dom} s^* \subseteq K^*$. However, one cannot conclude from here that $(s, v^*, y^*, v) \in \mathscr{B}_1^{G\mathscr{T}}$ implies $(v^*, y^*, v) \in \mathscr{B}_1^{G\mathscr{T}_l}$ because $v^* \in K^*$ in the former, while the v^* that is feasible to the vector dual problem to (PVG) that is obtained by means of the linear scalarization should belong to $K^* \setminus \{0\}$ and we have already seen in Remark 1.6 that the domain of the conjugate of a strictly K-increasing function is not necessarily a subset of $K^* \setminus \{0\}$. However, we noticed in Remarks 4.15 and 4.17, respectively, that at least the domains of the conjugates of the scalarization functions used in the maximum(-linear) scalarization and set scalarization are actually included in $K^* \setminus \{0\}$ and this guarantees that in the framework of Sect. 4.3.2 it holds $h_1^{G\mathscr{T}_l}\big(\mathscr{B}_1^{G\mathscr{T}_l}\big) \subseteq h_1^{G\mathscr{T}_{ml}}\big(\mathscr{B}_1^{G\mathscr{T}_{ml}}\big)$, while in the one of Sect. 4.3.3 one has $h_1^{G\mathscr{T}_l}\big(\mathscr{B}_1^{G\mathscr{T}_l}\big) \subseteq h_1^{G\mathscr{T}_s}\big(\mathscr{B}_1^{G\mathscr{T}_s}\big)$.

4.4 Vector Duality via Scalarization for Particular Vector Optimization Problems

In this section we particularize the duality investigations from Sect. 4.2 first for constrained vector optimization problems, then for unconstrained ones with the objective functions consisting in the sum of a vector function with the postcomposition of another with a linear continuous mapping.

4.4.1 Vector Duality via Scalarization for Constrained Vector Optimization Problems

Consider the nonempty convex set $S \subseteq X$ and the proper vector functions $f : X \to V^\bullet$ and $h : X \to Y^\bullet$ fulfilling $\operatorname{dom} f \cap S \cap h^{-1}(C) \neq \emptyset$. Let the primal vector optimization problem with geometric and cone constraints

$$(PVC) \qquad \underset{x \in \mathscr{A}}{\operatorname{Min}} f(x),$$

where

$$\mathscr{A} = \{x \in S : h(x) \in -C\}.$$

Since (PVC) is a special case of (PVG) obtained by taking

$$F : X \to V^\bullet, \ F(x) = \begin{cases} f(x), & \text{if } x \in \mathscr{A}, \\ \infty_K, & \text{otherwise,} \end{cases}$$

we use the approach developed in Sect. 4.2 in order to deal with it via duality.

Adapting the definition from the general case, an element $\bar{x} \in \mathscr{A}$ is said to be an $\mathscr{S}$*-properly efficient solution* to the vector optimization problem (PVC) if there exists a function $s \in \mathscr{S}$ such that $s(f(\bar{x})) \leq s(f(x))$ for all $x \in \mathscr{A}$. Here $\mathscr{S}$ is a set of functions $s : V^\bullet \to \overline{\mathbb{R}}$ that are proper, convex and strongly K-increasing on $f(\operatorname{dom} f \cap \mathscr{A}) + K$ fulfilling $f(\operatorname{dom} f \cap \mathscr{A}) + K \subseteq \operatorname{dom} s$ and $s(\infty_K) = +\infty$. Analogously, when $\operatorname{qi} K \neq \emptyset$ and $\mathscr{T}$ is a set of functions $s : V^\bullet \to \overline{\mathbb{R}}$ that are proper, convex and strictly K-increasing on $f(\operatorname{dom} f \cap \mathscr{A}) + K$ fulfilling $f(\operatorname{dom} f \cap \mathscr{A}) + K \subseteq \operatorname{dom} s$ and $s(\infty_K) = +\infty$, $\bar{x} \in \mathscr{A}$ is said to be a $\mathscr{T}$*-properly efficient solution* to the vector optimization problem (PVC) if there exists a function $s \in \mathscr{T}$ such that $s(f(\bar{x})) \leq s(f(x))$ for all $x \in \mathscr{A}$.

For convenient choices of the vector perturbation function Φ we obtain vector duals to (PVC) which are special cases of $(DVG_i^{\mathscr{S}})$ and $(DVG_i^{\mathscr{T}})$, $i \in \{1, 2\}$, respectively. Moreover, one can assign to (PVC) vector dual problems following from $(DVG_3^{\mathscr{S}})$ and $(DVG_3^{\mathscr{T}})$, too, where no perturbation functions are involved, namely

$$(DVC_3^{\mathscr{S}}) \qquad \underset{(s,v) \in \mathscr{B}_3^{C\mathscr{S}}}{\operatorname{Max}} h_3^{C\mathscr{S}}(s, v),$$

where

$$\mathscr{B}_3^{C\mathscr{S}} = \left\{(s, v) \in \mathscr{S} \times V : s(v) \leq \inf_{x \in \mathscr{A}} s(f(x))\right\}$$

and

$$h_3^{C\mathscr{S}}(s, v) = v,$$

and, respectively,

$$(DVC_3^{\mathscr{T}}) \qquad \underset{(s,v) \in \mathscr{B}_3^{C\mathscr{T}}}{\operatorname{WMax}} h_3^{C\mathscr{T}}(s, v),$$

where

$$\mathcal{B}_3^{C\mathcal{T}} = \left\{(s, v) \in \mathcal{T} \times V : s(v) \le \inf_{x\in\mathcal{A}} s(f(x))\right\}$$

and

$$h_3^{C\mathcal{T}}(s, v) = v.$$

The weak duality statements regarding these vector dual problems follow as special cases of the corresponding statements regarding (*PVG*) and its vector duals ($DVG_3^{\mathcal{S}}$) and ($DVG_3^{\mathcal{T}}$), namely Theorems 4.1 and 4.9, respectively.

Theorem 4.31. *(a) There are no* $x \in \mathcal{A}$ *and* $(s, v) \in \mathcal{B}_3^{C\mathcal{S}}$ *such that* $f(x) \le_K h_3^{C\mathcal{S}}(s, v)$.

(b) Assume that $\operatorname{qi} K \neq \emptyset$. *There are no* $x \in \mathcal{A}$ *and* $(s, v) \in \mathcal{B}_3^{C\mathcal{T}}$ *such that* $f(x) <_K h_3^{C\mathcal{T}}(s, v)$.

The strong duality statements regarding the vector optimization problems (*PVC*) and ($DVC_3^{\mathcal{S}}$), respectively ($DVC_3^{\mathcal{T}}$), follow automatically provided that the primal problem has at least a corresponding properly efficient solution, like in the general case, namely in Theorems 4.4 and 4.12, respectively. The same happens with the assertions delivering necessary and sufficient optimality conditions for these primal-dual pairs of problems, that are special cases of Theorems 4.6 and 4.14, respectively.

Theorem 4.32. *(a) If* $\bar{x} \in \mathcal{PE}_{\mathcal{S}}(PVC)$, *there exists an* $\bar{s} \in \mathcal{S}$ *such that* $(\bar{s}, f(\bar{x})) \in \mathcal{E}(DVC_3^{\mathcal{S}})$ *and* $f(\bar{x}) = h_3^{C\mathcal{S}}(\bar{s}, f(\bar{x}))$.

(b) If $\bar{x} \in \mathcal{PE}_{\mathcal{S}}(PVC)$, *there exists* $(\bar{s}, \bar{v}) \in \mathcal{E}(DVC_3^{\mathcal{S}})$ *such that*

(i) $f(\bar{x}) = \bar{v}$;

(ii) $\bar{s}(\bar{v}) = \bar{s}(f(\bar{x})) = \min_{x\in\mathcal{A}} \bar{s}(f(x))$.

(c) Assume that $\bar{x} \in \mathcal{A}$ *and* $(\bar{s}, \bar{v}) \in \mathcal{B}_3^{C\mathcal{S}}$ *fulfill the relations (i)–(ii) from (b). Then* $\bar{x} \in \mathcal{PE}_{\mathcal{S}}(PVC)$ *and* $(\bar{s}, \bar{v}) \in \mathcal{E}(DVC_3^{\mathcal{S}})$.

Theorem 4.33. *Assume that* $\operatorname{qi} K \neq \emptyset$.

(a) If $\bar{x} \in \mathcal{PE}_{\mathcal{T}}(PVC)$, *there exists an* $\bar{s} \in \mathcal{T}$ *such that* $(\bar{s}, f(\bar{x})) \in \mathcal{WE}(DVC_3^{\mathcal{T}})$ *and* $f(\bar{x}) = h_3^{C\mathcal{T}}(\bar{s}, f(\bar{x}))$.

(b) If $\bar{x} \in \mathcal{PE}_{\mathcal{T}}(PVC)$, *there exists* $(\bar{s}, \bar{v}) \in \mathcal{WE}(DVC_3^{\mathcal{T}})$ *such that*

(i) $f(\bar{x}) = \bar{v}$;

(ii) $\bar{s}(\bar{v}) = \bar{s}(f(\bar{x})) = \min_{x\in\mathcal{A}} \bar{s}(f(x))$.

(c) Assume that $\bar{x} \in \mathcal{A}$ *and* $(\bar{s}, \bar{v}) \in \mathcal{B}_3^{C\mathcal{T}}$ *fulfill the relations (i)–(ii) from (b). Then* $\bar{x} \in \mathcal{PE}_{\mathcal{T}}(PVC)$ *and* $(\bar{s}, \bar{v}) \in \mathcal{WE}(DVC_3^{\mathcal{T}})$.

Remark 4.28. The necessary and sufficient optimality conditions provided above for (*PVC*) and its vector duals can be rewritten by making use of subdifferentials like in Remark 4.6.

Remark 4.29. One can find in the literature the special cases of $(DVC_3^{L\mathscr{S}})$ and $(DVC_3^{L\mathscr{T}})$ obtained by means of linear scalarization in [48, 54, 55].

Now let us consider the Lagrange type vector perturbation function

$$\Phi_v^L : X \times Y \to V^\bullet,\ \Phi_v^L(x,z) = \begin{cases} f(x), & \text{if } x \in S, h(x) \in z - C, \\ \infty_K, & \text{otherwise,} \end{cases}$$

which is proper because f and h are proper and due to the fulfilment of the mentioned feasibility condition. For $v^* \in K^*$ and $z^* \in Y^*$ we have $(v^*\Phi_v^L)^*(0,z^*) = ((v^*f) - (z^*h) + \delta_S)^*(0) + \delta_{-C^*}(z^*)$, so the *Lagrange type vector duals* to (*PVC*) that follow from $(DVG_1^{\mathscr{S}})$ and $(DVG_1^{\mathscr{T}})$, respectively, are (note the change of sign of z^*)

$$(DVC_1^{L\mathscr{S}}) \qquad \operatorname*{Max}_{(s,v^*,z^*,v)\in\mathscr{B}_1^{L\mathscr{S}}} h_1^{L\mathscr{S}}(s,v^*,z^*,v),$$

where

$$\mathscr{B}_1^{L\mathscr{S}} = \left\{(s,v^*,z^*,v) \in \mathscr{S} \times K^* \times C^* \times V : s(v) \leq -s^*(v^*) - ((v^*f) + (z^*h) + \delta_S)^*(0)\right\}$$

and

$$h_1^{L\mathscr{S}}(s,v^*,z^*,v) = v,$$

and, when qi $K \neq \emptyset$,

$$(DVC_1^{L\mathscr{T}}) \qquad \operatorname*{WMax}_{(s,v^*,z^*,v)\in\mathscr{B}_1^{L\mathscr{T}}} h_1^{L\mathscr{T}}(s,v^*,z^*,v),$$

where

$$\mathscr{B}_1^{L\mathscr{T}} = \left\{(s,v^*,z^*,v) \in \mathscr{T} \times K^* \times C^* \times V : s(v) \leq -s^*(v^*) - ((v^*f) + (z^*h) + \delta_S)^*(0)\right\}$$

and

$$h_1^{L\mathscr{T}}(s,v^*,z^*,v) = v.$$

The other vector duals to (*PVC*) obtained via the Lagrange type vector perturbation are

$$(DVC_2^{L\mathscr{S}}) \qquad \underset{(s,z^*,v)\in\mathscr{B}_2^{L\mathscr{S}}}{\operatorname{Max}} h_2^{L\mathscr{S}}(s,z^*,v),$$

where

$$\mathscr{B}_2^{L\mathscr{S}} = \left\{(s,z^*,v) \in \mathscr{S} \times C^* \times V : s(v) \leq -(s \circ f + (z^*h) + \delta_S)^*(0)\right\}$$

and

$$h_2^{L\mathscr{S}}(s,z^*,v) = v,$$

and, when qi $K \neq \emptyset$,

$$(DVC_2^{L\mathscr{T}}) \qquad \underset{(s,z^*,v)\in\mathscr{B}_2^{L\mathscr{T}}}{\operatorname{WMax}} h_2^{L\mathscr{T}}(s,z^*,v),$$

where

$$\mathscr{B}_2^{L\mathscr{T}} = \left\{(s,z^*,v) \in \mathscr{T} \times C^* \times V : s(v) \leq -(s \circ f + (z^*h) + \delta_S)^*(0)\right\}$$

and

$$h_2^{L\mathscr{T}}(s,z^*,v) = v.$$

The weak duality statements regarding these vector dual problems follow as special cases of the corresponding statements regarding (PVG) and its duals, namely Theorems 4.2 and 4.3, respectively Theorems 4.10 and 4.11.

Theorem 4.34. *(a) There are no* $x \in \mathscr{A}$ *and* $(s,v^*,z^*,v) \in \mathscr{B}_1^{L\mathscr{S}}$ *such that* $f(x) \leq_K h_1^{L\mathscr{S}}(s,v^*,z^*,v)$.
(b) There are no $x \in \mathscr{A}$ *and* $(s,z^*,v) \in \mathscr{B}_2^{L\mathscr{S}}$ *such that* $f(x) \leq_K h_2^{L\mathscr{S}}(s,z^*,v)$.

Theorem 4.35. *Assume that* qi $K \neq \emptyset$.

(a) There are no $x \in \mathscr{A}$ *and* $(s,v^*,z^*,v) \in \mathscr{B}_1^{L\mathscr{T}}$ *such that* $f(x) <_K h_1^{L\mathscr{T}}(s,v^*,z^*,v)$.
(b) There are no $x \in \mathscr{A}$ *and* $(s,z^*,v) \in \mathscr{B}_2^{L\mathscr{T}}$ *such that* $f(x) <_K h_2^{L\mathscr{T}}(s,z^*,v)$.

The strong duality statements regarding the Lagrange type vector dual problems we assigned above to (PVC), as well as the corresponding necessary and sufficient optimality conditions regarding the mentioned pairs of primal-dual vector optimization problems require the fulfillment of some regularity conditions. The ones considered in Sect. 4.2 become

$$(RCV^{L\mathscr{S}}) \left| \begin{array}{l} \forall s \in \mathscr{S} \exists x' \in \operatorname{dom} f \cap S \text{ such that } h(x') \in -\operatorname{int} C \\ \text{and } s \text{ is continuous at } f(x'), \end{array} \right.$$

and, in case $\operatorname{int} K \neq \emptyset$,

$$(RCV_0^{L\mathscr{S}})\big|\ \exists x' \in \operatorname{dom} f \cap S \text{ such that } h(x') \in -\operatorname{int} C,$$

which is actually the classical *Slater constraint qualification* extended to the vector case.

Particularizing Theorem 4.5 for the situation considered here one obtains the following strong duality statement.

Theorem 4.36. *If f is a K-convex vector function, h is a C-convex vector function, the regularity condition $(RCV^{L\mathscr{S}})$ is fulfilled and $\bar{x} \in \mathscr{PE}_{\mathscr{S}}(PVC)$, there exist $\bar{s} \in \mathscr{S}$, $\bar{v}^* \in K^*$ and $\bar{z}^* \in Y^*$ such that $(\bar{s}, \bar{v}^*, \bar{z}^*, f(\bar{x})) \in \mathscr{E}(DVC_1^{L\mathscr{S}})$, $(\bar{s}, \bar{z}^*, f(\bar{x})) \in \mathscr{E}(DVC_2^{L\mathscr{S}})$ and $f(\bar{x}) = h_1^{L\mathscr{S}}(\bar{s}, \bar{v}^*, \bar{z}^*, f(\bar{x})) = h_2^{L\mathscr{S}}(\bar{s}, \bar{z}^*, f(\bar{x}))$.*

Employing now Theorems 4.7 and 4.8, respectively, the following necessary and sufficient optimality conditions statements involving (PVC) and its vector duals $(DVC_1^{L\mathscr{S}})$ and $(DVC_2^{L\mathscr{S}})$, respectively, are obtained.

Theorem 4.37. *(a) If f is a K-convex vector function, h is a C-convex vector function, the regularity condition $(RCV^{L\mathscr{S}})$ is fulfilled and $\bar{x} \in \mathscr{PE}_{\mathscr{S}}(PVC)$, there exists $(\bar{s}, \bar{v}^*, \bar{z}^*, \bar{v}) \in \mathscr{E}(DVC_1^{L\mathscr{S}})$ such that*

(i) $f(\bar{x}) = \bar{v}$;
(ii) $\bar{s}(\bar{v}) + \bar{s}^*(\bar{v}^*) = \langle \bar{v}^*, \bar{v} \rangle$;
(iii) $(\bar{v}^* f)(\bar{x}) + \big((\bar{v}^* f) + (\bar{z}^* h)\big)_S^*(0) = 0$;
(iv) $(\bar{z}^* h)(\bar{x}) = 0$.

(b) Assume that $\bar{x} \in X$ and $(\bar{s}, \bar{v}^*, \bar{z}^*, \bar{v}) \in \mathscr{B}_1^{L\mathscr{S}}$ fulfill the relations (i)–(iv). Then $\bar{x} \in \mathscr{PE}_{\mathscr{S}}(PVC)$ and $(\bar{s}, \bar{v}^*, \bar{z}^*, \bar{v}) \in \mathscr{E}(DVC_1^{L\mathscr{S}})$.

Theorem 4.38. *(a) If f is a K-convex vector function, h is a C-convex vector function, the regularity condition $(RCV^{L\mathscr{S}})$ is fulfilled and $\bar{x} \in \mathscr{PE}_{\mathscr{S}}(PVC)$, there exists $(\bar{s}, \bar{z}^*, \bar{v}) \in \mathscr{E}(DVC_2^{L\mathscr{S}})$ such that*

(i) $f(\bar{x}) = \bar{v}$;
(ii) $\bar{s}(f(\bar{x})) + (\bar{s} \circ f + (\bar{z}^* h))_S^*(0) = 0$;
(iii) $(\bar{z}^* h)(\bar{x}) = 0$.

(b) Assume that $\bar{x} \in \mathscr{A}$ and $(\bar{s}, \bar{z}^*, \bar{v}) \in \mathscr{B}_2^{L\mathscr{S}}$ fulfill the relations (i)–(iii). Then $\bar{x} \in \mathscr{PE}_{\mathscr{S}}(PVC)$ and $(\bar{s}, \bar{z}^*, \bar{v}) \in \mathscr{E}(DVC_2^{L\mathscr{S}})$.

Now let us give the similar strong duality and necessary and sufficient optimality conditions statements involving (PVC) and its Lagrange type vector duals with respect to $\mathscr{T}$-properly efficient solutions.

Theorem 4.39. *If $\operatorname{qi} K \neq \emptyset$, f is a K-convex vector function, h is a C-convex vector function, the regularity condition $(RCV^{L\mathscr{S}})$ is fulfilled and $\bar{x} \in \mathscr{PE}_{\mathscr{T}}(PVC)$, there exist $\bar{s} \in \mathscr{T}$, $\bar{v}^* \in K^*$ and $\bar{z}^* \in Y^*$ such that $(\bar{s}, \bar{v}^*, \bar{z}^*, f(\bar{x})) \in$*

$\mathscr{W}\mathscr{E}(DVC_1^{L\mathscr{T}})$, $(\bar{s}, \bar{z}^*, f(\bar{x})) \in \mathscr{W}\mathscr{E}(DVC_2^{L\mathscr{T}})$ *and* $f(\bar{x}) = h_1^{L\mathscr{T}}(\bar{s}, \bar{v}^*, \bar{z}^*, f(\bar{x})) = h_2^{L\mathscr{T}}(\bar{s}, \bar{z}^*, f(\bar{x}))$.

Theorem 4.40. *(a) If* $\operatorname{qi} K \neq \emptyset$, f *is a* K*-convex vector function,* h *is a* C*-convex vector function, the regularity condition* $(RCV^{L\mathscr{S}})$ *is fulfilled and* $\bar{x} \in \mathscr{P}\mathscr{E}_{\mathscr{T}}(PVC)$*, there exists* $(\bar{s}, \bar{v}^*, \bar{z}^*, \bar{v}) \in \mathscr{W}\mathscr{E}(DVC_1^{L\mathscr{T}})$ *such that conditions (i)–(iv) from Theorem 4.37 are fulfilled.*

(b) Assume that $\bar{x} \in X$ *and* $(\bar{s}, \bar{v}^*, \bar{z}^*, \bar{v}) \in \mathscr{B}_1^{L\mathscr{T}}$ *fulfill relations (i)–(iv) from Theorem 4.37. Then* $\bar{x} \in \mathscr{P}\mathscr{E}_{\mathscr{T}}(PVC)$ *and* $(\bar{s}, \bar{v}^*, \bar{z}^*, \bar{v}) \in \mathscr{W}\mathscr{E}(DVC_1^{L\mathscr{T}})$.

Theorem 4.41. *(a) If* $\operatorname{qi} K \neq \emptyset$, f *is a* K*-convex vector function,* h *is a* C*-convex vector function, the regularity condition* $(RCV^{L\mathscr{S}})$ *is fulfilled and* $\bar{x} \in \mathscr{P}\mathscr{E}_{\mathscr{T}}(PVC)$*, there exists* $(\bar{s}, \bar{z}^*, \bar{v}) \in \mathscr{W}\mathscr{E}(DVC_2^{L\mathscr{T}})$ *such that conditions (i)–(iii) from Theorem 4.38 are fulfilled.*

(b) Assume that $\bar{x} \in X$ *and* $(\bar{s}, \bar{z}^*, \bar{v}) \in \mathscr{B}_2^{L\mathscr{T}}$ *fulfill relations (i)–(iii) from Theorem 4.38. Then* $\bar{x} \in \mathscr{P}\mathscr{E}_{\mathscr{T}}(PVC)$ *and* $(\bar{s}, \bar{z}^*, \bar{v}) \in \mathscr{W}\mathscr{E}(DVC_2^{L\mathscr{T}})$.

Remark 4.30. The necessary and sufficient optimality conditions provided above for (PVC) and its Lagrange type vector duals can be rewritten by making use of subdifferentials like in Remarks 4.7 and 4.9, respectively.

Remark 4.31. The inclusions provided in Remarks 4.3, 4.4 and (4.2.1) can be particularized for (PVC) and its vector duals considered above, too.

Furthermore, one can particularize the scalarization functions as done in Sect. 4.3, obtaining different Lagrange type vector dual problems to (PVC).

Remark 4.32. One can find in the literature some special cases of the vector duals to (PVC) presented above, usually obtained for particular scalarizations. For instance, in [98] one can find $(DVC_2^{L\mathscr{S}})$ and in [192] $(DVC_2^{L\mathscr{T}})$, but with the scalarization functions assumed moreover continuous. In [102], in the same framework, a vector dual similar to $(DVC_2^{L\mathscr{S}})$ is considered, but with the inequality from the constraints replaced by the corresponding equality. In [48, 139, 140] the vector duals obtained in this framework from $(DVC_1^{L\mathscr{S}})$ and $(DVC_1^{L\mathscr{T}})$ by using the linear scalarization are treated.

Another vector perturbation function one can consider in order to assign vector dual problems to (PVC) is the Fenchel-Lagrange type vector perturbation function $\Phi_v^{FL} : X \times X \times Y \to V^{\bullet}$,

$$\Phi_v^{FL}(x, y, z) = \begin{cases} f(x + y), & \text{if } x \in S, h(x) \in z - C, \\ \infty_K, & \text{otherwise,} \end{cases}$$

which is proper since f and h are proper and due to the fulfilment of the mentioned feasibility condition. For $v^* \in K^*$, $y^* \in X^*$ and $z^* \in Y^*$ one has $(v^*\Phi_v^{CFL})^*(0, y^*, z^*) = (v^*f)^*(y^*) + (-(z^*h) + \delta_S)^*(-y^*) + \delta_{-C^*}(z^*)$, consequently, the *Fenchel-Lagrange type vector duals* to (PVC) obtained by making use of the vector perturbation function Φ_v^{FL} that follow from $(DVG_1^{\mathscr{S}})$ and $(DVG_1^{\mathscr{T}})$, respectively, are

$$(DVC_1^{FL\mathscr{S}}) \qquad \operatorname*{Max}_{(s,v^*,y^*,z^*,v)\in\mathscr{B}_1^{FL\mathscr{S}}} h_1^{FL\mathscr{S}}(s,v^*,y^*,z^*,v),$$

where

$$\mathscr{B}_1^{FL\mathscr{S}} = \Big\{(s,v^*,y^*,z^*,v) \in \mathscr{S} \times K^* \times X^* \times C^* \times V : \\ s(v) \leq -s^*(v^*) - (v^*f)^*(y^*) - ((z^*h) + \delta_S)^*(-y^*)\Big\}$$

and

$$h_1^{FL\mathscr{S}}(s,v^*,y^*,z^*,v) = v,$$

and, when qi $K \neq \emptyset$,

$$(DVC_1^{FL\mathscr{T}}) \qquad \operatorname*{Max}_{(s,v^*,y^*,z^*,v)\in\mathscr{B}_1^{FL\mathscr{T}}} h_1^{FL\mathscr{T}}(s,v^*,y^*,z^*,v),$$

where

$$\mathscr{B}_1^{FL\mathscr{T}} = \Big\{(s,v^*,y^*,z^*,v) \in \mathscr{T} \times K^* \times X^* \times C^* \times V : \\ s(v) \leq -s^*(v^*) - (v^*f)^*(y^*) - ((z^*h) + \delta_S)^*(-y^*)\Big\}$$

and

$$h_1^{FL\mathscr{T}}(s,v^*,y^*,z^*,v) = v.$$

The other vector duals to (PVC) obtained via the Fenchel-Lagrange type vector perturbation are

$$(DVC_2^{FL\mathscr{S}}) \qquad \operatorname*{Max}_{(s,y^*,z^*,v)\in\mathscr{B}_2^{FL\mathscr{S}}} h_2^{FL\mathscr{S}}(s,y^*,z^*,v),$$

where

$$\mathscr{B}_2^{FL\mathscr{S}} = \Big\{(s,y^*,z^*,v) \in \mathscr{S}\times X^*\times C^*\times V : s(v) \leq -(s\circ f)^*(y^*) - ((z^*h)+\delta_S)^*(-y^*)\Big\}$$

and

$$h_2^{FL\mathscr{S}}(s,y^*,z^*,v) = v,$$

and, respectively,

$$(DVC_2^{FL\mathcal{T}}) \qquad \underset{(s,y^*,z^*,v)\in\mathcal{B}_2^{FL\mathcal{T}}}{\text{WMax}} h_2^{FL\mathcal{T}}(s,y^*,z^*,v),$$

where

$$\mathcal{B}_2^{FL\mathcal{T}}=\Big\{(s,y^*,z^*,v)\in\mathcal{T}\times X^*\times C^*\times V : s(v)\leq-(s\circ f)^*(y^*)-((z^*h)+\delta_S)^*(-y^*)\Big\}$$

and

$$h_2^{FL\mathcal{T}}(s,y^*,z^*,v)=v.$$

The weak duality statements regarding these vector dual problems follow as special cases of the corresponding statements regarding (PVG) and its duals, namely Theorems 4.2 and 4.3, respectively Theorems 4.10 and 4.11.

Theorem 4.42. *(a) There are no* $x\in\mathcal{A}$ *and* $(s,v^*,y^*z^*,v)\in\mathcal{B}_1^{FL\mathcal{S}}$ *such that* $f(x)\leq_K h_1^{FL\mathcal{S}}(s,v^*,y^*,z^*,v)$.
(b) There are no $x\in\mathcal{A}$ *and* $(s,y^*,z^*,v)\in\mathcal{B}_2^{FL\mathcal{S}}$ *such that* $f(x)\leq_K h_2^{FL\mathcal{S}}(s,y^*,z^*,v)$.

Theorem 4.43. *Assume that* $\operatorname{qi} K\neq\emptyset$.

(a) There are no $x\in\mathcal{A}$ *and* $(s,v^*,y^*,z^*,v)\in\mathcal{B}_1^{FL\mathcal{T}}$ *such that* $f(x)<_K h_1^{FL\mathcal{T}}(s,v^*,y^*,z^*,v)$.
(b) There are no $x\in\mathcal{A}$ *and* $(s,y^*,z^*,v)\in\mathcal{B}_2^{FL\mathcal{T}}$ *such that* $f(x)<_K h_2^{FL\mathcal{T}}(s,y^*,z^*,v)$.

The strong duality statements regarding the Fenchel-Lagrange type vector dual problems to (PVC) introduced above, as well as the corresponding necessary and sufficient optimality conditions regarding the mentioned pairs of primal-dual vector optimization problems require the fulfillment of some regularity conditions. The ones considered in Sect. 4.2 become

$$(RCV^{FL\mathcal{S}})\left|\begin{array}{l}\forall s\in\mathcal{S}\,\exists x'\in\operatorname{dom} f\cap S \text{ such that } f \text{ is continuous at } x',\\ h(x')\in-\operatorname{int} C \text{ and } s \text{ is continuous at } f(x'),\end{array}\right.$$

and, in case $\operatorname{int} K\neq\emptyset$,

$$(RCV_0^{FL\mathcal{S}})\big|\exists x'\in\operatorname{dom} f\cap S \text{ such that } f \text{ is continuous at } x' \text{ and } h(x')\in-\operatorname{int} C.$$

Particularizing Theorem 4.5 for the situation considered here one obtains the following strong duality statement.

Theorem 4.44. *If* f *is a* K*-convex vector function,* h *is a* C*-convex vector function, the regularity condition* $(RCV^{FL\mathcal{S}})$ *is fulfilled and* $\bar{x}\in\mathcal{PE}_{\mathcal{S}}(PVC)$*, there exist* $\bar{s}\in$

$\mathcal{S}$, $\bar{v}^* \in K^*$, $\bar{y}^* \in X^*$ *and* $\bar{z}^* \in Y^*$ *such that* $(\bar{s}, \bar{v}^*, \bar{y}^*, \bar{z}^*, f(\bar{x})) \in \mathcal{E}(DVC_1^{FL_{\mathcal{S}}})$, $(\bar{s}, \bar{y}^*, \bar{z}^*, f(\bar{x})) \in \mathcal{E}(DVC_2^{FL_{\mathcal{S}}})$ *and* $f(\bar{x}) = h_1^{FL_{\mathcal{S}}}(\bar{s}, \bar{v}^*, \bar{y}^*, \bar{z}^*, f(\bar{x})) = h_2^{FL_{\mathcal{S}}}(\bar{s}, \bar{y}^*, \bar{z}^*, f(\bar{x}))$.

Employing now Theorems 4.7 and 4.8, respectively, the following necessary and sufficient optimality conditions statements involving (PVC) and its vector duals $(DVC_1^{FL_{\mathcal{S}}})$ and $(DVC_2^{FL_{\mathcal{S}}})$, respectively, are obtained.

Theorem 4.45. *(a) If f is a K-convex vector function, h is a C-convex vector function, the regularity condition $(RCV^{FL_{\mathcal{S}}})$ is fulfilled and $\bar{x} \in \mathcal{PE}_{\mathcal{S}}(PVC)$, there exists $(\bar{s}, \bar{v}^*, \bar{y}^*, \bar{z}^*, \bar{v}) \in \mathcal{E}(DVC_1^{FL_{\mathcal{S}}})$ such that*

(i) $f(\bar{x}) = \bar{v}$;
(ii) $\bar{s}(\bar{v}) + \bar{s}^*(\bar{v}^*) = \langle \bar{v}^*, \bar{v} \rangle$;
(iii) $(\bar{v}^* f)(\bar{x}) + (\bar{v}^* f)^*(\bar{y}^*) = \langle \bar{y}^*, \bar{x} \rangle$;
(iv) $(\bar{z}^* h)_S^*(-\bar{y}^*) = -\langle \bar{y}^*, \bar{x} \rangle$;
(v) $(\bar{z}^* h)(\bar{x}) = 0$.

(b) Assume that $\bar{x} \in X$ and $(\bar{s}, \bar{v}^*, \bar{y}^*, \bar{z}^*, \bar{v}) \in \mathcal{B}_1^{FL_{\mathcal{S}}}$ fulfill the relations (i)–(v). Then $\bar{x} \in \mathcal{PE}_{\mathcal{S}}(PVC)$ and $(\bar{s}, \bar{y}^*, \bar{v}^*, \bar{z}^*, \bar{v}) \in \mathcal{E}(DVC_1^{FL_{\mathcal{S}}})$.

Theorem 4.46. *(a) If f is a K-convex vector function, h is a C-convex vector function, the regularity condition $(RCV^{FL_{\mathcal{S}}})$ is fulfilled and $\bar{x} \in \mathcal{PE}_{\mathcal{S}}(PVC)$, there exists $(\bar{s}, \bar{y}^*, \bar{z}^*, \bar{v}) \in \mathcal{E}(DVC_2^{FL_{\mathcal{S}}})$ such that*

(i) $f(\bar{x}) = \bar{v}$;
(ii) $\bar{s}(f(\bar{x})) + (\bar{s} \circ f)^*(\bar{y}^*) = \langle \bar{y}^*, \bar{x} \rangle$;
(iii) $(\bar{z}^* h)_S^*(-\bar{y}^*) = -\langle \bar{y}^*, \bar{x} \rangle$;
(iv) $(\bar{z}^* h)(\bar{x}) = 0$.

(b) Assume that $\bar{x} \in X$ and $(\bar{s}, \bar{y}^*, \bar{z}^*, \bar{v}) \in \mathcal{B}_2^{FL_{\mathcal{S}}}$ fulfill the relations (i)–(iv). Then $\bar{x} \in \mathcal{PE}_{\mathcal{S}}(PVC)$ and $(\bar{s}, \bar{y}^*, \bar{z}^*, \bar{v}) \in \mathcal{E}(DVC_2^{FL_{\mathcal{S}}})$.

Now let us give the similar strong duality and necessary and sufficient optimality conditions statements involving (PVC) and its Fenchel-Lagrange type vector duals with respect to $\mathcal{T}$-properly efficient solutions.

Theorem 4.47. *If $\operatorname{qi} K \neq \emptyset$, f is a K-convex vector function, h is a C-convex vector function, the regularity condition $(RCV^{FL_{\mathcal{S}}})$ is fulfilled and $\bar{x} \in \mathcal{PE}_{\mathcal{T}}(PVC)$, there exist $\bar{s} \in \mathcal{T}$, $\bar{v}^* \in K^*$, $\bar{y}^* \in X^*$, and $\bar{z}^* \in Y^*$ such that $(\bar{s}, \bar{v}^*, \bar{y}^*, \bar{z}^*, f(\bar{x})) \in \mathcal{WE}(DVC_1^{FL_{\mathcal{T}}})$, $(\bar{s}, \bar{y}^*, \bar{z}^*, f(\bar{x})) \in \mathcal{WE}(DVC_2^{FL_{\mathcal{T}}})$ and $f(\bar{x}) = h_1^{FL_{\mathcal{T}}}(\bar{s}, \bar{v}^*, \bar{y}^*, \bar{z}^*, f(\bar{x})) = h_2^{FL_{\mathcal{T}}}(\bar{s}, \bar{y}^*, \bar{z}^*, f(\bar{x}))$.*

Theorem 4.48. *(a) If $\operatorname{qi} K \neq \emptyset$, f is a K-convex vector function, h is a C-convex vector function, the regularity condition $(RCV^{FL_{\mathcal{S}}})$ is fulfilled and $\bar{x} \in \mathcal{PE}_{\mathcal{T}}(PVC)$, there exists $(\bar{s}, \bar{v}^*, \bar{y}^*, \bar{z}^*, \bar{v}) \in \mathcal{WE}(DVC_1^{FL_{\mathcal{T}}})$ such that conditions (i)–(v) from Theorem 4.45 are fulfilled.*

(b) Assume that $\bar{x} \in X$ and $(\bar{s}, \bar{v}^*, \bar{y}^*, \bar{z}^*, \bar{v}) \in \mathcal{B}_1^{FL_{\mathcal{T}}}$ fulfill relations (i)–(v) from Theorem 4.45. Then $\bar{x} \in \mathcal{PE}_{\mathcal{T}}(PVC)$ and $(\bar{s}, \bar{v}^*, \bar{y}^*, \bar{z}^*, \bar{v}) \in \mathcal{WE}(DVC_1^{FL_{\mathcal{T}}})$.

Theorem 4.49. *(a) If f is a K-convex vector function, h is a C-convex vector function, the regularity condition $(RCV^{FL\mathscr{S}})$ is fulfilled and $\bar{x} \in \mathscr{PE}_{\mathscr{T}}(PVC)$, there exists $(\bar{s}, \bar{y}^*, \bar{z}^*, \bar{v}) \in \mathscr{WE}(DVC_2^{FL\mathscr{T}})$ such that conditions (i)–(iv) from Theorem 4.46 are fulfilled.*

(b) Assume that $\bar{x} \in X$ and $(\bar{s}, \bar{y}^*, \bar{z}^*, \bar{v}) \in \mathscr{B}_2^{FL\mathscr{T}}$ fulfill relations (i)–(iv) from Theorem 4.46. Then $\bar{x} \in \mathscr{PE}_{\mathscr{T}}(PVC)$ and $(\bar{s}, \bar{y}^*, \bar{z}^*, \bar{v}) \in \mathscr{WE}(DVC_2^{FL\mathscr{T}})$.

Remark 4.33. The necessary and sufficient optimality conditions provided above for (PVC) and its Fenchel-Lagrange type vector duals can be rewritten by making use of subdifferentials like in Remarks 4.7 and 4.9, respectively.

Remark 4.34. The inclusions provided in Remarks 4.3, 4.4 and (4.2.1) can be particularized for (PVC) and its vector duals considered above, too.

Furthermore, one can particularize the scalarization functions as done in Sect. 4.3, obtaining different Fenchel-Lagrange type vector dual problems to (PVC).

Remark 4.35. Note that $(DVC_1^{FL\mathscr{S}})$ and $(DVC_1^{FL\mathscr{T}})$ are actually the vector duals we introduced via the general scalarization in [37] in the finitely dimensional case and then extended to infinite dimensions in [48, Section 4.4]. In both these works the scalarization functions are then particularized, like in Sect. 4.3.

4.4.2 Vector Duality via Scalarization for Unconstrained Vector Optimization Problems

In this subsection we deal with unconstrained vector optimization problems whose objective functions consist of sums of functions. Let $f : X \to V^\bullet$ and $g : Y \to V^\bullet$ be given proper vector functions and $A : X \to Y$ a linear continuous mapping such that the feasibility condition $\text{dom}\, f \cap A^{-1}(\text{dom}\, g) \neq \emptyset$ is fulfilled.

The primal unconstrained vector optimization problem we work with is

$$(PVU) \qquad \underset{x \in X}{\text{Min}}[f(x) + g(Ax)].$$

Since (PVU) is a special case of (PVG) obtained by taking $F = f + g \circ A$, we use the approach developed in Sect. 4.2 in order to deal with it via duality.

Take $\mathscr{S}$ to be a set of functions $s : V^\bullet \to \overline{\mathbb{R}}$ that are proper, convex and strongly K-increasing on $(f + g \circ A)(\text{dom}\, f \cap A^{-1}(\text{dom}\, g)) + K$ that fulfill $(f + g \circ A)(\text{dom}\, f \cap A^{-1}(\text{dom}\, g)) + K \subseteq \text{dom}\, s$ and $s(\infty_K) = +\infty$. Adapting the definition from the general case, an element $\bar{x} \in X$ is said to be an *$\mathscr{S}$-properly efficient solution* to the vector optimization problem (PVU) if there exists a function $s \in \mathscr{S}$ such that $s((f + g \circ A)(\bar{x})) \leq s((f + g \circ A)(x))$ for all $x \in X$. Analogously, when $\text{qi}\, K \neq \emptyset$ and $\mathscr{T}$ is a set of functions $s : V^\bullet \to \overline{\mathbb{R}}$ that are proper, convex and strictly K-increasing on $(f + g \circ A)(\text{dom}\, f \cap A^{-1}(\text{dom}\, g)) + K$ fulfilling $(f + g \circ A)(\text{dom}\, f \cap A^{-1}(\text{dom}\, g)) + K \subseteq \text{dom}\, s$ and $s(\infty_K) = +\infty$, $\bar{x} \in X$ is said to be a $\mathscr{T}$*-properly efficient solution* to the vector optimization problem (PVU) if there exists a function $s \in \mathscr{T}$ such that $s((f + g \circ A)(\bar{x})) \leq s((f + g \circ A)(x))$ for all $x \in X$.

First, let us assign to (PVU) the vector dual problems following from $(DVG_3^{\mathscr{S}})$ and $(DVG_3^{\mathscr{T}})$, respectively, where no perturbation functions are involved, namely

$$(DVU_3^{\mathscr{S}}) \qquad \underset{(s,v)\in\mathscr{B}_3^{U\mathscr{S}}}{\operatorname{Max}} h_3^{U\mathscr{S}}(s,v),$$

where

$$\mathscr{B}_3^{U\mathscr{S}} = \left\{(s,v) \in \mathscr{S} \times V : s(v) \leq \inf_{x\in X} s((f + g \circ A)(x))\right\}$$

and

$$h_3^{U\mathscr{S}}(s,v) = v,$$

and, respectively,

$$(DVU_3^{\mathscr{T}}) \qquad \underset{(s,v)\in\mathscr{B}_3^{U\mathscr{T}}}{\operatorname{WMax}} h_3^{U\mathscr{T}}(s,v),$$

where

$$\mathscr{B}_3^{U\mathscr{T}} = \left\{(s,v) \in \mathscr{T} \times V : s(v) \leq \inf_{x\in X} s((f + g \circ A)(x))\right\}$$

and

$$h_3^{U\mathscr{T}}(s,v) = v.$$

The weak duality statements regarding these vector dual problems follow as special cases of the corresponding statements regarding (PVG) and its vector duals $(DVG_3^{\mathscr{S}})$ and $(DVG_3^{\mathscr{T}})$, namely Theorems 4.1 and 4.9, respectively.

Theorem 4.50. *(a) There are no $x \in X$ and $(s,v) \in \mathscr{B}_3^{U\mathscr{S}}$ such that $f(x) \leq_K h_3^{U\mathscr{S}}(s,v)$.*

(b) Assume that $\operatorname{qi} K \neq \emptyset$. *There are no $x \in X$ and $(s,v) \in \mathscr{B}_1^{U\mathscr{T}}$ such that $f(x) <_K h_1^{U\mathscr{T}}(s,v)$.*

The strong duality statements regarding the vector optimization problems (PVC) and $(DVU_3^{\mathscr{S}})$, respectively $(DVU_3^{\mathscr{T}})$ follow automatically provided that the primal problem has at least a corresponding properly efficient solution, like in the general case, namely in Theorems 4.4 and 4.12, respectively. The same happens with the assertions delivering necessary and sufficient optimality conditions for these primal-dual pairs of problems, that are special cases of Theorems 4.6 and 4.14, respectively.

Theorem 4.51. *(a) If $\bar{x} \in \mathscr{PE}_{\mathscr{S}}(PVU)$, there exists an $\bar{s} \in \mathscr{S}$ such that $(\bar{s}, (f + g \circ A)(\bar{x})) \in \mathscr{E}(DVU_3^{\mathscr{S}})$ and $(f + g \circ A)(\bar{x}) = h_3^{U_{\mathscr{S}}}(\bar{s}, (f + g \circ A)(\bar{x}))$.*
(b) If $\bar{x} \in \mathscr{PE}_{\mathscr{S}}(PVU)$, there exists $(\bar{s}, \bar{v}) \in \mathscr{E}(DVU_3^{\mathscr{S}})$ such that

(i) $(f + g \circ A)(\bar{x}) = \bar{v}$;
(ii) $\bar{s}(\bar{v}) = \bar{s}((f + g \circ A)(\bar{x})) = \min_{x \in X} \bar{s}((f + g \circ A)(x))$.

(c) Assume that $\bar{x} \in X$ and $(\bar{s}, \bar{v}) \in \mathscr{B}_3^{U_{\mathscr{S}}}$ fulfill the relations (i)–(ii) from (b). Then $\bar{x} \in \mathscr{PE}_{\mathscr{S}}(PVU)$ and $(\bar{s}, \bar{v}) \in \mathscr{E}(DVU_3^{\mathscr{S}})$.

Theorem 4.52. *Assume that* qi $K \neq \emptyset$.

(a) If $\bar{x} \in \mathscr{PE}_{\mathscr{T}}(PVU)$, there exists an $\bar{s} \in \mathscr{T}$ such that $(\bar{s}, (f + g \circ A)(\bar{x})) \in \mathscr{WE}(DVU_3^{\mathscr{T}})$ and $(f + g \circ A)(\bar{x}) = h_3^{U_{\mathscr{T}}}(\bar{s}, (f + g \circ A)(\bar{x}))$.
(b) If $\bar{x} \in \mathscr{PE}_{\mathscr{T}}(PVU)$, there exists $(\bar{s}, \bar{v}) \in \mathscr{WE}(DVU_3^{\mathscr{T}})$ such that

(i) $(f + g \circ A)(\bar{x}) = \bar{v}$;
(ii) $\bar{s}(\bar{v}) = \bar{s}((f + g \circ A)(\bar{x})) = \min_{x \in X} \bar{s}((f + g \circ A)(x))$.

(c) Assume that $\bar{x} \in X$ and $(\bar{s}, \bar{v}) \in \mathscr{B}_3^{U_{\mathscr{T}}}$ fulfill the relations (i)–(ii) from (b). Then $\bar{x} \in \mathscr{PE}_{\mathscr{T}}(PVU)$ and $(\bar{s}, \bar{v}) \in \mathscr{WE}(DVU_3^{\mathscr{T}})$.

Remark 4.36. The necessary and sufficient optimality conditions provided above for *(PVU)* and its vector duals can be rewritten by making use of subdifferentials like in Remark 4.6.

Like in the general case, one can assign vector duals to the primal vector optimization problem under consideration by making use of vector perturbation functions, too. Consider thus the vector perturbation function usually employed in the literature to approach *(PVU)*, namely

$$\Phi_v^U : X \times Y \to V^\bullet, \ \Phi_v^U(x, y) = f(x) + g(Ax + y),$$

which is proper because so are f and g and due to the fulfilment of the mentioned feasibility condition.

For $v^* \in K^*$ and $y^* \in Y^*$ one has $(v^*\Phi_v^U)^*(0, y^*) = (v^*f)^*(-A^*y^*) + (v^*g)^*(y^*)$. Now we are ready to formulate the vector duals to *(PVU)* that are special cases of $(DVG_1^{\mathscr{S}})$ and $(DVG_1^{\mathscr{T}})$, namely

$$(DVU_1^{\mathscr{S}}) \qquad \operatorname*{Max}_{(s,v^*,y^*,v) \in \mathscr{B}_1^{U_{\mathscr{S}}}} h_1^{U_{\mathscr{S}}}(s, v^*, y^*, v),$$

where

$$\mathscr{B}_1^{U_{\mathscr{S}}} = \Big\{(s, v^*, y^*, v) \in \mathscr{S} \times K^* \times Y^* \times V : \\ s(v) \leq -s^*(v^*) - (v^*f)^*(-A^*y^*) - (v^*g)^*(y^*)\Big\}$$

and

$$h_1^{U\mathscr{S}}(s, v^*, y^*, v) = v,$$

and, when $\operatorname{qi} K \neq \emptyset$,

$$(DVU_1^{\mathscr{T}}) \qquad \operatorname*{WMax}_{(s,v^*,y^*,v)\in\mathscr{B}_1^{U\mathscr{T}}} h_1^{U\mathscr{T}}(s, v^*, y^*, v),$$

where

$$\mathscr{B}_1^{U\mathscr{T}} = \Big\{(s, v^*, y^*, v) \in \mathscr{T} \times K^* \times Y^* \times V : \\ s(v) \leq -s^*(v^*) - (v^* f)^*(-A^* y^*) - (v^* g)^*(y^*)\Big\}$$

and

$$h_1^{U\mathscr{T}}(s, v^*, y^*, v) = v.$$

The other vector duals to (PVU) obtained via the considered vector perturbation function are

$$(DVU_2^{\mathscr{S}}) \qquad \operatorname*{Max}_{(s,y^*,v)\in\mathscr{B}_2^{U\mathscr{S}}} h_2^{U\mathscr{S}}(s, y^*, v),$$

where

$$\mathscr{B}_2^{U\mathscr{S}} = \Big\{(s, y^*, v) \in \mathscr{S} \times Y^* \times V : s(v) \leq -(s \circ \Phi_v^U)^*(0, y^*)\Big\}$$

and

$$h_2^{U\mathscr{S}}(s, y^*, v) = v,$$

and, when $\operatorname{qi} K \neq \emptyset$,

$$(DVU_2^{\mathscr{T}}) \qquad \operatorname*{WMax}_{(s,y^*,v)\in\mathscr{B}_2^{U\mathscr{T}}} h_2^{U\mathscr{T}}(s, y^*, v),$$

where

$$\mathscr{B}_2^{U\mathscr{T}} = \Big\{(s, y^*, v) \in \mathscr{T} \times Y^* \times V : s(v) \leq -(s \circ \Phi_v^U)^*(0, y^*)\Big\}$$

and

$$h_2^{U\mathscr{T}}(s, y^*, v) = v.$$

Remark 4.37. Regarding the definitions of $\mathscr{B}_2^{U\mathscr{S}}$ and $\mathscr{B}_2^{U\mathscr{T}}$, unfortunately we were not able to identify a formula for $(s \circ \Phi_v^U)^*(0, y^*)$ that uses only the functions s, f and g, similarly to the way the feasible sets of the vector duals of this type to (PVC) are defined. This can be done only under additional hypotheses, for instance via [49, Theorem 3.3] or when $\operatorname{int} K \neq \emptyset$.

The weak duality statements regarding these vector dual problems follow as special cases of the corresponding statements regarding (PVG) and its duals, namely Theorems 4.2 and 4.3, respectively Theorems 4.10 and 4.11.

Theorem 4.53. *(a) There are no* $x \in X$ *and* $(s, v^*, y^*, v) \in \mathscr{B}_1^{U\mathscr{S}}$ *such that* $(f + g \circ A)(x) \leq_K h_1^{U\mathscr{S}}(s, v^*, y^*, v)$.

(b) There are no $x \in X$ *and* $(s, y^*, v) \in \mathscr{B}_2^{U\mathscr{S}}$ *such that* $(f + g \circ A)(x) \leq_K h_2^{U\mathscr{S}}(s, y^*, v)$.

Theorem 4.54. *Assume that* $\operatorname{qi} K \neq \emptyset$.

(a) There are no $x \in X$ *and* $(s, v^*, y^*, v) \in \mathscr{B}_1^{U\mathscr{T}}$ *such that* $(f + g \circ A)(x) <_K h_1^{U\mathscr{T}}(s, v^*, y^*, v)$.

(b) There are no $x \in X$ *and* $(s, y^*, v) \in \mathscr{B}_2^{U\mathscr{T}}$ *such that* $(f + g \circ A)(x) <_K h_2^{U\mathscr{T}}(s, y^*, v)$.

The strong duality statements regarding the vector dual problems we assigned via perturbations to (PVU), as well as the corresponding necessary and sufficient optimality conditions regarding the mentioned pairs of primal-dual vector optimization problems require the fulfillment of some regularity conditions. The ones considered in Sect. 4.2 become

$$(RCV^{U\mathscr{S}}) \left| \begin{array}{l} \forall s \in \mathscr{S} \exists x' \in \operatorname{dom} f \cap A^{-1}(\operatorname{dom} g) \text{ such that } g \text{ is continuous} \\ \text{at } Ax' \text{ and } s \text{ is continuous at } f(x') + g(Ax') \end{array} \right.$$

and, in case $\operatorname{int} K \neq \emptyset$,

$$(RCV_0^{U\mathscr{S}}) \left| \ \exists x' \in \operatorname{dom} f \cap A^{-1}(\operatorname{dom} g) \text{ such that } g \text{ is continuous at } Ax'. \right.$$

Particularizing Theorem 4.5 for the situation considered here one obtains the following strong duality statement.

Theorem 4.55. *If* f *and* g *are* K*-convex vector functions, the regularity condition* $(RCV^{U\mathscr{S}})$ *is fulfilled and* $\bar{x} \in \mathscr{PE}_{\mathscr{S}}(PVU)$*, there exist* $\bar{s} \in \mathscr{S}$, $\bar{v}^* \in K^*$ *and* $\bar{y}^* \in Y^*$ *such that* $(\bar{s}, \bar{v}^*, \bar{y}^*, (f + g \circ A)(\bar{x})) \in \mathscr{E}(DVU_1^{\mathscr{S}})$, $(\bar{s}, \bar{y}^*, (f + g \circ A)(\bar{x})) \in \mathscr{E}(DVU_2^{\mathscr{S}})$ *and* $(f + g \circ A)(\bar{x}) = h_1^{U\mathscr{S}}(\bar{s}, \bar{v}^*, \bar{y}^*, (f + g \circ A)(\bar{x})) = h_2^{U\mathscr{S}}(\bar{s}, \bar{y}^*, (f + g \circ A)(\bar{x}))$.

Employing now Theorems 4.7 and 4.8, respectively, the following necessary and sufficient optimality conditions statements involving (PVU) and its vector duals $(DVU_1^{\mathscr{S}})$ and $(DVU_2^{\mathscr{S}})$, respectively, are obtained.

Theorem 4.56. *(a) If f and g are K-convex vector functions, the regularity condition $(RCV^{U\mathscr{S}})$ is fulfilled and $\bar{x} \in \mathscr{PE}_{\mathscr{S}}(PVU)$, there exists $(\bar{s}, \bar{v}^*, \bar{y}^*, \bar{v}) \in \mathscr{E}(DVU_1^{\mathscr{S}})$ such that*

(i) $(f + g \circ A)(\bar{x}) = \bar{v}$;
(ii) $\bar{s}(\bar{v}) + \bar{s}^*(\bar{v}^*) = \langle \bar{v}^*, \bar{v} \rangle$;
(iii) $(\bar{v}^* f)(\bar{x}) + (\bar{v}^* f)^*(-A^* \bar{y}^*) = -\langle \bar{y}^*, A\bar{x} \rangle$;
(iv) $(\bar{v}^* g)(A\bar{x}) + (\bar{v}^* g)^*(\bar{y}^*) = \langle \bar{y}^*, A\bar{x} \rangle$.

(b) Assume that $\bar{x} \in X$ and $(\bar{s}, \bar{v}^*, \bar{y}^*, \bar{v}) \in \mathscr{B}_1^{U\mathscr{S}}$ fulfill the relations (i)–(iv). Then $\bar{x} \in \mathscr{PE}_{\mathscr{S}}(PVU)$ and $(\bar{s}, \bar{v}^*, \bar{y}^*, \bar{v}) \in \mathscr{E}(DVU_1^{\mathscr{S}})$.

Theorem 4.57. *(a) If f and g are K-convex vector functions, the regularity condition $(RCV^{U\mathscr{S}})$ is fulfilled and $\bar{x} \in \mathscr{PE}_{\mathscr{S}}(PVU)$, there exists $(\bar{s}, \bar{y}^*, \bar{v}) \in \mathscr{E}(DVU_2^{\mathscr{S}})$ such that*

(i) $f(\bar{x}) = \bar{v}$;
(ii) $(\bar{v}^*(f + g \circ A))(\bar{x}) + (\bar{s} \circ \Phi_v^U)^*(0, \bar{y}^*) = 0$.

(b) Assume that $\bar{x} \in X$ and $(\bar{s}, \bar{y}^*, \bar{v}) \in \mathscr{B}_2^{U\mathscr{S}}$ fulfill the relations (i)–(ii). Then $\bar{x} \in \mathscr{PE}_{\mathscr{S}}(PVU)$ and $(\bar{s}, \bar{y}^*, \bar{v}) \in \mathscr{E}(DVU_2^{\mathscr{S}})$.

Now let us give the similar strong duality and necessary and sufficient optimality conditions statements involving *(PVU)* and its vector duals with respect to $\mathscr{T}$-properly efficient solutions obtained via perturbations.

Theorem 4.58. *If $\operatorname{qi} K \neq \emptyset$, f and g are K-convex vector functions, the regularity condition $(RCV^{U\mathscr{S}})$ is fulfilled and $\bar{x} \in \mathscr{PE}_{\mathscr{T}}(PVU)$, there exist $\bar{s} \in \mathscr{T}$, $\bar{v}^* \in K^*$ and $\bar{y}^* \in Y^*$ such that $(\bar{s}, \bar{v}^*, \bar{y}^*, (f + g \circ A)(\bar{x})) \in \mathscr{WE}(DVU_1^{\mathscr{T}})$, $(\bar{s}, \bar{y}^*, (f + g \circ A)(\bar{x})) \in \mathscr{WE}(DVU_2^{\mathscr{T}})$ and $(f + g \circ A)(\bar{x}) = h_1^{U\mathscr{T}}(\bar{s}, \bar{v}^*, \bar{y}^*, (f + g \circ A)(\bar{x})) = h_2^{U\mathscr{T}}(\bar{s}, \bar{y}^*, (f + g \circ A)(\bar{x}))$.*

Theorem 4.59. *(a) If $\operatorname{qi} K \neq \emptyset$, f and g are K-convex vector functions, the regularity condition $(RCV^{U\mathscr{S}})$ is fulfilled and $\bar{x} \in \mathscr{PE}_{\mathscr{T}}(PVU)$, there exists $(\bar{s}, \bar{v}^*, \bar{y}^*, \bar{v}) \in \mathscr{WE}(DVU_1^{\mathscr{T}})$ such that conditions (i)–(iv) from Theorem 4.56 are fulfilled.*
(b) Assume that $\bar{x} \in X$ and $(\bar{s}, \bar{v}^*, \bar{y}^*, \bar{v}) \in \mathscr{B}_1^{U\mathscr{T}}$ fulfill relations (i)–(iv) from Theorem 4.56. Then $\bar{x} \in \mathscr{PE}_{\mathscr{T}}(PVU)$ and $(\bar{s}, \bar{v}^*, \bar{y}^*, \bar{v}) \in \mathscr{WE}(DVU_1^{\mathscr{T}})$.

Theorem 4.60. *(a) If $\operatorname{qi} K \neq \emptyset$, f and g are K-convex vector functions, the regularity condition $(RCV^{U\mathscr{S}})$ is fulfilled and $\bar{x} \in \mathscr{PE}_{\mathscr{T}}(PVU)$, there exists $(\bar{s}, \bar{y}^*, \bar{v}) \in \mathscr{WE}(DVU_2^{\mathscr{T}})$ such that conditions (i)–(ii) from Theorem 4.57 are fulfilled.*
(b) Assume that $\bar{x} \in X$ and $(\bar{s}, \bar{y}^*, \bar{v}) \in \mathscr{B}_2^{U\mathscr{T}}$ fulfill relations (i)–(ii) from Theorem 4.57. Then $\bar{x} \in \mathscr{PE}_{\mathscr{T}}(PVU)$ and $(\bar{s}, \bar{y}^*, \bar{v}) \in \mathscr{WE}(DVU_2^{L\mathscr{T}})$.

Remark 4.38. The necessary and sufficient optimality conditions provided above for *(PVU)* and its vector duals can be rewritten by making use of subdifferentials like in Remarks 4.7 and 4.9, respectively.

Remark 4.39. The inclusions provided in Remarks 4.3, 4.4 and (4.2.1) can be particularized for (PVU) and its vector duals considered above, too.

Furthermore, one can particularize the scalarization functions as done in Sect. 4.3, obtaining different vector dual problems to (PVC).

Remark 4.40. One can find in the literature some special cases of the vector duals to (PVU) presented above, usually obtained for particular scalarizations. In [48] the vector duals obtained in this framework from $(DVU_1^{\mathscr{S}})$ and $(DVU_1^{\mathscr{T}})$ by using the linear scalarization are treated, while in [58, 59] a vector dual similar $(DVU_1^{\mathscr{S}})$ is obtained via linear scalarization, but with the inequality in the constraints replaced by an equality.

Remark 4.41. Valuable special cases of the vector optimization problem (PVU), met in the literature in various circumstances, can be obtained, for instance, by taking $X = Y$ and A to be the identity mapping on X or f to be a zero vector function, respectively. The vector duals assigned above to (PVU) and the corresponding duality and optimality conditions statements can be directly particularized for these problems, too.

Getting back to the constrained vector optimization problem considered in Sect. 4.4.1 and using the notations considered there, one can see (PVC) as an unconstrained vector optimization problem, namely

$$(PVC) \qquad \operatorname*{Min}_{x \in X} \left[f(x) + \delta^v_{\mathscr{A}}(x) \right],$$

Then, taking $A := \mathrm{id}_X$, $f := f$ and $g := \delta^v_{\mathscr{A}}$, the vector dual problems assigned to (PVU) with respect to $\mathscr{S}$-properly efficient solutions via perturbations turn into

$$(DVC_1^{F\mathscr{S}}) \qquad \operatorname*{Max}_{(s,v^*,y^*,v) \in \mathscr{B}_1^{F\mathscr{S}}} h_1^{F\mathscr{S}}(s, v^*, y^*, v),$$

where

$$\mathscr{B}_1^{F\mathscr{S}} = \left\{ (s, v^*, y^*, v) \in \mathscr{S} \times K^* \times Y^* \times V : s(v) \leq -s^*(v^*) - (v^* f)^*(y^*) - \sigma_{\mathscr{A}}(-y^*) \right\}$$

and

$$h_1^{F\mathscr{S}}(s, v^*, y^*, v) = v,$$

and, respectively,

$$(DVC_2^{F\mathscr{S}}) \qquad \operatorname*{Max}_{(s,y^*,v) \in \mathscr{B}_2^{F\mathscr{S}}} h_2^{F\mathscr{S}}(s, y^*, v),$$

where

$$\mathcal{B}_2^{F\mathcal{S}} = \left\{(s, y^*, v) \in \mathcal{S} \times Y^* \times V : s(v) \leq -(s \circ f)^*(y^*) - \sigma_{\mathcal{A}}(-y^*)\right\}$$

and

$$h_2^{F\mathcal{S}}(s, y^*, v) = v,$$

where the vector perturbation function Φ_v^U becomes actually the Fenchel type vector perturbation function for (PVC), namely

$$\Phi_v^F : X \times X \to V^\bullet,\ \Phi_v^F(x, y) = \begin{cases} f(x + y), & \text{if } x \in \mathcal{A}, \\ \infty_K, & \text{otherwise,} \end{cases}$$

which is proper because f and h are proper and due to the fulfilment of the mentioned feasibility condition. This is the reason why the vector dual problems assigned to (PVC) via the vector perturbation function Φ_v^F are said to be *Fenchel vector duals*.

When $\operatorname{qi} K \neq \emptyset$, vector dual problems can be assigned to (PVU) with respect to $\mathcal{T}$-properly efficient solutions, too, via the vector perturbation function Φ_v^F, namely

$$(DVC_1^{F\mathcal{T}}) \qquad \underset{(s,v^*,y^*,v)\in\mathcal{B}_1^{F\mathcal{T}}}{\operatorname{WMax}} h_1^{F\mathcal{T}}(s, v^*, y^*, v),$$

where

$$\mathcal{B}_1^{F\mathcal{T}} = \left\{(s, v^*, y^*, v) \in \mathcal{T} \times K^* \times Y^* \times V : s(v) \leq -s^*(v^*) - (v^* f)^*(y^*) - \sigma_{\mathcal{A}}(-y^*)\right\}$$

and

$$h_1^{F\mathcal{T}}(s, v^*, y^*, v) = v,$$

and, respectively,

$$(DVC_2^{F\mathcal{T}}) \qquad \underset{(s,y^*,v)\in\mathcal{B}_2^{F\mathcal{T}}}{\operatorname{WMax}} h_2^{F\mathcal{T}}(s, y^*, v),$$

where

$$\mathcal{B}_2^{F\mathcal{T}} = \left\{(s, y^*, v) \in \mathcal{S} \times Y^* \times V : s(v) \leq -(s \circ f)^*(y^*) - \sigma_{\mathcal{A}}(-y^*)\right\}$$

and

$$h_2^{F\mathcal{T}}(s, y^*, v) = v.$$

The weak duality statements regarding these vector dual problems follow as special cases of the corresponding statements regarding (PVU) and its vector duals or, alternatively from the ones regarding (PVG) and its vector duals, namely Theorems 4.2 and 4.3, respectively Theorems 4.10 and 4.11.

Theorem 4.61. *(a) There are no $x \in \mathscr{A}$ and $(s, v^*, y^*, v) \in \mathscr{B}_1^{F\mathscr{S}}$ such that $f(x) \leq_K h_1^{F\mathscr{S}}(s, v^*, y^*, v)$.*
(b) There are no $x \in \mathscr{A}$ and $(s, y^*, v) \in \mathscr{B}_2^{F\mathscr{S}}$ such that $f(x) \leq_K h_2^{F\mathscr{S}}(s, y^*, v)$.

Theorem 4.62. *Assume that* qi $K \neq \emptyset$.

(a) There are no $x \in \mathscr{A}$ and $(s, v^*, y^*, v) \in \mathscr{B}_1^{F\mathscr{T}}$ such that $f(x) <_K h_1^{F\mathscr{T}}(s, v^*, y^*, v)$.
(b) There are no $x \in \mathscr{A}$ and $(s, y^*, v) \in \mathscr{B}_2^{F\mathscr{T}}$ such that $f(x) <_K h_2^{F\mathscr{T}}(s, y^*, v)$.

The strong duality statements regarding the Fenchel vector dual problems we assigned above to (PVC), as well as the corresponding necessary and sufficient optimality conditions regarding the mentioned pairs of primal-dual vector optimization problems require the fulfillment of some regularity conditions. The ones considered for (PVU) become

$$(RCV^{F\mathscr{S}}) \left| \begin{array}{l} \forall s \in \mathscr{S} \exists x' \in \operatorname{dom} f \cap \mathscr{A} \text{ such that } f \text{ is continuous} \\ \text{at } x' \text{ and } s \text{ is continuous at } f(x'), \end{array} \right.$$

and, in case int $K \neq \emptyset$,

$$(RCV_0^{F\mathscr{S}}) \left| \ \exists x' \in \operatorname{dom} f \cap \mathscr{A} \text{ such that } f \text{ is continuous at } A'. \right.$$

Particularizing Theorem 4.55 for the situation considered here one obtains the following strong duality statement.

Theorem 4.63. *If f is a K-convex vector function, h is a C-convex vector function, the regularity condition $(RCV^{F\mathscr{S}})$ is fulfilled and $\bar{x} \in \mathscr{PE}_{\mathscr{S}}(PVC)$, there exist $\bar{s} \in \mathscr{S}$, $\bar{v}^* \in K^*$ and $\bar{y}^* \in Y^*$ such that $(\bar{s}, \bar{v}^*, \bar{y}^*, f(\bar{x})) \in \mathscr{E}(DVC_1^{F\mathscr{S}})$, $(\bar{s}, \bar{y}^*, f(\bar{x})) \in \mathscr{E}(DVC_2^{F\mathscr{S}})$ and $f(\bar{x}) = h_1^{F\mathscr{S}}(\bar{s}, \bar{v}^*, \bar{y}^*, f(\bar{x})) = h_2^{F\mathscr{S}}(\bar{s}, \bar{y}^*, f(\bar{x}))$.*

Employing now Theorems 4.56 and 4.57, respectively, the following necessary and sufficient optimality conditions statements involving (PVC) and its vector duals $(DVC_1^{F\mathscr{S}})$ and $(DVC_2^{F\mathscr{S}})$, respectively, are obtained.

Theorem 4.64. *(a) If f is a K-convex vector function, h is a C-convex vector function, the regularity condition $(RCV^{F\mathscr{S}})$ is fulfilled and $\bar{x} \in \mathscr{PE}_{\mathscr{S}}(PVC)$, there exists $(\bar{s}, \bar{v}^*, \bar{y}^*, \bar{v}) \in \mathscr{E}(DVC_1^{F\mathscr{S}})$ such that*

(i) $f(\bar{x}) = \bar{v}$;
(ii) $\bar{s}(\bar{v}) + \bar{s}^*(\bar{v}^*) = \langle \bar{v}^*, \bar{v} \rangle$;

(iii) $(\bar{v}^* f)(\bar{x}) + (\bar{v}^* f)(\bar{y}^*) = \langle \bar{y}^*, \bar{x} \rangle$;
(iv) $\min_{x \in \mathscr{A}} \langle \bar{y}^*, x \rangle = \langle \bar{y}^*, \bar{x} \rangle$.

(b) *Assume that* $\bar{x} \in X$ *and* $(\bar{s}, \bar{v}^*, \bar{y}^*, \bar{v}) \in \mathscr{B}_1^{F_{\mathscr{S}}}$ *fulfill the relations (i)–(iv). Then* $\bar{x} \in \mathscr{P}\mathscr{E}_{\mathscr{S}}(PVC)$ *and* $(\bar{s}, \bar{v}^*, \bar{y}^*, \bar{v}) \in \mathscr{E}(DVC_1^{F_{\mathscr{S}}})$.

Theorem 4.65. *(a) If f is a K-convex vector function, h is a C-convex vector function, the regularity condition $(RCV^{F_{\mathscr{S}}})$ is fulfilled and $\bar{x} \in \mathscr{P}\mathscr{E}_{\mathscr{S}}(PVC)$, there exists $(\bar{s}, \bar{y}^*, \bar{v}) \in \mathscr{E}(DVC_2^{F_{\mathscr{S}}})$ such that*

(i) $f(\bar{x}) = \bar{v}$;
(ii) $\bar{s}(f(\bar{x})) + (\bar{s} \circ f)^*(\bar{y}^*) = \langle \bar{y}^*, \bar{x} \rangle$;
(iii) $\min_{x \in \mathscr{A}} \langle \bar{y}^*, x \rangle = \langle \bar{y}^*, \bar{x} \rangle$.

(b) *Assume that* $\bar{x} \in \mathscr{A}$ *and* $(\bar{s}, \bar{y}^*, \bar{v}) \in \mathscr{B}_2^{F_{\mathscr{S}}}$ *fulfill the relations (i)–(iii). Then* $\bar{x} \in \mathscr{P}\mathscr{E}_{\mathscr{S}}(PVC)$ *and* $(\bar{s}, \bar{y}^*, \bar{v}) \in \mathscr{E}(DVC_2^{F_{\mathscr{S}}})$.

The similar strong duality and necessary and sufficient optimality conditions statements involving (*PVC*) and its Fenchel type vector duals with respect to $\mathscr{T}$-properly efficient solutions follow analogously.

Theorem 4.66. *If* $\operatorname{qi} K \neq \emptyset$, *$f$ is a K-convex vector function, h is a C-convex vector function, the regularity condition $(RCV^{F_{\mathscr{S}}})$ is fulfilled and $\bar{x} \in \mathscr{P}\mathscr{E}_{\mathscr{T}}(PVC)$, there exist $\bar{s} \in \mathscr{T}$, $\bar{v}^* \in K^*$ and $\bar{y}^* \in Y^*$ such that $(\bar{s}, \bar{v}^*, \bar{y}^*, f(\bar{x})) \in \mathscr{W}\mathscr{E}(DVC_1^{F_{\mathscr{T}}})$, $(\bar{s}, \bar{y}^*, f(\bar{x})) \in \mathscr{W}\mathscr{E}(DVC_2^{F_{\mathscr{T}}})$ and $f(\bar{x}) = h_1^{F_{\mathscr{T}}}(\bar{s}, \bar{v}^*, \bar{y}^*, f(\bar{x})) = h_2^{F_{\mathscr{T}}}(\bar{s}, \bar{y}^*, f(\bar{x}))$.*

Theorem 4.67. *(a) If* $\operatorname{qi} K \neq \emptyset$, *$f$ is a K-convex vector function, h is a C-convex vector function, the regularity condition $(RCV^{F_{\mathscr{S}}})$ is fulfilled and $\bar{x} \in \mathscr{P}\mathscr{E}_{\mathscr{T}}(PVC)$, there exists $(\bar{s}, \bar{v}^*, \bar{y}^*, \bar{v}) \in \mathscr{W}\mathscr{E}(DVC_1^{F_{\mathscr{T}}})$ such that conditions (i)–(iv) from Theorem 4.64 are fulfilled.*

(b) Assume that $\bar{x} \in X$ and $(\bar{s}, \bar{v}^*, \bar{y}^*, \bar{v}) \in \mathscr{B}_1^{F_{\mathscr{T}}}$ fulfill relations (i)–(iv) from Theorem 4.64. Then $\bar{x} \in \mathscr{P}\mathscr{E}_{\mathscr{T}}(PVC)$ and $(\bar{s}, \bar{v}^*, \bar{y}^*, \bar{v}) \in \mathscr{W}\mathscr{E}(DVC_1^{F_{\mathscr{T}}})$.

Theorem 4.68. *(a) If* $\operatorname{qi} K \neq \emptyset$, *$f$ is a K-convex vector function, h is a C-convex vector function, the regularity condition $(RCV^{F_{\mathscr{S}}})$ is fulfilled and $\bar{x} \in \mathscr{P}\mathscr{E}_{\mathscr{T}}(PVC)$, there exists $(\bar{s}, \bar{y}^*, \bar{v}) \in \mathscr{W}\mathscr{E}(DVC_2^{F_{\mathscr{T}}})$ such that conditions (i)–(iii) from Theorem 4.65 are fulfilled.*

(b) Assume that $\bar{x} \in X$ and $(\bar{s}, \bar{y}^*, \bar{v}) \in \mathscr{B}_2^{F_{\mathscr{T}}}$ fulfill relations (i)–(iii) from Theorem 4.65. Then $\bar{x} \in \mathscr{P}\mathscr{E}_{\mathscr{T}}(PVC)$ and $(\bar{s}, \bar{y}^*, \bar{v}) \in \mathscr{W}\mathscr{E}(DVC_2^{F_{\mathscr{T}}})$.

Remark 4.42. The necessary and sufficient optimality conditions provided above for (*PVC*) and its Lagrange type vector duals can be rewritten by making use of subdifferentials like in Remarks 4.7 and 4.9, respectively.

Remark 4.43. The inclusions provided in Remarks 4.3, 4.4 and (4.2.1) can be particularized for (*PVC*) and its Fenchel type vector duals, too.

Remark 4.44. The image sets of the vector duals assigned to (PVC) with respect to $\mathscr{S}$-properly efficient solutions and, respectively, the ones with respect to $\mathscr{T}$-properly efficient solutions can be compared, analogously to the investigations from their counterparts from Sect. 5.5.2. Using the properties of the conjugate functions, one obtains for $i \in \{1, 2\}$

$$h_i^{FL\mathscr{S}}(\mathscr{B}_i^{FL\mathscr{S}}) \subseteq \begin{array}{c} h_i^{L\mathscr{S}}(\mathscr{B}_i^{L\mathscr{S}}) \\ h_i^{F\mathscr{S}}(\mathscr{B}_i^{F\mathscr{S}}) \end{array} \subseteq h_3^{C\mathscr{S}}(\mathscr{B}_3^{C\mathscr{S}}),$$

and, respectively,

$$h_i^{FL\mathscr{T}}(\mathscr{B}_i^{FL\mathscr{T}}) \subseteq \begin{array}{c} h_i^{L\mathscr{T}}(\mathscr{B}_i^{L\mathscr{T}}) \\ h_i^{F\mathscr{T}}(\mathscr{B}_i^{F\mathscr{T}}) \end{array} \subseteq h_3^{C\mathscr{T}}(\mathscr{B}_3^{C\mathscr{T}}).$$

Carefully analyzing the differences between the way a Lagrange or Fenchel vector dual to (PVC) is defined and its Fenchel-Lagrange counterpart, one can derive sufficient conditions that guarantee their coincidence from the stable strong duality results from Chap. 2 or [48, Section 3.5].

Chapter 5
General Wolfe and Mond-Weir Duality

5.1 Historical Overview and Motivation

After having a solid duality theory for linear optimization problems, the next step was to extend it for more general problems. Following Dorn's successful generalization of duality for quadratic problems, the next step was to deal with convex optimization problems. In [215], Wolfe proposed a dual problem for a scalar convex optimization problem in which the involved functions were taken differentiable, too. Then, Schechter extended in [186] Wolfe's duality for convex nondifferentiable functions, by replacing the gradients with (convex) subdifferentials. Then it was noticed that this duality approach can be extended for some classes of nonconvex optimization problems where the involved functions have certain generalized convexity properties. On the other hand, Mond and Weir proposed in [169] other dual problems to a constrained optimization problem where the involved functions were taken pseudoconvex and quasiconvex, respectively. For both these duals, in order to achieve strong duality a known optimal solution of the primal problem is required, extending thus somehow faithfully the duality approach for linear optimization problems. This is actually the main difference between the mentioned duality approaches and the classical conjugate one. Unlike Wolfe's duality approach, the Mond-Weir one has proven to be useful also when dealing with fractional problems, leading to the achievement of strong duality even for problems with fractions as objective functions where other duals proposed in the literature (by Bector or Schaible, for instance) failed. Afterwards, both Wolfe and Mond-Weir duality approaches were employed, separately, parallelly or even combined, for different classes of optimization problems, like the ones involving second order convex or invex functions. Moreover, they were used for developing symmetric duality for certain classes of problems and for constructing primal-dual pairs of problems where strong duality occurs without assuming the satisfaction of any constraint qualification. The rich literature on Wolfe and Mond-Weir duality concepts has developed in the last decades especially in the differentiable case.

S.-M. Grad, *Vector Optimization and Monotone Operators via Convex Duality*,
Vector Optimization, DOI 10.1007/978-3-319-08900-3_5

The main direction followed in this research was the one of relaxing the convexity assumptions on the functions involved, the connections of these duality concepts to other duality types based on convex functions remaining somehow neglected. On the other hand, in most of the papers dealing with these duality concepts only finitely dimensional spaces were considered.

A natural step was to extend the Wolfe and Mond-Weir duality approaches from scalar to vector optimization problems, too. The investigations begun by Mond, Weir, Craven and Egudo in papers like [82, 83, 205–207, 209, 211, 212] presented Wolfe and Mond-Weir type vector duals, respectively, to a constrained vector optimization problem obtained by retaining the objective functions of the primal problem into the objective functions of the vector duals. They quickly continued in various directions that led to a large number of papers where the Wolfe and Mond-Weir duality approaches were employed for constrained vector optimization problems with the objective functions usually containing (generalized convex) differentiable vector functions, vectors of fractions or other combinations of differentiable functions. On the other hand, Chien proposed in [71] another approach to construct a Mond-Weir type vector dual for a constrained vector minimization problem, namely by employing an idea considered in [58, 59, 140] for Fenchel and respectively Lagrange type vector duals, which consists in using the objective function of the primal problem only in the constraints of the dual. This duality approach also been employed for both Wolfe type and Mond-Weir type duality for fractional vector optimization problems.

Motivated by the huge amount of works where the classical Wolfe and Mond-Weir duality concepts are considered in various circumstances, we present in Sect. 5.2 a more general approach by embedding them into a larger class of dual problems obtained via perturbation theory, following our works [29, 48]. In this way these duality concepts can be applied to unconstrained optimization problems, too, and on the other hand one can deliver other Wolfe and Mond-Weir type duals for constrained optimization problems, too. Moreover, the functions involved in our investigations are defined on Hausdorff locally convex vector spaces and for achieving strong duality different weak regularity conditions are proposed. In Sects. 5.3 and 5.4 we extend our investigations from the scalar case also for vector optimization problems embedding the Wolfe and Mond-Weir duality concepts in classes of dual vector optimization problems attached via perturbations to a general vector optimization problem with respect to its properly efficient solutions, by exploiting the two mentioned main directions from the literature, respectively, following our recent papers [32, 108]. Like in the scalar case, our work was performed in the very general setting of Hausdorff locally convex vector spaces. Then the primal problem is specialized to be unconstrained, respectively constrained, and vector duals of both Wolfe and Mond-Weir types for it are obtained via different perturbation functions. Afterwards, we present in Sect. 5.5 some comparisons involving the image sets of different vector dual problems attached to the same primal problem.

5.2 Wolfe and Mond-Weir Type Duality for Scalar Optimization Problems

We begin our investigations with a general scalar optimization problem, to which dual problems of both Wolfe and Mond-Weir types are assigned. Then we particularize the primal problem to be constrained and unconstrained, respectively, and the corresponding dual problems are derived.

5.2.1 General Scalar Optimization Problems

Consider two Hausdorff locally convex vector spaces X and Y, the proper function $F : X \to \overline{\mathbb{R}}$ and the general optimization problem

$$(PG) \qquad \inf_{x \in X} F(x),$$

Making use of a proper perturbation function $\Phi : X \times Y \to \overline{\mathbb{R}}$, fulfilling $\Phi(x, 0) = F(x)$ for all $x \in X$, a hypothesis that guarantees that $0 \in \Pr_Y \operatorname{dom} \Phi$, the problem (PG) means nothing but

$$(PG) \qquad \inf_{x \in X} \Phi(x, 0).$$

To it one can attach besides the conjugate dual problem (DG) introduced in Sect. 1.2.2 also the *Wolfe type* dual problem

$$(DG_W) \qquad \sup_{\substack{u \in X, y \in Y, y^* \in Y^*, \\ (0, y^*) \in \partial \Phi(u, y)}} \big\{ -\Phi^*(0, y^*) \big\},$$

and the *Mond-Weir type* one

$$(DG_M) \qquad \sup_{\substack{u \in X, y^* \in Y^*, \\ (0, y^*) \in \partial \Phi(u, 0)}} \Phi(u, 0).$$

As we shall see later, when one takes as primal an optimization problem consisting in minimizing a function subject to both geometric and cone-inequality constraints, an appropriate perturbation function is employed and all the functions involved are Gâteaux differentiable and convex, the two dual problems introduced above lead to the classical Wolfe and Mond-Weir duals to the mentioned problem, respectively.

Next we show that weak duality holds for (PG) and these two duals, too.

Theorem 5.1. *One has*

$$-\infty \le v(DG_M) \le v(DG_W) \le v(PG) \le +\infty.$$

Proof. Noting that (DG_M) can be obtained from (DG_W) by taking $y = 0$ and using the constraint involving the subdifferential, it follows that $-\infty \le v(DG_M) \le v(DG_W)$. On the other hand, (DG_W) is actually the problem (DG) with an additional constraint. Consequently, $v(DG_W) \le v(DG)$ and, taking into account the weak duality statement for (DG) and (PG), we are done. □

Remark 5.1. As a byproduct of the proof of Theorem 5.1 one obtains the inequality $v(DG_W) \le v(DG)$.

Remark 5.2. Situations where the last inequality in Theorem 5.1 is strict are widely known in the literature, while for an example to have $-\infty < v(DG_M)$ it suffices to have this dual problem feasible. Later, in Examples 5.2 and 5.3 we bring into attention two situations where $v(DG_M) < v(DG_W)$ and $v(DG_W) < v(DG)$, respectively.

One of the directions in which both Wolfe and Mond-Weir duality concepts were developed in the literature is towards introducing dual problems for which strong duality holds without asking the fulfillment of any regularity condition. As it can be noticed in the following observation, (DG_M) is such a dual problem, provided the nonemptiness of its feasible set.

Remark 5.3. If the feasible set of (DG_M) is nonempty, containing for instance the element $(\bar{u}, \bar{y}^*)$, then $(0, \bar{y}^*) \in \partial\Phi(\bar{u}, 0)$ yields via Corollary 2.8(b) that $\bar{u}$ is an optimal solution to (PG), $\bar{y}^*$ is an optimal solution to (DG) and $v(DG) = v(PG) = \Phi(\bar{u}, 0) \le v(DG_M)$. Via Theorem 5.1 and Remark 5.1 we obtain $v(DG_M) = v(DG_W) = v(DG) = v(PG)$ and moreover that $(\bar{u}, \bar{y}^*)$ is an optimal solution to (DG_M) and $(\bar{u}, 0, \bar{y}^*)$ is one to (DG_W). Consequently, in this case we have strong duality for all three dual problems we assigned to (PG), namely (DG_M), (DG_W) and (DG).

However, in order to guarantee strong duality for the two duals to (PG) introduced above one needs not necessarily directly verify the nonemptiness of the feasible set of (DG_M), since this can be guaranteed by classical weak hypotheses, as the following statement shows. The regularity conditions used below are actually the ones considered in Sect. 1.2.2 for ensuring strong duality for (PG) and (DG).

Theorem 5.2. *Assume that Φ is a convex function. Let $\bar{x} \in X$ be an optimal solution to (PG) and assume that one of the regularity conditions (RC_i^G), $i \in \{1,2,3,4\}$, is fulfilled. Then $v(PG) = v(DG_W) = v(DG_M)$ and there exists a $\bar{y}^* \in Y^*$ for which $(\bar{x}, \bar{y}^*)$ is an optimal solution to (DG_M) and $(\bar{x}, 0, \bar{y}^*)$ is one to (DG_W).*

Proof. Corollary 2.8(a) guarantees that under the present hypotheses there exists a $\bar{y}^* \in Y^*$, which is an optimal solution to (DG), such that $(0, \bar{y}^*) \in \partial\Phi(\bar{x}, 0)$.

Thus the feasible set of (DG_M) is nonempty, containing at least the element $(\bar{x}, \bar{y}^*)$. The conclusion follows via Remark 5.3. □

Remark 5.4. Other regularity conditions can be used in order to guarantee strong duality for (DG_M) and (DG_W), too, as long as they ensure the stability of the problem (PG) with respect to the perturbation function Φ.

Remark 5.5. One can note that there is strong duality for (PG) and (DG_M) if and only if there is strong duality for (PG) and (DG), while the strong duality for (PG) and (DG) implies the same thing for (PG) and (DG_W). However, the strong duality for (PG) and (DG_W) implies in general only that for (PG) and (DG) there is zero duality gap. Similar observations can be made for the special cases of these problems that will be treated further within this section, too.

Let us see now how do the duals arising from (DG_W) and (DG_M) look when the primal problem takes several classical particular formulations and the perturbation functions are carefully chosen. More precisely, the primal is taken first to mean finding the infimum of a function subject to both geometric and cone-inequality constraints, and afterwards to consist in the unconstrained minimization of a sum of a function with the composition of another function with a linear continuous mapping.

5.2.2 Constrained Scalar Optimization Problems

The first class of particular optimization problems for which we particularize the investigations from Sect. 5.2.1 is the one of the constrained optimization problems. Consider the nonempty set $S \subseteq X$ and let the nonempty convex cone $C \subseteq Y$ induce a partial ordering on Y. Take the proper functions $f : X \to \overline{\mathbb{R}}$ and $h : X \to Y^\bullet$, fulfilling the feasibility condition $\operatorname{dom} f \cap S \cap h^{-1}(-C) \neq \emptyset$. The primal problem we treat further is

$$(PC) \qquad \inf_{x \in \mathscr{A}} f(x),$$

where

$$\mathscr{A} = \{x \in S : h(x) \in -C\}.$$

Like in Sect. 2.2.2, we consider two perturbation functions which will lead to two different types of dual problems to (PC) that arise from (DG_W) and (DG_M), respectively. Concerning the Lagrange perturbation function Φ^L, one has $(0, z^*) \in \partial \Phi^L(u, z)$ if and only if $u \in S$, $h(u) \in z - C$ and $(f - (z^* h) + \delta_S)^*(0) + \delta_{C^*}(-z^*) + (f + \delta_S)(u) + \delta_{-C}(h(u) - z) = \langle z^*, z \rangle$. Using the fact that $\delta_{-C}^* = \delta_{C^*}$, the latter can be equivalently rewritten as

$\big((f - (z^*h) + \delta_S)^*(0) + (f - (z^*h) + \delta_S)(u)\big) + \big(\delta^*_{-C}(-z^*) + \delta_{-C}(h(u) - z) + \langle z^*, h(u) - z\rangle\big) = 0$. By the Young-Fenchel inequality it follows that this equality is nothing but $(f - (z^*h) + \delta_S)^*(0) + (f - (z^*h) + \delta_S)(u) = 0$ and $\delta^*_{-C}(-z^*) + \delta_{-C}(h(u) - z) + \langle z^*, h(u) - z\rangle = 0$, i.e. $0 \in \partial(f - (z^*h) + \delta_S)(u)$, $z^* \in -C^*$ and $\delta_{-C}(h(u) - z) - \langle -z^*, h(u) - z\rangle = 0$. Thus we obtain from (DG_W) the following dual problem to (PC)

$$(DC^L_W) \qquad \sup_{\substack{u\in S, z\in Y, z^*\in -C^*,\\ h(u)-z\in -C, (z^*h)(u)=\langle z^*, z\rangle,\\ 0\in\partial(f-(z^*h)+\delta_S)(u)}} \{f(u) - \langle z^*, z\rangle\},$$

which can be equivalently rewritten as

$$(DC^L_W) \qquad \sup_{\substack{u\in S, z^*\in C^*,\\ 0\in\partial(f+(z^*h)+\delta_S)(u)}} \{f(u) + (z^*h)(u)\}.$$

We call this the *Wolfe dual of Lagrange type* to (PC). We shall see later that, in the particular instance where the classical Wolfe duality was considered, this dual turns into the well-known Wolfe dual problem to (PC).

Analogously we get a dual problem to (PC) arising from (DG_M), namely

$$(DC^L_M) \qquad \sup_{\substack{u\in S, z^*\in C^*,\\ h(u)\in -C, (z^*h)(u)\geq 0,\\ 0\in\partial(f+(z^*h)+\delta_S)(u)}} f(u).$$

Note that in the constraints of this dual one can replace $(z^*h)(u) \geq 0$ by $(z^*h)(u) = 0$ without altering anything. Because the classical *Mond-Weir dual* to (PC) can be obtained, in the corresponding framework, as a special case to (DC^L_M) by removing the constraint $h(u) \in -C$, we consider here also the *Mond-Weir dual problem of Lagrange type* to (PC)

$$(DC^L_{MW}) \qquad \sup_{\substack{u\in S, z^*\in C^*, (z^*h)(u)\geq 0,\\ 0\in\partial(f+(z^*h)+\delta_S)(u)}} f(u).$$

By construction it is clear that $v(DC^L_M) \leq v(DC^L_{MW})$. On the other hand, whenever (u, z^*) is feasible to (DC^L_{MW}) it is feasible to (DC^L_W), too, and moreover $(z^*h)(u) \geq 0$. This yields $f(u) \leq f(u) + (z^*h)(u) \leq v(DC^L_W)$. Considering the supremum over all the pairs (u, z^*) feasible to (DC^L_{MW}) we obtain $v(DC^L_{MW}) \leq v(DC^L_W)$. Applying the weak duality statement Theorem 5.1, we get

$$v(DC^L_M) \leq v(DC^L_{MW}) \leq v(DC^L_W) \leq v(DC^L) \leq v(PC). \tag{5.2.1}$$

As can be seen in the following situations, the Wolfe dual has sometimes an indeed larger optimal objective value than the Mond-Weir one, while the classical conjugate dual can have a strictly greater one than the other mentioned two duals. Moreover, the first example we give below exhibits a situation where the first inequality in (5.2.1) is strictly fulfilled.

Example 5.1. Let $X = \mathbb{R}$, $Y = \mathbb{R}$, $C = \mathbb{R}_+$, $Y^\bullet = \mathbb{R} \cup \{\infty_{\mathbb{R}_+}\}$, $S = \mathbb{R}_+$, $f : \mathbb{R} \to \mathbb{R}$, $f(x) = x$, and $h : \mathbb{R} \to \mathbb{R} \cup \{\infty_{\mathbb{R}_+}\}$,

$$h(x) = \begin{cases} -x, & \text{if } x > 0, \\ 2, & \text{if } x = 0, \\ \infty_{\mathbb{R}_+}, & \text{if } x < 0, \end{cases}$$

where we note that $\infty_{\mathbb{R}_+}$ can be actually identified with $+\infty$.

We have $0 \in \partial(f + (0h) + \delta_S)(0) = (-\infty, 1]$ and $(0h)(0) = 0$, thus $(0, 0)$ is feasible to (DC^L_{MW}). So $v(DC^L_{MW}) \geq 0$ and since $v(PC) = 0$ one gets $v(DC^L_{MW}) = 0$. Employing (5.2.1), we obtain that $v(DC^L_W) = 0$, too.

On the other hand, $h(0) = 2 > 0$, thus there is no $z^* \in \mathbb{R}_+$ for which $(0, z^*)$ is feasible to (DC^L_M). Noting that $h(u) \neq 0$ for all $u \in \mathbb{R}$, from the constraint $(z^*h)(u) = 0$ (see the discussion after introducing (DC^L_M)) one obtains that whenever (u, z^*) were feasible to (DC^L_M) there should be $z^* = 0$. Since for every $u > 0$ we have $\partial(f + (0h) + \delta_S)(u) = \{1\}$, it follows that (DC^L_M) has no feasible points, consequently $v(DC^L_M) = -\infty$.

Therefore, $v(DC^L_M) < v(DC^L_{MW}) = v(DC^L_W)$ in this case. Employing also (5.2.1), one can see that $v(DG_M)$ can in general be smaller than $v(DG_W)$.

Example 5.2. Let $X = \mathbb{R}$, $Y = \mathbb{R}^2$, $C = \mathbb{R}^2_+$, $S = \mathbb{R}_+$, $f : \mathbb{R} \to \overline{\mathbb{R}}$,

$$f(x) = \begin{cases} x, & \text{if } x > 0, \\ +\infty, & \text{otherwise,} \end{cases}$$

and $h : \mathbb{R} \to \mathbb{R}^2$, $h(x) = (x - 1, -x)^\top$.

When, for $u > 0$ and $z^* = (z_1^*, z_2^*)^\top \geqq 0$ it holds $0 \in \partial(f + (z^*h) + \delta_S)(u)$, one obtains $z_1^* - z_2^* = -1$. Thus (DC^L_W) has feasible points and, for some $u > 0$ we obtain $v(PC) = 0 \geq v(DC^L_W) \geq \sup\{f(u) + (z^*h)(u) : z^* = (z_1^*, z_2^*)^\top \in \mathbb{R}^2_+, z_1^* - z_2^* = -1\} = \sup\{u + z_1^*(u-1) - z_2^* u : z^* = (z_1^*, z_2^*)^\top \in \mathbb{R}^2_+, z_1^* - z_2^* = -1\} = \sup\{u(1 + z_1^* - z_2^*) - z_2^* : z^* = (z_1^*, z_2^*)^\top \in \mathbb{R}^2_+, z_1^* - z_2^* = -1\} = \sup_{z_2^* \geq 0}\{-z_2^*\} = 0$. Then obviously $v(DC^L_W) = 0$.

On the other hand, $(z^*h)(u) \geq 0$ yields $u(z_1^* - z_2^*) - z_1^* \geq 0$, i.e. $-u - z_1^* \geq 0$. But $u > 0$ and $z_1^* \geq 0$, thus we obtained a contradiction, consequently (DC^L_{MW}) has no feasible points. Employing also (5.2.1), we obtain $v(DC^L_M) = v(DC^L_{MW}) = -\infty$.

Therefore, $v(DC^L_M) = v(DC^L_{MW}) < v(DC^L_W)$ in this case.

Example 5.3. Let $X = \mathbb{R}^2$, $Y = \mathbb{R}$, $C = \mathbb{R}_+$, $S = \mathbb{R}^2$, $f : \mathbb{R}^2 \to \mathbb{R}$, $f(x, y) = y$, and $h : \mathbb{R}^2 \to \mathbb{R}$, $h(x, y) = e^x - y$.

For $u = (u_1, u_2)^\top \in \mathbb{R}^2$ and $z^* \geq 0$ we have, taking into account the continuity of f and h, $\partial(f + (z^*h) + \delta_S)(u) = \partial f(u) + \partial(z^*h)(u)$, hence

$$\partial(f + (z^*h) + \delta_S)(u) = \left\{ \begin{pmatrix} 0 \\ 1 \end{pmatrix} \right\} + \left\{ \begin{pmatrix} z^* e^{u_1} \\ -z^* \end{pmatrix} \right\} = \left\{ \begin{pmatrix} z^* e^{u_1} \\ 1 - z^* \end{pmatrix} \right\}.$$

Then $(0,0)^\top \in \partial(f + (z^*h) + \delta_S)(u)$ if and only if concomitantly $z^* = 0$ and $z^* = 1$, that is impossible. Consequently, via (5.2.1), $v(DC_M^L) = v(DC_{MW}^L) = v(DC_W^L) = -\infty$.

On the other hand,

$$v(DC^L) = \sup_{z^* \geq 0} \inf_{(u_1,u_2) \in \mathbb{R}^2} \left[u_2 + z^* e^{u_1} - z^* u_2 \right] = \sup_{z^* \geq 0} \left\{ \inf_{u_1 \in \mathbb{R}} z^* e^{u_1} + \inf_{u_2 \in \mathbb{R}} u_2 (1 - z^*) \right\} = 0.$$

Therefore, $v(DC_M^L) = v(DC_{MW}^L) = v(DC_W^L) < v(DC^L)$ in this case. Employing also (5.2.1), one can see that $v(DG_W)$ can in general be smaller than $v(DG)$.

For strong duality, which follows directly from Theorem 5.2, besides convexity assumptions which guarantee the convexity of the perturbation function Φ^L we use regularity conditions, too, obtained in Sect. 2.2.2 by particularizing (RC_i^G), $i \in \{1,2,3,4\}$. Specializing Theorem 5.2 for the present context we obtain strong duality statements for (PC) and (DC_W^L) and (DC_M^L), respectively, while the one concerning (DC_{MW}^L) follows analogously or via (5.2.1).

Theorem 5.3. *Assume that S is a convex set, f is a convex function and h is a C-convex vector function. Let $\bar{x} \in X$ be an optimal solution to (PC) and assume that one of the regularity conditions (RC_i^L), $i \in \{1,2,3,4\}$, is fulfilled. Then $v(PC) = v(DC_W^L) = v(DC_M^L) = v(DC_{MW}^L)$ and there exists a $\bar{z}^* \in C^*$ for which $(\bar{x}, \bar{z}^*)$ is an optimal solution to all three duals.*

Remark 5.6. One can notice that in the situation considered in Example 5.3 the convexity hypotheses of Theorem 5.3 and the Slater constraint qualification (RC_1^L) are valid, but strong duality fails for the Wolfe and Mond-Weir type duals. This happens because the infimal objective value of (PC) is not attained, the primal problem having no optimal solutions.

Remark 5.7. Assume that S is a convex set, f is a convex function and h is a C-convex vector function. When one of the following conditions

(i) f and h are continuous at a point in $\operatorname{dom} f \cap \operatorname{dom} h \cap S$;
(ii) $\operatorname{dom} f \cap \operatorname{int} S \cap \operatorname{dom} h \neq \emptyset$ and f or h is continuous at a point in $\operatorname{dom} f \cap \operatorname{dom} h$;
(iii) X is a Fréchet space, S is closed, f is lower semicontinuous, h is star C-lower semicontinuous and $0 \in \operatorname{sqri}(\operatorname{dom} f \times S \times \operatorname{dom} h - \Delta_{X^3})$;
(iv) $\dim \operatorname{lin}(\operatorname{dom} f \times S \times \operatorname{dom} h - \Delta_{X^3}) < +\infty$ and $0 \in \operatorname{ri}(\operatorname{dom} f \times S \times \operatorname{dom} h - \Delta_{X^3})$;

is satisfied, then (see [21, 48, 221])

$$\partial f(x) + \partial(z^*h)(x) + N_S(x) = \partial(f + (z^*h) + \delta_S)(x) \ \forall x \in S \ \forall z^* \in C^*$$

Consequently, when one of these situations occurs, the constraint involving the subdifferential in (DC_W^L), (DC_M^L) and (DC_{MW}^L) can be correspondingly modified. Moreover, in order to split $\partial(f + (z^*h) + \delta_S)(x)$ into a sum of only two subdifferentials, one can apply [48, Theorem 3.5.6].

Remark 5.8. If $X = \mathbb{R}^n$, $Y = \mathbb{R}^m$, $C = \mathbb{R}_+^m$, $f : \mathbb{R}^n \to \mathbb{R}$ and $h = (h_1, \ldots, h_m)^\top : \mathbb{R}^n \to \mathbb{R}^m$, and the functions f and h_j, $j = 1, \ldots, m$, are convex, then these are also continuous, hence the condition (i) in Remark 5.7 is fulfilled and the subdifferentials in the constraints of the duals we assigned above to (PC) can be split. Then (DC_W^L) turns out to be the classical *nondifferentiable Wolfe dual problem* mentioned in the literature (see for instance [145, 186]). Meanwhile, (DC_{MW}^L) is the classical *nondifferentiable Mond-Weir dual problem* to (PC).

Remark 5.9. If in addition to the hypotheses of Remark 5.8 the set S is open and the functions f and h_j, $j = 1, \ldots, m$, are moreover Gâteaux differentiable on it, the subdifferentials in the constraints can be replaced by the corresponding gradients, (DC_W^L) turns out to be the classical *Wolfe dual problem* (see [215]), while (DC_{MW}^L) is nothing but the classical *Mond-Weir dual problem* from [169].

Another perturbation function employed to assign conjugate dual problems to (PC) is the Fenchel-Lagrange dual perturbation function Φ^{FL} for which one has $(0, y^*, z^*) \in \partial\Phi^{FL}(u, y, z)$ if and only if $u \in S$, $h(u) \in z - C$ and $f^*(y^*) + (-(z^*h) + \delta_S)^*(-y^*) + \delta_{-C^*}(z^*) + f(u + y) + \delta_{-C}(h(u) - z) + \delta_S(u) = \langle y^*, y\rangle + \langle z^*, z\rangle$, which is nothing but $u \in S$, $h(u) \in z - C$ and $\big(f^*(y^*) + f(u+y) - \langle y^*, u+y\rangle\big) + \big((-(z^*h) + \delta_S)^*(-y^*) + (-(z^*h) + \delta_S)(u) - \langle -y^*, u\rangle\big) + \big(\sigma_{-C}(-z^*) + \delta_{-C}(h(u) - z) + \langle z^*, h(u) - z\rangle\big) = 0$, i.e. $u \in S$, $-z^* \in C^*$, $h(u) - z \in -C$, $y^* \in \partial f(u + y)$, $-y^* \in \partial(-(z^*h) + \delta_S)(u)$ and $(z^*h)(u) = \langle z^*, z\rangle$. Thus we obtain from (DG_W) the following dual problem to (PC)

$$(DC_W^{FL}) \qquad \sup_{\substack{u\in S, y\in X, z\in Y, y^*\in X^*, z^*\in -C^*,\\ h(u)-z\in -C, (z^*h)(u)=\langle z^*, z\rangle,\\ y^*\in\partial f(u+y)\cap(-\partial((z^*h)+\delta_S)(u))}} \{f(u+y) - \langle y^*, y\rangle - \langle z^*, z\rangle\},$$

which can be equivalently turned into

$$(DC_W^{FL}) \qquad \sup_{\substack{u\in S, y\in X, y^*\in X^*, z^*\in C^*,\\ y^*\in\partial f(u+y)\cap(-\partial((z^*h)+\delta_S)(u))}} \{\langle y^*, u\rangle + (z^*h)(u) - f^*(y^*)\},$$

further referred to as the *Wolfe dual of Fenchel-Lagrange type* to (PC). Analogously, the dual problem to (PC) arising from (DG_M) is

$$(DC_M^{FL}) \qquad \sup_{\substack{u\in S, z^*\in C^*,\\ (z^*h)(u)\geq 0, h(u)\in -C,\\ 0\in\partial f(u)+\partial((z^*h)+\delta_S)(u)}} f(u).$$

Note that in the constraints of this dual one can replace $(z^*h)(u) \geq 0$ by $(z^*h)(u) = 0$ without altering anything. Removing, like in the Lagrange case, from it the constraint $h(u) \in -C$, we obtain the *Mond-Weir dual problem of Fenchel-Lagrange type* to (PC)

$$(DC_{MW}^{FL}) \qquad \sup_{\substack{u \in S, z^* \in C^*, (z^*h)(u) \geq 0, \\ 0 \in \partial f(u) + \partial((z^*h) + \delta_S)(u)}} f(u).$$

Applying Theorem 5.1 and Remark 5.1 and using similar arguments to the ones used concerning (DC_{MW}^{L}), we obtain the following weak duality inequality

$$v(DC_M^{FL}) \leq v(DC_{MW}^{FL}) \leq v(DC_W^{FL}) \leq v(DC^{FL}) \leq v(PC). \tag{5.2.2}$$

As can be seen in the following situations, the Wolfe dual has sometimes an indeed larger optimal objective value than the Mond-Weir one, while the classical conjugate dual can have a strictly greater one than the other mentioned two duals.

Example 5.4. Consider again the situation from Example 5.1. One can analogously show that $v(DC_{MW}^{FL}) = v(DC_W^{FL}) = 0$, while (DC_M^{FL}) has no feasible points, consequently $v(DC_M^{FL}) = -\infty$.

Therefore, $v(DC_M^{FL}) < v(DC_{MW}^{FL}) = v(DC_W^{FL})$ in this case.

Example 5.5. Consider again the situation from Example 5.2. One can verify that for $u > 0$ and $z^* = (z_1^*, z_2^*)^\top \geqq 0$ fulfilling $z_1^* - z_2^* = -1$ it holds $1 \in \partial f(u)$ and $-1 \in \partial((z^*h) + \delta_S)(u)$, so $(u, 0, 1y^*, z^*)$ is feasible to (DC_W^{FL}). Moreover, $f(u) = u - f^*(1)$, so one obtains like in Example 5.2 that $v(DC_W^{FL}) = 0$, while both (DC_M^{FL}) and (DC_{MW}^{FL}) are unfeasible.

Therefore, $v(DC_M^{FL}) = v(DC_{MW}^{FL}) < v(DC_W^{FL})$ in this case.

Example 5.6. Consider again the situation from Example 5.3. One can analogously show that $v(DC_M^{FL}) = v(DC_{MW}^{FL}) = v(DC_W^{FL}) = -\infty$, while $v(DC^{FL}) = 0$.

Therefore, $v(DC_M^{FL}) = v(DC_{MW}^{FL}) = v(DC_W^{FL}) < v(DC^{FL})$ in this case.

For strong duality, which follows directly from Theorem 5.2, besides convexity assumptions which guarantee the convexity of the perturbation function Φ^{FL} we use regularity conditions, too, obtained in Sect. 2.2.2 by particularizing (RC_i^G), $i \in \{1, 2, 3, 4\}$. Specializing Theorem 5.2 for the present context we obtain strong duality statements for (PC) and (DC_W^{FL}) and (DC_M^{FL}), respectively, while the one concerning (DC_{MW}^{FL}) follows analogously or via (5.2.2).

Theorem 5.4. *Assume that S is a convex set, f is a convex function and h is C-convex vector function. Let $\bar{x} \in X$ be an optimal solution to (PC) and assume that one of the regularity conditions (RC_i^{FL}), $i \in \{1, 2, 3, 4\}$, is fulfilled. Then $v(PC) = v(DC_W^{FL}) = v(DC_M^{FL}) = v(DC_{MW}^{FL})$ and there exist $\bar{y}^* \in X^*$ and $\bar{z}^* \in C^*$ for which $(\bar{x}, 0, \bar{y}^*, \bar{z}^*)$ is an optimal solution to (DC_W^{FL}) and $(\bar{x}, \bar{z}^*)$ is an optimal solution to (DC_M^{FL}) and (DC_{MW}^{FL}).*

Remark 5.10. Results analogous to the one from Remark 5.7 can be given for the Fenchel-Lagrange type duals, too. Assume that S is a convex set and h is a C-convex vector function, the satisfaction of any of the following conditions (cf. [48, Theorem 3.5.6])

(i) h is continuous at a point in $S \cap \operatorname{dom} h$;
(ii) $\operatorname{int} S \cap \operatorname{dom} h \neq \emptyset$;
(iii) X is a Fréchet space, S is closed, h is star C-lower semicontinuous and $0 \in \operatorname{sqri}(S - \operatorname{dom} h)$;
(iv) $\dim \operatorname{lin}(S - \operatorname{dom} h) < +\infty$ and $\operatorname{ri} S \cap \operatorname{ri} \operatorname{dom} h \neq \emptyset$;

ensures the fulfillment of the formula

$$\partial((z^*h) + \delta_S)(u) = \partial(z^*h)(u) + N_S(u) \ \forall u \in X \ \forall z^* \in C^*,$$

in which case the constraint involving $\partial((z^*h) + \delta_S)(u)$ in (DC_W^{FL}), (DC_M^{FL}) and (DC_{MW}^{FL}) can be correspondingly modified.

5.2.3 Unconstrained Scalar Optimization Problems

Consider now the unconstrained optimization problem

$$(PU) \qquad \inf_{x \in X} \big[f(x) + g(Ax)\big],$$

where $A : X \to Y$ is a linear continuous mapping and $f : X \to \overline{\mathbb{R}}$ and $g : Y \to \overline{\mathbb{R}}$ are proper functions fulfilling the feasibility condition $\operatorname{dom} f \cap A^{-1}(\operatorname{dom} g) \neq \emptyset$. The perturbation function considered for assigning the Wolfe type and Mond-Weir type dual problems to (PU) is the Fenchel type one Φ^U already considered in Sect. 2.2.3, for which one has

$$(0, y^*) \in \partial\Phi(u, y) \Leftrightarrow A^*y^* \in -\partial f(u) \text{ and } y^* \in \partial g(Au + y).$$

The duals we assign to (PU) by using Φ^U turn out to be

$$(DU_W) \qquad \sup_{\substack{u \in X, y \in Y, y^* \in Y^*, \\ y^* \in (A^*)^{-1}(-\partial f(u)) \cap \partial g(Au+y)}} \{-f^*(-A^*y^*) - g^*(y^*)\},$$

which is a *Wolfe type dual*, and, respectively the *Mond-Weir type dual*,

$$(DU_M) \qquad \sup_{\substack{u \in X, \\ 0 \in (A^*)^{-1}(-\partial f(u)) - \partial g(Au)}} \{f(u) + g(Au)\}.$$

Remark 5.11. When the primal problem is taken to be more particular, as happens for instance when f, respectively takes everywhere the value 0, or when $X = Y$ and A is the identity mapping of X, duals correspondingly obtained from (DU_W) and (DU_M) can be easily assigned to it.

Employing Theorem 5.1 and Remark 5.1, one obtains the weak duality statements corresponding to the just introduced dual problems to (PU), namely

$$v(DU_M) \leq v(DU_W) \leq v(DU) \leq v(PU).$$

For strong duality, which follows directly from Theorem 5.2, besides convexity assumptions which guarantee the convexity of the perturbation function Φ^U, we use the regularity conditions considered in Sect. 2.2.3.

Theorem 5.5. *Assume that f and g are convex functions. Let $\bar{x} \in X$ be an optimal solution to (PU) and assume that one of the regularity conditions (RC_i^U), $i \in \{1,2,3,4\}$, is fulfilled. Then $v(PU) = v(DU_W) = v(DU_M)$ and there exists a $\bar{y}^* \in Y^*$ for which $(\bar{x}, \bar{y}^*)$ is an optimal solution to (DG_M) and $(\bar{x}, 0, \bar{y}^*)$ is one to (DG_W).*

As noted in Sect. 2.2.3, one can see (PC) as an unconstrained optimization problem, too. Using the notations considered in Sect. 5.2.2, the duals (DU_W) and (DU_M) turn into

$$(DC_W^F) \qquad \sup_{\substack{u \in S, y \in X, y^* \in X^*, \\ y^* \in \partial f(u+y) \cap (-N_{\mathscr{A}}(u))}} \{\langle y^*, u\rangle - f^*(y^*)\},$$

the *Wolfe dual of Fenchel type* to (PC), and, respectively

$$(DC_M^F) \qquad \sup_{\substack{u \in S, \\ 0 \in \partial f(u) + N_{\mathscr{A}}(u)}} f(u).$$

From the weak duality statement involving (PU) and its duals, or alternatively, by employing Theorem 5.1 and Remark 5.1, one obtains weak duality assertions for the just introduced duals to (PC), namely

$$v(DC_M^F) \leq v(DC_W^F) \leq v(DC^F) \leq v(PC).$$

For strong duality, which follows directly from either Theorems 5.2 or 5.5, besides convexity assumptions which guarantee the convexity of the corresponding perturbation function one can use the regularity conditions introduced in Sect. 2.2.3.

Theorem 5.6. *Assume that $\mathscr{A}$ is a convex set and f is a convex function. Let $\bar{x} \in X$ be an optimal solution to (PC) and assume that one of the regularity conditions (RC_i^F), $i \in \{1,2,3,4\}$, is fulfilled. Then $v(PC) = v(DC_W^F) = v(DC_M^F)$, $\bar{x}$ is an* *optimal solution to* (DC_M^F) *and there exist* $(\bar{y}, \bar{y}^*) \in X \times X^*$ *for which* $(\bar{x}, \bar{y}, \bar{y}^*)$ *is an optimal solution to* (DC_W^F).

Remark 5.12. In order to ensure the convexity of the set $\mathscr{A}$ it is sufficient to take the set S convex and h to be a C-convex vector function. To guarantee that the set $\mathscr{A}$ is closed it is enough to assume that S is a closed set and h a C-epi-closed vector function.

Remark 5.13. As one can see in Sect. 5.2.4, the dual problems we assigned to (PC) do not coincide in general. Sufficient conditions that ensure the equivalence of the corresponding duals of Lagrange type and Fenchel-Lagrange type, respectively, can be easily obtained via [48, Theorem 3.5.6], while for the equivalence of the corresponding duals of Fenchel type and Fenchel-Lagrange type, respectively, one can apply [48, Theorem 3.5.13].

5.2.4 Comparisons Between the Duals for Constrained Scalar Optimization Problems

From [21, 48] it is known that the optimal objective values of the conjugate duals attached to (PC) we mentioned before fulfill the following inequality

$$v(DC^{FL}) \leq \begin{matrix} v(DC^L) \\ v(DC^F) \end{matrix} \leq v(PC). \tag{5.2.3}$$

A natural question is if similar inequalities exist also for the dual problems introduced in Sect. 5.2.3. First we deal with the ones that are particular instances of (DG_M), where the answer is positive, as the following statement shows.

Proposition 5.1. *One has*

$$v(DC_M^{FL}) \leq \begin{matrix} v(DC_M^L) \\ v(DC_M^F) \end{matrix} \leq v(PC).$$

Proof. If the feasible set of (DC_M^{FL}) is empty, there is nothing to prove. Let (u, z^*) be feasible to (DC_M^{FL}). Then $u \in S$, $z^* \in C^*$, $(z^*h)(u) \geq 0$, $h(u) \in -C$ and $0 \in \partial f(u) + \partial((z^*h) + \delta_S)(u)$. The last relation implies $0 \in \partial(f + (z^*h) + \delta_S)(u)$, consequently (u, z^*) is feasible to (DC_M^L), too. As both (DC_M^{FL}) and (DC_M^L) have f as objective function, it is clear that $v(DC_M^{FL}) \leq v(DC_M^L)$.

On the other hand, $\partial((z^*h) + \delta_S)(u) \subseteq N_{\mathscr{A}}(u)$ since $u \in \mathscr{A}$, thus u is feasible to (DC_M^F). Since both (DC_M^{FL}) and (DC_M^F) have f as objective function, it is clear that $v(DC_M^{FL}) \leq v(DC_M^F)$. □

Analogously one can show that for the Mond-Weir dual problems to (PC) there is a similar inequality.

Proposition 5.2. *One has*

$$v(DC_{MW}^{FL}) \leq v(DC_{MW}^{L}).$$

Remark 5.14. When the feasible set of (DC_M^L) or (DC_M^F) is empty, then so is the feasible set of (DC_M^{FL}), too. In general, the fact that (DC_M^{FL}) has no feasible points does not imply the emptiness of any of the feasible sets of (DC_M^L) and (DC_M^F), as it can be seen in Examples 5.7 and 5.8.

Situations where the inequalities from Propositions 5.1 and 5.2 are strictly fulfilled can be found below. In the view of Remark 5.3, it is clear that if this is the case the dual (DC_{MW}^{FL}) is infeasible and (DC_M^L) or (DC_M^F), respectively, is feasible and there is strong duality for it. As a byproduct we obtain that, like in the conjugate case, in general there cannot be established an inequality involving the optimal objective values of the dual problems of Lagrange and Fenchel types to (PC). First we deal with a problem whose Lagrange dual derived as a special case of (DG_M) has a larger optimal objective value than both its Fenchel and Fenchel-Lagrange type duals obtained from (DG_M).

Example 5.7. Let $X = \mathbb{R}^2$, $Y = \mathbb{R}$, $C = \mathbb{R}_+$,

$$S = \left\{ (x_1, x_2)^\top \in \mathbb{R}^2 : 0 \leq x_1 \leq 2, \begin{array}{l} 3 \leq x_2 \leq 4, \text{ if } x_1 = 0, \\ 1 \leq x_2 \leq 4, \text{ if } x_1 \in (0, 2] \end{array} \right\},$$

$$f : \mathbb{R}^2 \to \overline{\mathbb{R}},\ f(x_1, x_2) = \begin{cases} x_2, & \text{if } x_1 \leq 0, \\ +\infty, & \text{otherwise}, \end{cases}$$

and $h : \mathbb{R}^2 \to \mathbb{R}$, $h(x_1, x_2) = 0$.

Note first that because $h(x_1, x_2) = 0 \in \mathbb{R}_+$ for all $(x_1, x_2)^\top \in \mathbb{R}^2$, it follows that the dual problem (DC_M^L) is equivalent to (DC_{MW}^L) and, respectively, that (DC_M^{FL}) and (DC_{MW}^{FL}) are equivalent, too. One has $\mathscr{A} = S$ and since $(z^*h)(u) = 0$ for all $u \in S$, it follows that (DC_M^F) is equivalent to (DC_M^{FL}) and (DC_{MW}^{FL}), too. One has

$$(f + \delta_S)(u_1, u_2) = \begin{cases} u_2, & \text{if } u_1 = 0, u_2 \in [3, 4], \\ +\infty, & \text{otherwise}. \end{cases}$$

For any $z^* \in \mathbb{R}_+$ we get $(0, 0) \in \partial(f + (z^*h) + \delta_S)(0, 3)$, thus $v(DC_M^L) = v(DC_{MW}^L) \geq 3$. Since it can be seen that $v(PC) = 3$, we get $v(DC_M^L) = v(DC_{MW}^L) = 3$.

On the other hand, taking without loss of generality $z^* = 1$, to have, for some $u = (u_1, u_2)^\top \in S$, $0 \in \partial f(u) + \partial((z^*h) + \delta_S)(u)$ means actually that there exists a $y^* \in \partial f(u) \cap (-N_S(u))$. From $y^* \in \partial f(u)$ we obtain that $y^* = (y_1^*, y_2^*)^\top \in \mathbb{R}_+ \times \{1\}$ and $u_1 = 0$. Consequently, $y_2^* = 1$. Let us see now for what $y_1^* \in \mathbb{R}_+$ does one obtain $(-y_1^*, -1) \in N_S(0, u_2)$. We have $(-y_1^*, -1) \in N_S(0, u_2)$ if and only if

$\sigma_S(-y_1^*, -1) = -u_2$. This yields $u_2 = 1$, but $(0,1) \notin S$, consequently (DC_M^{FL}) is infeasible. Hence, one has $v(DC_M^F) = v(DC_M^{FL}) = v(DC_{MW}^{FL}) = -\infty$.

Therefore, $v(DC_M^F) = v(DC_M^{FL}) = v(DC_{MW}^{FL}) < v(DC_M^L) = v(DC_{MW}^L)$ in this case.

The second example we present brings into attention an optimization problem for which the Fenchel dual derived as a special case of (DG_M) has a larger optimal objective value than both its Lagrange and Fenchel-Lagrange type duals obtained from (DG_M).

Example 5.8. Let $X = \mathbb{R}^2$, $Y = \mathbb{R}$, $C = \mathbb{R}_+$,

$$S = \left\{ (x_1, x_2)^\top \in \mathbb{R}^2 : 0 \le x_1 \le 2, \begin{array}{l} 3 \le x_2 \le 4, \text{ if } x_1 = 0, \\ 1 \le x_2 \le 4, \text{ if } x_1 \in (0,2] \end{array} \right\},$$

$f : \mathbb{R}^2 \to \mathbb{R}$, $f(x_1, x_2) = x_2$ and $h : \mathbb{R}^2 \to \mathbb{R}$, $h(x_1, x_2) = x_1$.

Then $\mathscr{A} = \{0\} \times [3,4]$. As the functions f and h are continuous, the condition (ii) in Remark 5.7 is fulfilled, hence the subdifferentials from the dual constraints can be split and it follows that for the optimization problem we are dealing with the dual problems (DC_M^{FL}) and (DC_M^L) are equivalent. Moreover, (DC_{MW}^{FL}) and (DC_{MW}^L) are equivalent, too. Since $(0,1)^\top \in \partial f(0,3) \cap (-N_{\mathscr{A}}(0,3))$, $v(DC_M^F) \ge 3$. But $v(PC) = 3$, consequently $v(DC_M^F) = 3$.

On the other hand, for $u \in S$, $y^* \in \mathbb{R}^2$ and $z^* \ge 0$, $0 \in \partial f(u) + \partial((z^* h) + \delta_S)(u)$ if and only if one concomitantly has $(0,1)^\top \in \partial f(u) \cap (-\partial((z^* h) + \delta_S)(u))$, $z^* = 0$ and $u \in (0,2] \times \{1\}$. But then $h(u) > 0$, so (DC_M^{FL}) is infeasible, thus $v(DC_M^{FL}) = v(DC_M^L) = -\infty$. Moreover, $v(DC_{MW}^L) = v(DC_{MW}^{FL}) = \sup\{u_2 : (u_1, u_2)^\top \in (0,2] \times \{1\}\} = 1$.

Therefore, $v(DC_M^{FL}) = v(DC_M^L) < v(DC_{MW}^L) = v(DC_{MW}^{FL}) < v(DC_M^F)$ in this case.

Another observation that can be drawn after analyzing the two examples from above is that for $v(DC_{MW}^{FL})$ and $v(DC_M^L)$ no generally valid order can be established. Moreover, we have seen that in Example 5.8 $v(DC_{MW}^{FL})$ is strictly less than $v(DC_M^F)$. Though, this inequality is not valid in general, as the following situation shows.

Example 5.9. Consider again the situation from Example 5.1. As $\mathscr{A} = (0, +\infty)$, $N_{\mathscr{A}}(u) = \{0\}$ for all $u \in \mathscr{A}$ and $\partial f(u) = \{1\}$ for all $u \in \mathbb{R}$, it follows that $\partial f(u) \cap (-N_{\mathscr{A}}(u)) = \emptyset$ for all $u \in S$. Consequently, $v(DC_M^F) = -\infty$.

On the other hand, taking $z^* = 0$ we get $\partial((z^* h) + \delta_S)(u) = N_{\mathbb{R}_+}(u)$ for all $u \ge 0$ and it can be shown that $N_{\mathbb{R}_+}(0) = \mathbb{R}_-$. Thus $1 \in \partial f(0) \cap \partial(-((z^* h) + \delta_S)(0))$. As $z^* h(u) = 0$, the element $(0,0)$ is feasible to (DC_{MW}^{FL}). This yields $v(DC_{MW}^{FL}) \ge 0 = v(PC)$, thus $v(DC_{MW}^{FL}) = 0$.

Therefore, $v(DC_M^F) < v(DC_{MW}^{FL})$ in this case.

Remark 5.15. Since the constraint $h(u) \in -C$ does not explicitly appear in the definition of the feasible set of the problem (DC_M^F), one may assume that this can be considered per se the Mond-Weir dual problem of Fenchel type to (PC). However, as seen in Example 5.9, a situation analogous to the one in (5.2.3) or Proposition 5.1 would not hold for the Mond-Weir duals, despite Proposition 5.2. That is why we chose not to introduce a Mond-Weir dual problem of Fenchel type to (PC). Note also that, on the other hand, $h(u) \in -C$ is "hidden" in the constraint containing $N_{\mathscr{A}}(u)$, which cannot be fulfilled if $h(u) \notin -C$.

However, a result similar to (5.2.3) or Proposition 5.1 does not hold for the Wolfe type duals to (PC). Even if the primal problem is convex, the optimal objective values of (DC_W^{FL}), (DC_W^L) and (DC_W^F) cannot be ordered in general. Because of (5.2.3), this fact was expected to happen for the Fenchel and Lagrange type duals, but, surprisingly, the optimal objective value of the Wolfe dual of Fenchel-Lagrange type is not always smaller than them. In the following we sustain this claim by several examples. First we deal with the Wolfe duals of types Lagrange and Fenchel-Lagrange, respectively.

Example 5.10. Consider again the situation from Example 5.7. Since $v(DC_M^L) = v(DC_{MW}^L) = v(PC) = 3$, it follows via (5.2.1) that $v(DC_W^L) = 3$, too.

On the other hand, as $\mathscr{A} = S$ and since $(z^*h)(u) = 0$ for all $u \in S$, it follows that (DC_W^F) is equivalent to (DC_W^{FL}) for the optimization problem in discussion. Because of the investigations in Example 5.7 and (5.2.2), we know that $1 \leq v(DC_W^F) = v(DC_W^{FL}) \leq 3$, so these dual problems are feasible. Taking without loss of generality $z^* = 1$, we have $y^* \in \partial f(u+y) \cap (-N_S(u))$ for some $u = (u_1, u_2) \in S$ and $y \in \mathbb{R}^2$. From $y^* \in \partial f(u+y)$ we obtain that $y^* = (y_1^*, y_2^*)^\top \in \mathbb{R}_+ \times \{1\}$. Consequently, $y_2^* = 1$. Let us see now for what $y_1^* \in \mathbb{R}_+$ does one obtain $(-y_1^*, -1) \in N_S(u_1, u_2)$. We have $(-y_1^*, -1) \in N_S(u_1, u_2)$ if and only if $\sigma_S(-y_1^*, -1) = -y_1^* u_1 - u_2$. Since this can take place only if $y_1^* = 0$, it follows that $u_1 \in (0, 2]$, $u_2 = 1$ and $(0, 1)^\top$ is the only possible value for y^*. Consequently, $v(DC_W^F) = v(DC_W^{FL}) = \sup\{u_2 : u_1 \in (0, 2], u_2 = 1\} = 1$.

Therefore, $v(DC_W^F) = v(DC_W^{FL}) < v(DC_W^L)$ in this case.

Example 5.11. Let $X = \mathbb{R}$, $Y = \mathbb{R}$, $C = \mathbb{R}_+$, $S = \mathbb{R}$,

$$f : \mathbb{R} \to \overline{\mathbb{R}},\ f(x) = \begin{cases} x, & \text{if } x > 0, \\ +\infty, & \text{otherwise,} \end{cases}$$

and

$$h : \mathbb{R} \to \mathbb{R},\ h(x) = \begin{cases} -x, & \text{if } x \leq 0, \\ 0, & \text{otherwise.} \end{cases}$$

Note that $\mathscr{A} = \mathbb{R}_+$, $v(PC) = 0$ and for all $z^* \geq 0$ one has $f + (z^*h) + \delta_S \equiv f$. Thus, for all $z^* \geq 0$,

$$\partial(f + (z^*h) + \delta_S)(u) = \partial f(u) = \begin{cases} \{1\}, & \text{if } u > 0, \\ \emptyset, & \text{otherwise.} \end{cases}$$

Consequently, (DC_W^L) has no feasible points, therefore $v(DC_W^L) = -\infty$.

On the other hand, taking $u = 0$, $y = 1$, $y^* = 1$ and $z^* = 1$, we have $(z^*h) + \delta_S \equiv h$, $1 \in \partial f(1)$ and $-1 \in \partial h(0)$. Thus $(0, 1, 1, 1)$ is feasible to (DC_W^{FL}). Moreover, $-1 \in N_{\mathscr{A}}$, so $(0, 1, 1)$ is feasible to (DC_W^F). Then $v(DC_W^{FL}) \geq 0 \leq v(DC_W^F)$, but as $v(PC) = 0$ it follows $v(DC_W^{FL}) = v(DC_W^F) = 0$.

Therefore, $v(DC_W^L) < v(DC_W^F) = v(DC_W^{FL})$ in this case.

For both these problems we had $v(DC_W^F) = v(DC_W^{FL})$. But each of these optimal objective values can be larger than the other in a specific situation, as one can see in the following two examples.

Example 5.12. Consider again the situation from Example 5.8. As the condition (ii) in Remark 5.7 is fulfilled, it follows that for the optimization problem we are dealing with the dual problems (DC_W^{FL}) and (DC_W^L) are equivalent. Since $v(DC_M^F) = v(PC) = 3$, it follows via (5.2.1) that $v(DC_W^F) = 3$.

On the other hand, for $u \in S$, $y, y^* \in \mathbb{R}^2$ and $z^* \geq 0$, $y^* \in \partial f(u + y) \cap (-\partial((z^*h) + \delta_S)(u))$ if and only if $y^* = (0, 1)^\top$, $z^* = 0$ and $u \in (0, 2] \times \{1\}$. Then $v(DC_W^L) = v(DC_W^{FL}) = \sup\{u_2 : (u_1, u_2)^\top \in (0, 2] \times \{1\}\} = 1$.

Therefore, $v(DC_W^L) = v(DC_W^{FL}) < v(DC_W^F)$ in this case.

Example 5.13. Consider again the situation from Examples 5.1 and 5.9. We have $v(DC_{MW}^L) = v(DC_{MW}^{FL}) = v(PC) = 0$, thus, via (5.2.1), $v(DC_W^{FL}) = v(DC_W^L) = 0$, too.

As $\mathscr{A} = (0, +\infty)$, $N_{\mathscr{A}}(u) = \{0\}$ for all $u \in \mathscr{A}$ and $\partial f(u) = \{1\}$ for all $u \in \mathbb{R}$, it follows that $\partial f(u+y) \cap (-N_{\mathscr{A}}(u)) = \emptyset$ for all $u \in S$ and all $y \in \mathbb{R}$. Consequently, $v(DC_W^F) = -\infty$.

Therefore, $v(DC_W^F) < v(DC_{MW}^{FL}) = v(DC_W^{FL}) = v(DC_{MW}^L) = v(DC_W^L)$ in this case.

In the last two examples one can see as a byproduct that no order can be established in general between $v(DC_{MW}^{FL})$ and $v(DC_W^F)$. A natural question is whether the same conclusion can be drawn for $v(DC_{MW}^{FL})$ and $v(DC_W^L)$, too. The next statement answers to it negatively and completes the investigations on the optimal objective values of the duals we introduced to (PC).

Proposition 5.3. *One has*

$$v(DC_{MW}^{FL}) \leq v(DC_W^L).$$

Proof. Let (u, z^*) be feasible to (DC_{MW}^{FL}). Then $u \in S$, $z^* \in C^*$, $(z^*h)(u) \geq 0$ and $0 \in \partial f(u) + \partial((z^*h) + \delta_S)(u)$. The last relation implies $0 \in \partial(f + (z^*h) + \delta_S)(u)$, consequently (u, z^*) is feasible to (DC_W^L), too. Since in this case $f(u) \leq f(u) + (z^*h)(u)$, it follows that $v(DC_{MW}^{FL}) \leq v(DC_W^L)$. □

Remark 5.16. To show that the inequality provided in Proposition 5.3 can in general be strictly fulfilled one can use for instance (5.2.2) and Example 5.10 or (5.2.1) and Example 5.7, respectively.

5.2.5 A Glimpse into Generalized Convexity

A characteristic of many results in the literature concerning Wolfe duality and Mond-Weir duality in the differential case is the usage of different generalized convexity hypotheses, like quasiconvexity, pseudoconvexity or invexity, for the functions involved in order to achieve weak and strong duality. However, generalized convexity hypotheses can be taken into account in the nondifferentiable case, too, as it can be seen in the following. In this subsection we assume that $X = \mathbb{R}^n$, $Y = \mathbb{R}^m$ and $C \subseteq \mathbb{R}^m$. We consider here only the notions of *almost convexity* (for properties we refer to [9, 39, 53, 75] and the references therein) and *near convexity* (see [39, 41, 52] and the references therein) for both sets and functions. Other generalized convexity notions successfully used in conjugate duality, for instance the *convexlikeness* or the *even convexity* can be employed here, too.

A set $U \subseteq \mathbb{R}^n$ is called *almost convex* if $\operatorname{cl} U$ is convex and $\operatorname{ri}\operatorname{cl} U \subseteq U$ and, respectively *nearly convex* if there is a constant $\alpha \in]0,1[$ such that for any x and y belonging to U one has $\alpha x + (1-\alpha)y \in U$. An example of an almost convex but not convex set is $([0,1]\times[0,1]) \setminus \{(0,y) : y \in \mathbb{R}\setminus\mathbb{Q}\} \subseteq \mathbb{R}^2$, while $\mathbb{Q} \subseteq \mathbb{R}$ is nearly convex but not convex.

A function $f : \mathbb{R}^n \to \overline{\mathbb{R}}$ is said to be *almost convex* if $\bar{f}$ is convex and $\operatorname{ri}\operatorname{epi} \bar{f} \subseteq \operatorname{epi} f$ and, respectively, *nearly convex* if $\operatorname{epi} f$ is a nearly convex set. Moreover, a vector function $g : \mathbb{R}^n \to \mathbb{R}^m$ is said to be *C-almost convex* if $\operatorname{epi}_C g$ is an almost convex set and, respectively, *C-nearly convex* if $\operatorname{epi}_C g$ is a nearly convex set. Each convex set or function is both almost convex and nearly convex, too, while the C-convex vector functions are both C-almost convex and C-nearly convex. Note also that a nearly convex set with nonempty relative interior is almost convex and a nearly convex function $f : \mathbb{R}^n \to \overline{\mathbb{R}}$ whose epigraph has a nonempty relative interior is almost convex, too.

As the weak duality holds for all the primal-dual pairs of problems considered in this section in the most general case, thus without any additional hypotheses, we focus in the following on strong duality. First we give the corresponding statements involving (PG) and the duals we considered for it, where the just introduced generalized convexity concepts play an important role.

Theorem 5.7. *Assume that $\Phi : \mathbb{R}^n \times \mathbb{R}^m \to \overline{\mathbb{R}}$ is a proper and almost convex function, with its domain fulfilling $0 \in \operatorname{Pr}_{\mathbb{R}^m}(\operatorname{dom}\Phi)$. Let $\bar{x} \in \mathbb{R}^n$ be an optimal solution to (PG) and assume that the regularity condition*

$$0 \in \operatorname{ri}\operatorname{Pr}_Y(\operatorname{dom}\Phi)$$

is fulfilled. Then $v(PG) = v(DG_W) = v(DG_M)$ and there exists a $\bar{y}^* \in \mathbb{R}^m$ for which $(\bar{x}, \bar{y}^*)$ is an optimal solution to (DG_M) and $(\bar{x}, 0, \bar{y}^*)$ is one to (DG_W).

Proof. From [53, Corollary 3.1] it is known that under these hypotheses one has $v(PG) = v(DG)$ with the latter attained at some $\bar{y}^* \in \mathbb{R}^m$. Then the optimality condition (2.2.10) holds for $\bar{x}$ and $\bar{y}^*$ and this means that $(\bar{x}, 0, \bar{y}^*)$ is feasible to (DG_W) and $(\bar{x}, \bar{y}^*)$ is feasible to (DG_M). The conclusion follows via Remark 5.3. □

Corollary 5.1. *Assume that $\Phi : \mathbb{R}^n \times \mathbb{R}^m \to \overline{\mathbb{R}}$ is a proper and nearly convex function, with its domain fulfilling $0 \in \Pr_{\mathbb{R}^m}(\operatorname{dom} \Phi)$ and with the relative interior of its epigraph nonempty. Let $\bar{x} \in \mathbb{R}^n$ be an optimal solution to (PG) and assume that the regularity condition*

$$0 \in \operatorname{ri} \Pr_Y(\operatorname{dom} \Phi)$$

is fulfilled. Then $v(PG) = v(DG_W) = v(DG_M)$ and there exists a $\bar{y}^* \in \mathbb{R}^m$ for which $(\bar{x}, \bar{y}^*)$ is an optimal solution to (DG_M) and $(\bar{x}, 0, \bar{y}^*)$ is one to (DG_W).

Note that the regularity condition used in Theorem 5.7 and Corollary 5.1 is nothing but (RC_3^G) written in the framework considered in this section. Of course this statement can be particularized for the duals considered in Sects. 5.2.2 and 5.2.3, too, as follows. First we deal with constrained optimization problems.

Theorem 5.8. *Assume that S is a nonempty and almost convex set, $C \subseteq \mathbb{R}^m$ is a nonempty convex cone, $f : \mathbb{R}^n \to \overline{\mathbb{R}}$ is a proper and almost convex function and $h : \mathbb{R}^n \to \mathbb{R}^m$ is a C-almost convex vector function fulfilling the feasibility condition* $\operatorname{dom} f \cap S \cap h^{-1}(-C) \neq \emptyset$. *Let $\bar{x} \in \mathscr{A}$ be an optimal solution to (PC) and assume that the regularity condition*

$$0 \in \operatorname{ri}\big(h(\operatorname{dom} f \cap S) + C\big)$$

is fulfilled. Then $v(PC) = v(DC_W^L) = v(DC_M^L) = v(DC_{MW}^L)$ and there exists a $\bar{z}^* \in C^*$ for which $(\bar{x}, \bar{z}^*)$ is an optimal solution to all three duals.

Corollary 5.2. *Assume that S is a nonempty and nearly convex set with a nonempty relative interior, $C \subseteq \mathbb{R}^m$ is a nonempty convex cone, $f : \mathbb{R}^n \to \overline{\mathbb{R}}$ is a proper and nearly convex function with* $\operatorname{ri}\operatorname{epi} f \neq \emptyset$ *and $h : \mathbb{R}^n \to \mathbb{R}^m$ is a C-nearly convex vector function with* $\operatorname{ri}\operatorname{epi}_C h \neq \emptyset$ *fulfilling the feasibility condition* $\operatorname{dom} f \cap S \cap h^{-1}(-C) \neq \emptyset$. *Let $\bar{x} \in \mathscr{A}$ be an optimal solution to (PC) and assume that the regularity condition*

$$0 \in \operatorname{ri}\big(h(\operatorname{dom} f \cap S) + C\big)$$

is fulfilled. Then $v(PC) = v(DC_W^L) = v(DC_M^L) = v(DC_{MW}^L)$ and there exists a $\bar{z}^* \in C^*$ for which $(\bar{x}, \bar{z}^*)$ is an optimal solution to all three duals.

Theorem 5.9. *Assume that S is a nonempty and almost convex set, $C \subseteq \mathbb{R}^m$ is a nonempty convex cone, $f : \mathbb{R}^n \to \overline{\mathbb{R}}$ is a proper and almost convex function and $h : \mathbb{R}^n \to \mathbb{R}^m$ is a C-almost convex vector function fulfilling the feasibility condition* $\operatorname{dom} f \cap S \cap h^{-1}(-C) \neq \emptyset$. *Let $\bar{x} \in \mathscr{A}$ be an optimal solution to (PC) and assume that the regularity condition*

$$0 \in \operatorname{ri}\big(\operatorname{dom} f \times C - \operatorname{epi}_{(-C)}(-h) \cap (S \times \mathbb{R}^m)\big)$$

is fulfilled. Then $v(PC) = v(DC_W^{FL}) = v(DC_M^{FL}) = v(DC_{MW}^{FL})$ *and there exist* $\bar{y}^* \in \mathbb{R}^n$ *and* $\bar{z} \in C^*$ *for which* $(\bar{x}, 0, \bar{y}^*, \bar{z}^*)$ *is an optimal solution to* (DC_W^{FL}) *and* $(\bar{x}, \bar{z}^*)$ *is an optimal solution to* (DC_M^{FL}) *and* (DC_{MW}^{FL}).

Corollary 5.3. *Assume that* S *is a nonempty and nearly convex set with a nonempty relative interior,* $C \subseteq \mathbb{R}^m$ *is a nonempty convex cone,* $f : \mathbb{R}^n \to \overline{\mathbb{R}}$ *is a proper and nearly convex function with* $\operatorname{ri}\operatorname{epi} f \neq \emptyset$ *and* $h : \mathbb{R}^n \to \mathbb{R}^m$ *is a* C*-nearly convex vector function with* $\operatorname{ri}\operatorname{epi}_C h \neq \emptyset$ *fulfilling the feasibility condition* $\operatorname{dom} f \cap S \cap h^{-1}(-C) \neq \emptyset$. *Let* $\bar{x} \in \mathscr{A}$ *be an optimal solution to (PC) and assume that the regularity condition*

$$0 \in \operatorname{ri}\big(\operatorname{dom} f \times C - \operatorname{epi}_{(-C)}(-h) \cap (S \times \mathbb{R}^m)\big)$$

is fulfilled. Then $v(PC) = v(DC_W^{FL}) = v(DC_M^{FL}) = v(DC_{MW}^{FL})$ *and there exist* $\bar{y}^* \in \mathbb{R}^n$ *and* $\bar{z} \in C^*$ *for which* $(\bar{x}, 0, \bar{y}^*, \bar{z}^*)$ *is an optimal solution to* (DC_W^{FL}) *and* $(\bar{x}, \bar{z}^*)$ *is an optimal solution to* (DC_M^{FL}) *and* (DC_{MW}^{FL}).

Theorem 5.10. *Assume that* $f : \mathbb{R}^n \to \overline{\mathbb{R}}$ *is a proper and almost convex function and* $\mathscr{A}$ *is a nonempty and almost convex set fulfilling the feasibility condition* $\operatorname{dom} f \cap \mathscr{A} \neq \emptyset$. *Let* $\bar{x} \in \mathscr{A}$ *be an optimal solution to (PC) and assume that the regularity condition*

$$\operatorname{ri}\operatorname{dom} f \cap \operatorname{ri}\mathscr{A} \neq \emptyset$$

is fulfilled. Then $v(PC) = v(DC_W^F) = v(DC_M^F)$ *and there exist* $\bar{y}, \bar{y}^* \in \mathbb{R}^n$ *for which* $\bar{x}$ *is an optimal solution to* (DC_M^F) *and* $(\bar{x}, \bar{y}, \bar{y}^*)$ *is one to* (DC_W^F).

Corollary 5.4. *Assume that* $f : \mathbb{R}^n \to \overline{\mathbb{R}}$ *is a proper and nearly convex function with* $\operatorname{ri}\operatorname{epi} f \neq \emptyset$ *and* $\mathscr{A}$ *is a nonempty and nearly convex set with a nonempty relative interior fulfilling the feasibility condition* $\operatorname{dom} f \cap \mathscr{A} \neq \emptyset$. *Let* $\bar{x} \in \mathscr{A}$ *be an optimal solution to (PC) and assume that the regularity condition*

$$\operatorname{ri}\operatorname{dom} f \cap \operatorname{ri}\mathscr{A} \neq \emptyset$$

is fulfilled. Then $v(PC) = v(DC_W^F) = v(DC_M^F)$ *and there exist* $\bar{y}, \bar{y}^* \in \mathbb{R}^n$ *for which* $\bar{x}$ *is an optimal solution to* (DC_M^F) *and* $(\bar{x}, \bar{y}, \bar{y}^*)$ *is one to* (DC_W^F).

Remark 5.17. Other hypotheses requesting the (C-)near convexity of the involved functions that guarantee the strong duality statements in Corollaries 5.2–5.4 can be found in [52, Theorem 3.3].

Now let us deal with unconstrained optimization problems, as well.

Theorem 5.11. *Assume that* $f : \mathbb{R}^n \to \overline{\mathbb{R}}$ *and* $g : \mathbb{R}^m \to \overline{\mathbb{R}}$ *are proper and almost convex functions and* $A : \mathbb{R}^n \to \mathbb{R}^m$ *is a linear mapping fulfilling the feasibility* *condition* $\operatorname{dom} f \cap A^{-1}(\operatorname{dom} g) \neq \emptyset$. *Let* $\bar{x} \in \mathbb{R}^n$ *be an optimal solution to (PU) and assume that the regularity condition*

$$A(\operatorname{ri}\operatorname{dom} f) \cap \operatorname{ri}\operatorname{dom} g \neq \emptyset$$

is fulfilled. Then $v(PU) = v(DU_W) = v(DU_M)$ *and there exists a* $\bar{y}^* \in \mathbb{R}^m$ *for which* $(\bar{x}, \bar{y}^*)$ *is an optimal solution to* (DU_M) *and* $(\bar{x}, 0, \bar{y}^*)$ *is one to* (DU_W).

Corollary 5.5. *Assume that* $f : \mathbb{R}^n \to \overline{\mathbb{R}}$ *and* $g : \mathbb{R}^m \to \overline{\mathbb{R}}$ *are proper and nearly convex functions with the relative interiors of their epigraphs nonempty, and* $A : \mathbb{R}^n \to \mathbb{R}^m$ *is a linear mapping fulfilling the feasibility condition* $\operatorname{dom} f \cap A^{-1}(\operatorname{dom} g) \neq \emptyset$. *Let* $\bar{x} \in \mathbb{R}^n$ *be an optimal solution to (PU) and assume that the regularity condition*

$$A(\operatorname{ri}\operatorname{dom} f) \cap \operatorname{ri}\operatorname{dom} g \neq \emptyset$$

is fulfilled. Then $v(PU) = v(DU_W) = v(DU_M)$ *and there exists a* $\bar{y}^* \in \mathbb{R}^m$ *for which* $(\bar{x}, \bar{y}^*)$ *is an optimal solution to* (DU_M) *and* $(\bar{x}, 0, \bar{y}^*)$ *is one to* (DU_W).

5.3 General Wolfe and Mond-Weir Vector Duality of Classical Type

As mentioned in Sect. 5.1, we present two ways of assigning vector dual problems of Wolfe and Mond-Weir type to vector optimization problems. In this section we deal with the so-called *classical* approach, where the objective vector function of the dual vector optimization problems contains its counterpart from the primal vector optimization problem. We begin our investigations with a general vector optimization problem, to which vector duals of both Wolfe and Mond-Weir types are assigned. Then we particularize the primal problem to be constrained and unconstrained, respectively, and the corresponding vector dual problems are derived, following the scalar case.

5.3.1 General Vector Optimization Problems

Let X, Y and V be Hausdorff locally convex vector spaces, with V partially ordered by the nontrivial pointed convex cone $K \subseteq V$. Let $F : X \to V^\bullet$ be a proper vector function and consider the general vector-minimization problem

$$(PVG) \qquad \operatorname*{Min}_{x \in X} F(x).$$

As solution concepts for (PVG) we consider the efficient solutions (cf. Definition 3.13) and the properly efficient solutions in the sense of linear scalarization (cf. Sect. 4.3), respectively. Recall that an element $\bar{x} \in X$ is said to be a *properly efficient solution* to the vector optimization problem (PVG) in the sense of linear scalarization if there exists a $v^* \in K^{*0}$ such that $(v^*F)(\bar{x}) \leq (v^*F)(x)$ for all $x \in X$. The set of all properly efficient solutions to (PVG) in the sense of linear scalarization is denoted by $\mathscr{P}\mathscr{E}_{LS}(PVG)$.

Remark 5.18. Since within this chapter the properly efficient solutions in the sense of linear scalarization are the only type of properly efficient solutions assigned to (PVG) we will call them here simply properly efficient. Every properly efficient solution to (PVG) belongs to $\operatorname{dom} F$ and it is also an efficient solution to the same vector optimization problem.

Consider now the proper vector perturbation function $\Phi : X \times Y \to V^\bullet$ which fulfills $\Phi(x, 0) = F(x)$ for all $x \in X$. Then $0 \in \Pr_Y(\operatorname{dom} \Phi)$. The primal vector optimization problem introduced above can be reformulated as

$$(PVG) \qquad \operatorname*{Min}_{x \in X} \Phi(x, 0).$$

Inspired by the way conjugate dual problems are attached to a given primal problem via perturbations in the scalar case and by the investigations from Sect. 5.2, where we embedded the classical Wolfe and Mond-Weir duality concepts into classes of scalar dual problems obtained via perturbation theory, and incorporating also ideas from different papers on Wolfe and Mond-Weir vector duality like [66, 82, 83, 205–207, 209, 211, 212], we attach to (PVG) the following vector dual problems with respect to properly efficient solutions

$$(DVG_W) \qquad \operatorname*{Max}_{(v^*, y^*, u, y, r) \in \mathscr{B}_W^G} h_W^G(v^*, y^*, u, y, r),$$

where

$$\mathscr{B}_W^G = \Big\{(v^*, y^*, u, y, r) \in K^{*0} \times Y^* \times X \times Y \times (K \setminus \{0\}) : (0, y^*) \in \partial(v^*\Phi)(u, y)\Big\}$$

and

$$h_W^G(v^*, y^*, u, y, r) = \Phi(u, y) - \frac{\langle y^*, y \rangle}{\langle v^*, r \rangle} r,$$

further referred to as the *Wolfe type vector dual* to (PVG), and, respectively, the *Mond-Weir type vector dual* to it

$$(DVG_M) \qquad \operatorname*{Max}_{(v^*, y^*, u) \in \mathscr{B}_M^G} h_M^G(v^*, y^*, u),$$

where

$$\mathscr{B}_M^G = \left\{(v^*, y^*, u) \in K^{*0} \times Y^* \times X : (0, y^*) \in \partial(v^*\Phi)(u, 0)\right\}$$

and

$$h_M^G(v^*, y^*, u) = \Phi(u, 0).$$

Remark 5.19. If $(v^*, y^*, u, y, r) \in \mathscr{B}_W^G$, one can immediately notice that $h_W^G(v^*, y^*, u, y, r) = h_W^G(v^*, y^*, u, y, \alpha r)$ for all $\alpha > 0$ and $(v^*, y^*, u, y, s) \in \mathscr{B}_W^G$ for all $s \in K \setminus \{0\}$.

Remark 5.20. Fixing $r \in K \setminus \{0\}$, we can construct, starting from (DVG_W), another dual problem to (PVG), namely

$$(DVG_{W^r}) \qquad \underset{(v^*, y^*, u, y) \in \mathscr{B}_{W^r}^G}{\text{Max}} h_{W^r}^G(v^*, y^*, u, y),$$

where

$$\mathscr{B}_{W^r}^G = \left\{(v^*, y^*, u, y) \in K^{*0} \times Y^* \times X \times Y : (0, y^*) \in \partial(v^*\Phi)(u, y), \langle v^*, r\rangle = 1\right\}$$

and

$$h_{W^r}^G(v^*, y^*, u, y) = \Phi(u, y) - \langle y^*, y\rangle r.$$

In this way one introduces a whole family of vector duals to (PVG). Moreover, one can consider other such families of vector dual problems to (PVG) by taking in (DVG_W) $\langle v^*, r\rangle$ equal to a given positive constant.

For these vector-maximization problems we consider efficient solutions, defined below for (DVG_W) and analogously for the others.

Definition 5.1. An element $(\bar{v}^*, \bar{y}^*, \bar{u}, \bar{y}, \bar{r}) \in \mathscr{B}_W^G$ is said to be an *efficient solution* to the vector optimization problem (DVG_W) if $(\bar{v}^*, \bar{y}^*, \bar{u}, \bar{y}, \bar{r}) \in \operatorname{dom} h_W^G$ and for all $(v^*, y^*, u, y, r) \in \mathscr{B}_W^G$ from $h_W^G(\bar{v}^*, \bar{y}^*, \bar{u}, \bar{y}, \bar{r}) \leqq_K h_W^G(v^*, y^*, u, y, r)$ follows $h_W^G(\bar{v}^*, \bar{y}^*, \bar{u}, \bar{y}, \bar{r}) = h_W^G(v^*, y^*, u, y, r)$. The set of all efficient solutions to (DVG_W) is called the *efficiency set of* (DVG_W), being denoted by $\mathscr{E}(DVG_W)$.

From the way the vector duals are defined above one can obtain the following results involving the images of their feasible sets via their objective functions.

Proposition 5.4. *It holds*

$$h_M^G(\mathscr{B}_M^G) \subseteq \bigcup_{r \in K \setminus \{0\}} h_{W^r}^G(\mathscr{B}_{W^r}^G) = h_W^G(\mathscr{B}_W^G).$$

Proof. Take $(v^*, y^*, u) \in \mathscr{B}_M^G$. Then $v^* \in K^{*0}$ and there exists an $r \in K \setminus \{0\}$ such that $\langle v^*, r\rangle = 1$. Thus $(v^*, y^*, u, 0) \in \mathscr{B}_{W^r}^G$ and $h_{W^r}^G(v^*, y^*, u, 0) = h_M^G(v^*, y^*, u) = \Phi(u, 0) = F(u)$.

Let now $r \in K \setminus \{0\}$ and $(v^*, y^*, u, y) \in \mathscr{B}_{W^r}^G$. It is obvious that $(v^*, y^*, u, y, r) \in \mathscr{B}_W^G$ and $h_W^G(v^*, y^*, u, y, r) = h_{W^r}^G(v^*, y^*, u, y) = \Phi(u, y) - \langle y^*, y\rangle r$.

Finally, if $(v^*, y^*, u, y, r) \in \mathscr{B}_W^G$, then taking $s = (1/\langle v^*, r\rangle)r \in K \setminus \{0\}$, it follows $\langle v^*, s\rangle = 1$ and, consequently, $(v^*, y^*, u, y) \in \mathscr{B}_{W^s}^G$. Moreover, $h_W^G(v^*, y^*, u, y, r) = h_{W^s}^G(v^*, y^*, u, y) = \Phi(u, y) - \langle y^*, y\rangle s$. □

Remark 5.21. Situations where the inclusion in Proposition 5.4 is strictly fulfilled will be presented later, in Examples 5.14 and 5.15.

Remark 5.22. It is a simple verification to show that if $(v^*, y^*, u, y, r) \in \mathscr{B}_W^G$ and $s \in K \setminus \{0\}$ such that $\langle y^*, y\rangle \neq 0$ and $h_W^G(v^*, y^*, u, y, r) - h_W^G(v^*, y^*, u, y, s) \in K$, then $r = s$ (see also Remark 5.19). Moreover, if $(v^*, y^*, u) \in \mathscr{B}_M^G$, then whenever $\alpha > 0$ one has $(\alpha v^*, \alpha y^*, u) \in \mathscr{B}_M^G$, while $(v^*, y^*, u, y, r) \in \mathscr{B}_W^G$ yields $(\alpha v^*, \alpha y^*, u, y, r) \in \mathscr{B}_W^G$ for all $\alpha > 0$. Similarly, for $r \in K \setminus \{0\}$ and $(v^*, y^*, u, y) \in \mathscr{B}_{W^r}^G$ one has $(\alpha v^*, \alpha y^*, u, y) \in \mathscr{B}_{W^r}^G$ for all $\alpha > 0$.

Remark 5.23. If $(\bar{v}^*, \bar{y}^*, \bar{u}, \bar{y}, \bar{r}) \in \mathscr{E}(DVG_W)$ and $\langle \bar{y}^*, \bar{y}\rangle = 0$, then $(\bar{v}^*, \bar{y}^*, \bar{u}, \bar{y}, \bar{s}) \in \mathscr{E}(DVG_W)$ and $(\bar{v}^*, \bar{y}^*, \bar{u}, \bar{y}) \in \mathscr{E}(DVG_{W^{\bar{s}}})$ for all $\bar{s} \in K \setminus \{0\}$. However, it is still an open problem whether the latter assertions remain valid after removing the hypothesis $\langle \bar{y}^*, \bar{y}\rangle = 0$.

Let us prove now that for the just introduced vector dual problems there is weak duality.

Theorem 5.12. *There are no $x \in X$ and $(v^*, y^*, u, y, r) \in \mathscr{B}_W^G$ such that $F(x) \leq_K h_W^G(v^*, y^*, u, y, r)$.*

Proof. Assume to the contrary that there exist $x \in X$ and $(v^*, y^*, u, y, r) \in \mathscr{B}_W^G$ fulfilling $F(x) \leq_K h_W^G(v^*, y^*, u, y, r)$. Then $x \in \operatorname{dom} F$ and it follows

$$\left\langle v^*, \Phi(u, y) - \frac{\langle y^*, y\rangle}{\langle v^*, r\rangle} r - \Phi(x, 0)\right\rangle > 0.$$

On the other hand, from the feasibility of (v^*, y^*, u, y, r) to (DVG_W), it follows $(v^*\Phi)(x, 0) - (v^*\Phi)(u, y) \geq \langle y^*, 0 - y\rangle$, from which

$$\left\langle v^*, \Phi(u, y) - \frac{\langle y^*, y\rangle}{\langle v^*, r\rangle} r - \Phi(x, 0)\right\rangle \leq \langle y^*, y\rangle - \left\langle v^*, \frac{\langle y^*, y\rangle}{\langle v^*, r\rangle} r\right\rangle = 0.$$

This leads to a contradiction to the strict inequality proven above. □

By making use of Theorem 5.12 and Proposition 5.4, one can prove also the following two weak duality statements involving the other vector duals to (PVG) introduced above.

Theorem 5.13. *There are no $x \in X$ and $(v^*, y^*, u) \in \mathscr{B}_M^G$ such that $F(x) \leq_K h_M^G(v^*, y^*, u)$.*

Theorem 5.14. *Let $r \in K \setminus \{0\}$. Then there are no $x \in X$ and $(v^*, y^*, u, y) \in \mathscr{B}_{W^r}^G$ such that $F(x) \leq_K h_{W^r}^G(v^*, y^*, u, y)$.*

One of the directions in which both Wolfe and Mond-Weir duality concepts were developed is towards introducing dual problems for which strong duality holds without asking the fulfillment of a regularity condition (see [84, 208, 210]). Like in the scalar case, (DVG_M) can be considered as such a vector dual problem to (PVG).

Proposition 5.5. *One always has $\mathscr{B}_M^G = \mathscr{E}(DVG_M)$ and $h_M^G(\mathscr{B}_M^G) = \operatorname{Max}(h_M^G(\mathscr{B}_M^G), K) \subseteq \operatorname{PMin}_{LS}(F(\operatorname{dom} F), K)$.*

Proof. If $\mathscr{B}_M^G = \emptyset$ there is nothing to prove. Assume thus that there is some $(v^*, y^*, u) \in \mathscr{B}_M^G$. Then $(v^*\Phi)^*(0, y^*) + (v^*\Phi)(u, 0) = 0$, which implies

$$(v^*\Phi)(u,0) = \inf_{x \in X, y \in Y} \left[(v^*\Phi)(x,y) - \langle y^*, y\rangle\right] \leq \inf_{x \in X}(v^*\Phi)(x,0).$$

Hence $u \in \mathscr{PE}_{LS}$ (PVG) and $\Phi(u,0) = F(u)$ is a value taken by the objective functions of both (PVG) and (DVG_M). Assuming that $(v^*, y^*, u) \notin \mathscr{E}(DVG_M)$, a contradiction is immediately obtained by employing Theorem 5.13. Consequently, $\mathscr{B}_M^G = \mathscr{E}(DVG_M)$ and using that $u \in \mathscr{PE}_{LS}(PVG)$ we obtain also that $h_M^G(\mathscr{B}_M^G) = \operatorname{Max}(h_M^G(\mathscr{B}_M^G), K) \subseteq \operatorname{PMin}_{LS}(F(\operatorname{dom} F), K)$. □

Two immediate consequences of this assertion follow.

Corollary 5.6. *If $(\bar{v}^*, \bar{y}^*, \bar{u}, 0, \bar{r}) \in \mathscr{B}_W^G$, then $(\bar{v}^*, \bar{y}^*, \bar{u}) \in \mathscr{E}(DVG_M)$, $(\bar{v}^*, \bar{y}^*, \bar{u}, 0, \bar{r}) \in \mathscr{E}(DVG_W)$, $(\bar{v}^*, \bar{y}^*, \bar{u}, 0) \in \mathscr{E}(DVG_{W^{\bar{r}}})$, $\bar{u} \in \mathscr{PE}_{LS}(PVG)$ and $F(\bar{u}) = h_M^G(\bar{v}^*, \bar{y}^*, \bar{u}) = h_W^G(\bar{v}^*, \bar{y}^*, \bar{u}, 0, \bar{r}) = h_{W^{\bar{r}}}^G(\bar{v}^*, \bar{y}^*, \bar{u}, 0)$.*

Proof. If $(\bar{v}^*, \bar{y}^*, \bar{u}, 0, \bar{r}) \in \mathscr{B}_W^G$, then it can be immediately verified that $F(\bar{u}) = h_W^G(\bar{v}^*, \bar{y}^*, \bar{u}, 0, \bar{r})$ and $(\bar{v}^*, \bar{y}^*, \bar{u}) \in \mathscr{B}_M^G$. By Proposition 5.5 it follows that $(\bar{v}^*, \bar{y}^*, \bar{u}) \in \mathscr{E}(DVG_M)$ and, consequently, $\bar{u} \in \mathscr{PE}_{LS}(PVG)$. Knowing these, the efficiency of $(\bar{v}^*, \bar{y}^*, \bar{u}, 0, \bar{r})$ to (DVG_W) follows by employing Theorem 5.12, while $(\bar{v}^*, \bar{y}^*, \bar{u}, 0) \in \mathscr{E}(DVG_{W^{\bar{r}}})$ follows via Theorem 5.14. □

Corollary 5.7. *Let $\bar{r} \in K \setminus \{0\}$. If $(\bar{v}^*, \bar{y}^*, \bar{u}, 0) \in \mathscr{B}_{W^{\bar{r}}}^G$, then $(\bar{v}^*, \bar{y}^*, \bar{u}) \in \mathscr{E}(DVG_M)$, $(\bar{v}^*, \bar{y}^*, \bar{u}, 0, \bar{r}) \in \mathscr{E}(DVG_W)$, $(\bar{v}^*, \bar{y}^*, \bar{u}, 0) \in \mathscr{E}(DVG_{W^{\bar{r}}})$, $\bar{u} \in \mathscr{PE}_{LS}$ (PVG) and $F(\bar{u}) = h_M^G(\bar{v}^*, \bar{y}^*, \bar{u}) = h_W^G(\bar{v}^*, \bar{y}^*, \bar{u}, 0, \bar{r}) = h_{W^{\bar{r}}}^G(\bar{v}^*, \bar{y}^*, \bar{u}, 0)$.*

Next we give some results involving the maximal sets of the vector duals introduced above. Combining Propositions 5.4 and 5.5, we obtain the following statement.

Proposition 5.6. *It holds*

$$\begin{aligned} h_M^G(\mathscr{B}_M^G) = \operatorname{Max}(h_M^G(\mathscr{B}_M^G), K) &\subseteq \operatorname{Max}(h_W^G(\mathscr{B}_W^G), K) \\ &\subseteq \bigcup_{r \in K \setminus \{0\}} \operatorname{Max}(h_{W^r}^G(\mathscr{B}_{W^r}^G), K). \end{aligned}$$

Proof. From Propositions 5.4 and 5.5 it is known that $h_M^G(\mathscr{B}_M^G) = \operatorname{Max}(h_M^G(\mathscr{B}_M^G), K) \subseteq \operatorname{PMin}_{LS}(F(\operatorname{dom} F), K) \cap h_W^G(\mathscr{B}_W^G)$. On the other hand, Theorem 5.12 yields that $\operatorname{PMin}_{LS}(F(\operatorname{dom} F), K) \cap h_W^G(\mathscr{B}_W^G) \subseteq \operatorname{Max}(h_W^G(\mathscr{B}_W^G), K)$ and the first inclusion is proven.

To demonstrate the second one, let $\bar{d} \in (h_W^G(\mathscr{B}_W^G), K)$. This means that there exists $(\bar{v}^*, \bar{y}^*, \bar{u}, \bar{y}, \bar{r}) \in \mathscr{E}(DVG_W)$ such that $h_W^G(\bar{v}^*, \bar{y}^*, \bar{u}, \bar{y}, \bar{r}) = \bar{d}$. For $\bar{s} = (1/\langle \bar{v}^*, \bar{r}\rangle)\bar{r}$, we obtain that $(\bar{v}^*, \bar{y}^*, \bar{u}, \bar{y}) \in \mathscr{B}_{W\bar{s}}^G$ and $h_{W\bar{s}}^G(\bar{v}^*, \bar{y}^*, \bar{u}, \bar{y}) = \bar{d}$.

Assuming that $(\bar{v}^*, \bar{y}^*, \bar{u}, \bar{y})$ were not efficient to $(DVG_{W\bar{s}})$ would bring, via Proposition 5.4, a contradiction to the efficiency of $(\bar{v}^*, \bar{y}^*, \bar{u}, \bar{y}, \bar{r})$ to (DVG_W). □

Now we turn our attention to strong duality for the vector duals introduced above, for whose attainment one needs, besides convexity hypotheses, the fulfillment of certain regularity conditions. Inspired by the ones considered in Sect. 2.2.1 in the scalar case and following [48], we consider four types of regularity conditions, namely a classical one involving continuity,

$$(RCV_1^G) \mid \exists x' \in X \text{ such that } (x', 0) \in \operatorname{dom} \Phi \text{ and } \Phi(x', \cdot) \text{ is continuous at } 0,$$

a weak interiority type one,

$$(RCV_2^G) \left| \begin{array}{l} X \text{ and } Y \text{ are Fréchet spaces, } \Phi \text{ is } C\text{-lower semicontinuous} \\ \text{and } 0 \in \operatorname{sqri} \operatorname{Pr}_Y(\operatorname{dom} \Phi), \end{array} \right.$$

a generalized interiority type one which works in finitely dimensional spaces,

$$(RCV_3^G) \mid \dim \operatorname{lin}(\operatorname{Pr}_Y(\operatorname{dom} \Phi)) < +\infty \text{ and } 0 \in \operatorname{ri} \operatorname{Pr}_Y(\operatorname{dom} \Phi),$$

and finally a closedness type one,

$$(RCV_4^G) \left| \begin{array}{l} \Phi \text{ is } C\text{-lower semicontinuous and for any } v^* \in K^{*0} \operatorname{Pr}_{X^* \times \mathbb{R}}(\operatorname{epi}(v^* \Phi)^*) \\ \text{is closed in the topology } \omega(X^*, X) \times \mathscr{R}. \end{array} \right.$$

Theorem 5.15. *Let $\bar{r} \in K \setminus \{0\}$. Assume that Φ is a K-convex function and one of the regularity conditions (RCV_i^G), $i \in \{1, 2, 3, 4\}$, is fulfilled. If $\bar{x} \in \mathscr{P}\mathscr{E}_{LS}(PVG)$, then there exist $\bar{v}^* \in K^{*0}$ and $\bar{y}^* \in Y^*$ such that $(\bar{v}^*, \bar{y}^*, \bar{x}, 0, \bar{r}) \in \mathscr{E}(DVG_W)$, $(\bar{v}^*, \bar{y}^*, \bar{x}, 0) \in \mathscr{E}(DVG_{W\bar{r}})$, $(\bar{v}^*, \bar{y}^*, \bar{x}) \in \mathscr{E}(DVG_M)$ and $F(\bar{x}) = h_W^G(\bar{v}^*, \bar{y}^*, \bar{x}, 0, \bar{r}) = h_{W\bar{r}}^G(\bar{v}^*, \bar{y}^*, \bar{x}, 0) = h_M^G(\bar{v}^*, \bar{y}^*, \bar{x})$.*

Proof. Since $\bar{x} \in \mathscr{P}\mathscr{E}_{LS}(PVG)$, there exists a $\bar{v}^* \in K^{*0}$ such that $\langle \bar{v}^*, F(\bar{x})\rangle \leq \langle \bar{v}^*, F(x)\rangle$ for all $x \in X$. As $\bar{r} \in K \setminus \{0\}$ assuming that $\langle \bar{v}^*, \bar{r}\rangle = 1$ does not imply losing the generality. From [48] (see also Corollary 2.6 and Remark 2.5) it is known that each of the regularity conditions (RCV_i^G), $i \in \{1, 2, 3, 4\}$, ensures the stability of the scalar optimization problem

$$\inf_{x \in X} (\bar{v}^* F)(x),$$

with respect to the perturbation function Φ. Then, via Theorem 5.2 (see also Remark 5.4), there is strong duality for it and its Wolfe type dual

$$\sup_{\substack{u \in X, y \in Y, y^* \in Y^*, \\ (0, y^*) \in \partial(\bar{v}^* \Phi)(u, y)}} \left\{ -(\bar{v}^* \Phi)^*(0, y^*) \right\},$$

i.e. there exists a $\bar{y}^* \in Y^*$ such that

$$-(\bar{v}^* \Phi)^*(0, \bar{y}^*) = \sup_{y^* \in Y^*} \{-(\bar{v}^* \Phi)^*(0, y^*)\} = \inf_{x \in X} \langle \bar{v}^*, F(x) \rangle = \langle \bar{v}^*, F(\bar{x}) \rangle,$$

and $(0, \bar{y}^*) \in \partial(\bar{v}^* \Phi)(\bar{x}, 0)$. Then $(\bar{v}^*, \bar{y}^*, \bar{x}) \in \mathscr{B}_M^G$ and, moreover, $(\bar{v}^*, \bar{y}^*, \bar{x}, 0) \in \mathscr{B}_{W^{\bar{r}}}^G$. The conclusion follows by using Proposition 5.5 and Corollary 5.7. □

Remark 5.24. In case $V = \mathbb{R}$ and $K = \mathbb{R}_+$, identifying $V^\bullet$ with $\mathbb{R} \cup \{+\infty\}$ and $\infty_{\mathbb{R}_+}$ with $+\infty$, and taking the function $F : X \to \overline{\mathbb{R}}$ proper we rediscover the Wolfe and Mond-Weir type scalar duality schemes from the scalar case presented in Sect. 5.2.1. More precisely the problem (*PVG*) becomes then the general scalar optimization problem (*PG*), while the duals (DVG_W) and (DVG_{W^r}), $r > 0$, turn out to coincide with (DG_W), the scalar Wolfe type dual to (*PG*), and (DVG_M) is nothing but the scalar Mond-Weir type dual (DG_M).

Remark 5.25. Other regularity conditions can be used in order to guarantee strong duality for (DVG_M) and (DVG_W), too, as long as they guarantee the stability of the scalar optimization problem $\inf_{x \in X}(v^* F)(x)$ with respect to the perturbation function Φ for all $v^* \in K^{*0}$.

Remark 5.26. Besides the properly efficient solutions in the sense of linear scalarization, one can use for (*PVG*) the solution concepts considered in Sect. 4.2, too. In this case one can assign to (*PVG*) vector duals of Wolfe and Mond-Weir types with respect to these types of solutions, defined by making use of the corresponding optimality conditions.

Remark 5.27. Another interesting vector duality approach for (*PVG*) can be developed starting from the observation that for a fixed $v^* \in K^{*0}$ one can show that an element $\bar{x} \in X$ is efficient to (*PVG*) if and only if it is an optimal solution of the scalar optimization problem

$$(EP) \qquad \inf_{\substack{F(\bar{x}) - F(x) \in K, \\ x \in X}} (v^* F)(x).$$

Having different scalar duals assigned to this scalar optimization problem, one can use them to formulate vector optimization dual problems with respect to efficient solutions to (*PVG*). More precisely, the strong duality statements regarding (*PVG*) and these new vector duals would ask the existence of an efficient solution to (*PVG*), besides convexity hypotheses and regularity conditions, in order to obtain efficient solutions to the vector duals.

Remark 5.28. It can also be interesting to study how can one give weak and strong duality statements for the primal-dual pairs of vector optimization problems considered in this section when the functions involved are differentiable on an open set S and the subdifferentials are replaced by gradients in the duals by using generalized convexity notions like quasiconvexity, pseudoconvexity, even invexity.

In the next subsections we consider like in Chap. 3 as special instances of (PVG) the two main classes of vector optimization problems, namely we work with a constrained and an unconstrained vector optimization problem, respectively. To these problems we attach vector duals that are special cases of (DVG_M), (DVG_W) and (DVG_{W^r}), $r > 0$, respectively, obtained for different choices of the vector perturbation function Φ.

5.3.2 Wolfe and Mond-Weir Vector Duals of Classical Type for Constrained Vector Optimization Problems

Besides the framework defined in the beginning of the section, consider that the space Y is partially ordered by the nonempty convex cone $C \subseteq Y$. Let the nonempty set $S \subseteq X$ and the proper vector functions $f : X \to V^\bullet$ and $h : X \to Y^\bullet$ fulfilling the feasibility condition $\operatorname{dom} f \cap S \cap h^{-1}(C) \neq \emptyset$. Let the primal vector optimization problem with geometric and cone constraints

$$(PVC) \qquad \operatorname*{Min}_{x \in \mathscr{A}} f(x),$$

where

$$\mathscr{A} = \{x \in S : h(x) \in -C\}.$$

Since (PVC) is a special case of (PVG) obtained by taking

$$F : X \to V^\bullet,\ F(x) = \begin{cases} f(x), & \text{if } x \in \mathscr{A}, \\ \infty_K, & \text{otherwise,} \end{cases}$$

we use the approach developed in Sect. 5.3.1 in order to deal with it via duality. More precisely, for convenient choices of the vector perturbation function Φ we obtain vector duals to (PVC) which are special cases of (DVG_M) and (DVG_W), respectively. Following the investigations from Sect. 5.3.1, we work with properly efficient solutions to (PVC), while for the vector dual we assign to it in this subsection we consider efficient solutions.

Consider first the Lagrange type vector perturbation function used in Sect. 4.4.1

$$\Phi_v^L : X \times Y \to V^\bullet,\ \Phi_v^L(x,z) = \begin{cases} f(x), & \text{if } x \in S, h(x) \in z - C, \\ \infty_K, & \text{otherwise,} \end{cases}$$

which is proper due to the fulfilment of the mentioned feasibility condition. For $u \in X$, $z \in Y$, $v^* \in K^{*0}$ and $z^* \in Y^*$ we have $(0, z^*) \in \partial(v^*\Phi_v^L)(u, z)$ if and only if $(v^*\Phi_v^L)^*(0, z^*) + (v^*\Phi_v^L)(u, z) = \langle z^*, z\rangle$, i.e. $((v^*f) - (z^*h) + \delta_S)^*(0) + \delta_{C^*}(-z^*) + f(u) + \delta_S(u) + \delta_{-C}(h(u) - z) = \langle z^*, z\rangle$. Using that $\delta^*_{-C} = \delta_{C^*}$, this can be rewritten as $\big(((v^*f) - (z^*h) + \delta_S)^*(0) + ((v^*f) - (z^*h) + \delta_S)(u)\big) + \big(\delta^*_{-C}(-z^*) + \delta_{-C}(h(u) - z) - \langle -z^*, h(u) - z\rangle\big) = 0$. Having the Young-Fenchel inequality and the characterization of the subdifferential by its equality case, it follows that $(0, z^*) \in \partial(v^*\Phi_v^L)(u, z)$ if and only if $0 \in \partial((v^*f) - (z^*h) + \delta_S)(u)$, $z^* \in -C^*$ and $\delta_{-C}(h(u) - z) - \langle -z^*, h(u) - z\rangle = 0$. Thus, from (DVG_W) we obtain the following vector dual to (PVC)

$$(DVC^L_{\tilde{W}}) \qquad \operatorname*{Max}_{(v^*,z^*,u,z,r)\in\mathscr{B}^L_{\tilde{W}}} h^L_{\tilde{W}}(v^*, z^*, u, z, r),$$

where

$$\mathscr{B}^L_{\tilde{W}} = \Big\{(v^*, z^*, u, z, r) \in K^{*0} \times C^* \times S \times Y \times (K \setminus \{0\}) : h(u) - z \in -C,$$
$$(z^*h)(u) = \langle z^*, z\rangle, 0 \in \partial((v^*f) + (z^*h) + \delta_S)(u)\Big\}$$

and

$$h^L_{\tilde{W}}(v^*, z^*, u, z, r) = f(u) + \frac{\langle z^*, z\rangle}{\langle v^*, r\rangle} r,$$

which can be equivalently rewritten as

$$(DVC^L_W) \qquad \operatorname*{Max}_{(v^*,z^*,u,r)\in\mathscr{B}^L_W} h^L_W(v^*, z^*, u, r),$$

where

$$\mathscr{B}^L_W = \Big\{(v^*, z^*, u, r) \in K^{*0} \times C^* \times S \times (K \setminus \{0\}) : 0 \in \partial((v^*f) + (z^*h) + \delta_S)(u)\Big\}$$

and

$$h^L_W(v^*, z^*, u, r) = f(u) + \frac{\langle z^*, h(u)\rangle}{\langle v^*, r\rangle} r,$$

further referred to as the *vector Wolfe dual of Lagrange type*, while the vector dual to (PVC) that results from (DVG_M) is

$$(DVC^L_M) \qquad \operatorname*{Max}_{(v^*,z^*,u)\in\mathscr{B}^L_M} h^L_M(v^*, z^*, u),$$

where

$$\mathscr{B}_M^L = \Big\{(v^*, z^*, u) \in K^{*0} \times C^* \times S : (z^*h)(u) \geq 0, h(u) \in -C,$$
$$0 \in \partial((v^* f) + (z^*h) + \delta_S)(u)\Big\}$$

and

$$h_M^L(v^*, z^*, u, r) = f(u).$$

Note that in the constraints of this dual one can replace $(z^*h)(u) \geq 0$ by $(z^*h)(u) = 0$ without altering anything since $h(u) \in -C$ and $z^* \in C^*$. Removing like in the scalar case from $\mathscr{B}_M^L$ the constraint $h(u) \in -C$, we obtain another vector dual to (*PVC*), namely

$$(DVC_{MW}^L) \qquad \operatorname*{Max}_{(v^*,z^*,u)\in\mathscr{B}_{MW}^L} h_{MW}^L(v^*, z^*, u),$$

where

$$\mathscr{B}_{MW}^L = \Big\{(v^*, z^*, u) \in K^{*0} \times C^* \times S : (z^*h)(u) \geq 0, 0 \in \partial((v^* f) + (z^*h) + \delta_S)(u)\Big\}$$

and

$$h_{MW}^L(v^*, z^*, u, r) = f(u),$$

further called the *vector Mond-Weir dual of Lagrange type* to (*PVC*). We can consider also the particularizations of the family of vector duals introduced in Remark 5.20. For each $r \in K \setminus \{0\}$ we have the vector dual

$$(DVC_{Wr}^L) \qquad \operatorname*{Max}_{(v^*,z^*,u)\in\mathscr{B}_{Wr}^L} h_{Wr}^L(v^*, z^*, u),$$

where

$$\mathscr{B}_{Wr}^L = \Big\{(v^*, z^*, u) \in K^{*0} \times C^* \times X : \langle v^*, r\rangle = 1, 0 \in \partial((v^* f) + (z^*h) + \delta_S)(u)\Big\}$$

and

$$h_{Wr}^L(v^*, z^*, u) = f(u) + \langle z^*, h(u)\rangle r.$$

Remark 5.29. Due to the way the vector duals we assigned above to (*PVC*) are constructed it is clear that $h_M^L(\mathscr{B}_M^L) \subseteq h_{MW}^L(\mathscr{B}_{MW}^L)$ and, via Proposition 5.4, it holds $h_M^L(\mathscr{B}_M^L) \subseteq \cup_{r\in K\setminus\{0\}} h_{Wr}^L(\mathscr{B}_{Wr}^L) = h_W^L(\mathscr{B}_W^L)$. The following examples show that there are situations when these inclusion are strictly fulfilled. Moreover, in both of them we see that $h_W^L(\mathscr{B}_W^L)$ is strictly larger than $h_{MW}^L(\mathscr{B}_{MW}^L)$, but it is not known whether the latter is in general a subset of the former image set.

Example 5.14. Let $X = \mathbb{R}$, $Y = \mathbb{R}$, $C = \mathbb{R}_+$, $Y^\bullet = \mathbb{R} \cup \{+\infty\}$, $V = \mathbb{R}^2$, $K = \mathbb{R}_+^2$, $S = \mathbb{R}_+$, $f : \mathbb{R} \to \mathbb{R}^2$, $f(x) = (x, x)^\top$, and $h : \mathbb{R} \to \mathbb{R} \cup \{+\infty\}$,

$$h(x) = \begin{cases} -x, & \text{if } x > 0, \\ 2, & \text{if } x = 0, \\ +\infty, & \text{if } x < 0. \end{cases}$$

For $v^* = (v_1^*, v_2^*)^\top$ we have $0 \in \partial((v^* f) + (0h) + \delta_S)(0) = (-\infty, v_1^* + v_2^*]$ and $(0h)(0) = 0$, thus $(v^*, 0, 0) \in \mathscr{B}_{MW}^L$, therefore $(0, 0)^\top \in h_{MW}^L(\mathscr{B}_{MW}^L)$.

If $u \geq 0$ and $z^* > 0$ it holds $(z^* h)(u) < 0$, while when $u \geq 0$ and $z^* = 0$ one gets $0 \notin \partial((v^* f) + (0h) + \delta_S)(u) = \{v_1^* + v_2^*\}$ for all $v^* = (v_1^*, v_2^*)^\top \in \operatorname{int} \mathbb{R}_+^2$. Consequently, $h_{MW}^L(\mathscr{B}_{MW}^L) = \{(0, 0)^\top\}$.

However, when $u > 0$ and $z^* > 0$ it follows that $\partial((v^* f) + (z^* h) + \delta_S)(u) = \{v_1^* + v_2^* - z^*\}$ for all $v^* = (v_1^*, v_2^*)^\top \in \operatorname{int} \mathbb{R}_+^2$, so $\partial((v^* f) + (z^* h) + \delta_S)(u) = \{0\}$ if $v_1^* + v_2^* = z^*$. Then, for $v^* = (v_1^*, v_2^*)^\top \in \operatorname{int} \mathbb{R}_+^2$, $u > 0$ and $r = (r_1, r_2)^\top \in \mathbb{R}_+^2 \setminus \{0\}$, one has $(v^*, v_1^* + v_2^*, u, r) \in \mathscr{B}_W^L$ and $h_W^L(v^*, v_1^* + v_2^*, u, r) = (u, u)^\top - \big((v_1^* + v_2^*)u/(v_1^* r_1 + v_2^* r_2)\big)(r_1, r_2)^\top \in h_W^L(\mathscr{B}_W^L)$. Taking $v^* = (1/2, 1/2)^\top$ and $r = (0, 1)^\top$, it follows that $(u, -u)^\top \in h_W^L(\mathscr{B}_W^L)$ for all $u > 0$.

On the other hand it can be shown like in Example 5.1 that $\mathscr{B}_M^L = \emptyset$.

Therefore, $h_M^L(\mathscr{B}_M^L) \subsetneq h_{MW}^L(\mathscr{B}_{MW}^L) \subsetneq h_W^L(\mathscr{B}_W^L)$ in this case. This shows that in general one has $h_M^G(\mathscr{B}_M^G) \subsetneq \cup_{r \in K \setminus \{0\}} h_{Wr}^G(\mathscr{B}_{Wr}^G) = h_W^G(\mathscr{B}_W^G)$.

Example 5.15. Let $X = \mathbb{R}$, $Y = \mathbb{R}^2$, $C = \mathbb{R}_+^2$, $V = \mathbb{R}^2$, $K = \mathbb{R}_+^2$, $V^\bullet = (\mathbb{R}^2)^\bullet = \mathbb{R}^2 \cup \{\infty_{\mathbb{R}_+^2}\}$, $S = \mathbb{R}_+$, $f : \mathbb{R} \to (\mathbb{R}^2)^\bullet$,

$$f(x) = \begin{cases} \begin{pmatrix} 1 \\ 1 \end{pmatrix} x & \text{if } x > 0, \\ \infty_{\mathbb{R}_+^2}, & \text{otherwise,} \end{cases}$$

and $h : \mathbb{R} \to \mathbb{R}^2$, $h(x) = (x - 1, -x)^\top$.

Like in Example 5.2, it can be shown that $h_M^L(\mathscr{B}_M^L) = h_{MW}^L(\mathscr{B}_{MW}^L) = \emptyset$, while for $r = (1, 1)^\top$, one has $\big((1/2, 1/2)^\top, (2, 3)^\top, 1\big) \in \mathscr{B}_{Wr}^L$, consequently, $(-2, -2)^\top \in h_{Wr}^L(\mathscr{B}_{Wr}^L) \subseteq h_W^L(\mathscr{B}_W^L)$.

Therefore, $h_M^L(\mathscr{B}_M^L) = h_{MW}^L(\mathscr{B}_{MW}^L) \subsetneq h_W^L(\mathscr{B}_W^L)$ in this case.

Remark 5.30. Assume that S is a convex set, f is a K-convex vector function and h is a C-convex vector function. Then it is a simple verification to see that the vector perturbation function Φ_v^L is K-convex. When one of the following conditions

(i) f and h are continuous at a point in $\operatorname{dom} f \cap \operatorname{dom} h \cap S$;
(ii) $\operatorname{dom} f \cap \operatorname{int} S \cap \operatorname{dom} h \neq \emptyset$ and f or h is continuous at a point in $\operatorname{dom} f \cap \operatorname{dom} h$;

(iii) X is a Fréchet space, S is closed, f is K-lower semicontinuous, h is C-lower semicontinuous and $0 \in \operatorname{sqri}(\operatorname{dom} f \times S \times \operatorname{dom} h - \Delta_{X^3})$;
(iv) $\dim \operatorname{lin}(\operatorname{dom} f \times S \times \operatorname{dom} h - \Delta_{X^3}) < +\infty$ and $\operatorname{ri} \operatorname{dom} f \cap \operatorname{ri} S \cap \operatorname{ri} \operatorname{dom} h \neq \emptyset$;

is satisfied, then (see [21,48,221]) for all $v^* \in K^{*0}$ and all $z^* \in C^*$, it holds

$$\partial((v^* f) + (z^* h) + \delta_S)(x) = \partial(v^* f)(x) + \partial(z^* h)(x) + N_S(x) \ \forall x \in X.$$

Consequently, when one of these situations occurs, the constraint involving the subdifferential in (DVC_W^L), $(DVC_{W^r}^L)$, for any $r \in K \setminus \{0\}$, (DVC_M^L) and, respectively, (DVC_{MW}^L) can be correspondingly modified. Moreover, in order to split $\partial((v^* f) + (z^* h) + \delta_S)(x)$ into a sum of only two subdifferentials, one can apply [48, Theorem 3.5.6].

Remark 5.31. If $X = \mathbb{R}^n$, $Y = \mathbb{R}^m$, $C = \mathbb{R}_+^m$, $V = \mathbb{R}^k$, $K = \mathbb{R}_+^k$, S is convex, $f = (f_1, \dots, f_k)^\top : \mathbb{R}^n \to \mathbb{R}^k$ and $h = (h_1, \dots, h_m)^\top : \mathbb{R}^n \to \mathbb{R}^m$, and the functions f_i, $i = 1, \dots, k$, and h_j, $j = 1, \dots, m$, are convex, then $(DVC_{W^e}^L)$, turns out to be the *nondifferentiable vector Wolfe dual* problem mentioned in the literature (see [84,134,210]), while (DVC_{MW}^L) is the *nondifferentiable vector Mond-Weir dual* problem to (PVC).

Remark 5.32. If, in addition to the hypotheses of Remark 5.31, the set S is open and the functions f_i, $i = 1, \dots, k$, and h_j, $j = 1, \dots, m$, are moreover Gâteaux differentiable on it, the subdifferentials in the constraints can be replaced by the corresponding gradients, $(DVC_{W^r}^L)$ turns out to be, after fixing r, the classical *vector Wolfe dual* problem from the literature (see [212] and, for the case $r = e$, [83, 205,206,209]), while (DVC_{MW}^L) is the classical *vector Mond-Weir dual* problem to (PVC) considered in papers like [82,83,205,207–209]).

Like in the previous section, the results involving (PVG) and its vector duals can be particularized for the problems introduced above, however we give here only the weak and strong duality statements involving (PVC) and its vector duals of Lagrange type.

Theorem 5.16. *There are no $x \in \mathscr{A}$ and $(v^*, z^*, u, r) \in \mathscr{B}_W^L$ such that $f(x) \leq_K h_W^L(v^*, z^*, u, r)$.*

Theorem 5.17. *There are no $x \in \mathscr{A}$ and $(v^*, z^*, u) \in \mathscr{B}_M^L$ such that $f(x) \leq_K h_M^L(v^*, z^*, u)$.*

Theorem 5.18. *Let $r \in K \setminus \{0\}$. Then there are no $x \in \mathscr{A}$ and $(v^*, z^*, u) \in \mathscr{B}_{W^r}^L$ such that $f(x) \leq_K h_{W^r}^L(v^*, z^*, u)$.*

Analogously, one can prove also the following weak duality statement involving (PVC) and (DVC_{MW}^L).

Theorem 5.19. *There are no $x \in \mathscr{A}$ and $(v^*, z^*, u) \in \mathscr{B}_{MW}^L$ such that $f(x) \leq_K h_{MW}^L(v^*, z^*, u)$.*

For strong duality, which follows directly from Theorem 5.15, besides convexity assumptions which guarantee the K-convexity of the vector perturbation function Φ_v^L we use regularity conditions, too, obtained by particularizing (RCV_i^G), $i \in \{1,2,3,4\}$, namely (cf. [21,48])

$$(RCV_1^L) \big| \; \exists x' \in \operatorname{dom} f \cap S \text{ such that } h(x') \in -\operatorname{int} C,$$

which is the classical *Slater constraint qualification* extended to the vector case,

$$(RCV_2^L) \left| \begin{array}{l} X \text{ and } Y \text{ are Fréchet spaces, } S \text{ is closed, } f \text{ is } K\text{-lower semicontinuous,} \\ h \text{ is } C\text{-epi-closed and } 0 \in \operatorname{sqri}\big(h(\operatorname{dom} f \cap S \cap \operatorname{dom} h) + C\big), \end{array} \right.$$

$$(RCV_3^L) \left| \begin{array}{l} \dim \operatorname{lin}\left(h(\operatorname{dom} f \cap S \cap \operatorname{dom} h) + C\right) < +\infty \text{ and} \\ 0 \in \operatorname{ri}\left(h(\operatorname{dom} f \cap S \cap \operatorname{dom} h) + C\right), \end{array} \right.$$

and, respectively,

$$(RCV_4^L) \left| \begin{array}{l} S \text{ is closed, } f \text{ is } K\text{-lower semicontinuous, } h \text{ is } C\text{-epi-closed and} \\ \text{for any } v^* \in K^{*0} \bigcup\limits_{z^* \in C^*} \operatorname{epi}((v^* f) + (z^* h) + \delta_S)^* \text{ is closed} \\ \text{in the topology } \omega(X^*, X) \times \mathscr{R}. \end{array} \right.$$

The strong duality assertions concerning (DVC_W^L) and (DVC_M^L), respectively, follow via Theorem 5.15, while their counterparts for (DVC_{MW}^L) and $(DVC_{W\bar{r}}^L)$, where $\bar{r} \in K \setminus \{0\}$, can be proven analogously.

Theorem 5.20. *Let $\bar{r} \in K \setminus \{0\}$. Assume that S is a convex set, f is a K-convex vector function, h is a C-convex vector function and one of the regularity conditions (RCV_i^L), $i \in \{1,2,3,4\}$, is fulfilled. If $\bar{x} \in \mathscr{P}\mathscr{E}_{LS}(PVC)$, then there exist $\bar{v}^* \in K^{*0}$ and $\bar{z}^* \in C^*$ such that $(\bar{v}^*, \bar{z}^*, \bar{x}, \bar{r}) \in \mathscr{E}(DVC_W^L)$, $(\bar{v}^*, \bar{z}^*, \bar{x}) \in \mathscr{E}(DVC_{W\bar{r}}^L) \cap \mathscr{E}(DVC_M^L) \cap \mathscr{E}(DVC_{MW}^L)$ and $f(\bar{x}) = h_W^L(\bar{v}^*, \bar{z}^*, \bar{x}, \bar{r}) = h_{W\bar{r}}^L(\bar{v}^*, \bar{z}^*, \bar{x}) = h_M^L(\bar{v}^*, \bar{z}^*, \bar{x}) = h_{MW}^L(\bar{v}^*, \bar{z}^*, \bar{x})$.*

Like in the scalar case and Sect. 4.4.1, one can consider a Fenchel-Lagrange type vector perturbation function in order to assign vector dual problems to (PVC), too, namely $\Phi_v^{FL} : X \times X \times Y \to V^\bullet$,

$$\Phi_v^{FL}(x,y,z) = \begin{cases} f(x+y), & \text{if } x \in S, h(x) \in z - C, \\ \infty_K, & \text{otherwise.} \end{cases}$$

For $v^* \in K^{*0}$, $z^* \in Y^*$, $y^* \in X^*$, $z \in Y$ and $y \in X$, one has $(0, y^*, z^*) \in \partial \Phi_v^{FL}(u,y,z)$ if and only if $u \in S$, $h(u) \in z - C$ and $(v^* f)^*(y^*) + (-(z^* h) + \delta_S)^*(-y^*) + \delta_{-C^*}(z^*) + f(u+y) + \delta_{-C}(h(u)-z) + \delta_S(u) = \langle z^*, z\rangle + \langle y^*, y\rangle$, which is nothing but $u \in S$, $h(u) \in z - C$ and $\big((v^* f)^*(y^*) + (v^* f)(u+y) - \langle y^*, u+y\rangle\big) + \big((-(z^* h) + \delta_S)^*(-y^*) + (-(z^* h) + \delta_S)(u) - \langle -y^*, u\rangle\big) + \big(\delta_{-C}^*(-z^*) + \delta_{-C}(h(u) - z) - \langle -z^*, h(u) - z\rangle\big) = 0$. Consequently, $(0, y^*, z^*) \in \partial \Phi_v^{FL}(u,y,z)$ if and only if

$u \in S$, $z^* \in -C^*$, $h(u) - z \in -C$, $y^* \in \partial f(u+y) \cap (-\partial(-(z^*h) + \delta_S)(u))$ and $(z^*h)(u) = \langle z^*, z\rangle$. Therefore, the vector duals to (PVC) obtained, by making use of the vector perturbation function Φ_v^{FL}, from the vector duals introduced in Sect. 5.3.1 are

$$(DVC_{\tilde{W}}^{FL}) \qquad \operatorname*{Max}_{(v^*,y^*,z^*,u,y,z,r)\in\mathscr{B}_{\tilde{W}}^{FL}} h_{\tilde{W}}^{FL}(v^*, y^*, z^*, u, y, z, r),$$

where

$$\mathscr{B}_{\tilde{W}}^{FL} = \Big\{(v^*, y^*, z^*, u, y, z, r) \in K^{*0} \times X^* \times C^* \times S \times X \times Y \times (K \setminus \{0\}) : \\ h(u) - z \in -C, (z^*h)(u) = \langle z^*, z\rangle, \\ y^* \in \partial(v^*f)(u+y) \cap (-\partial((z^*h) + \delta_S))(u)\Big\}$$

and

$$h_{\tilde{W}}^{FL}(v^*, y^*, z^*, u, y, z, r) = f(u+y) - \frac{\langle z^*, z\rangle + \langle y^*, y\rangle}{\langle v^*, r\rangle} r,$$

which can be equivalently rewritten as

$$(DVC_{W}^{FL}) \qquad \operatorname*{Max}_{(v^*,y^*,z^*,u,y,r)\in\mathscr{B}_{W}^{FL}} h_{W}^{FL}(v^*, y^*, z^*, u, y, r),$$

where

$$\mathscr{B}_{W}^{FL} = \Big\{(v^*, y^*, z^*, u, y, r) \in K^{*0} \times X^* \times C^* \times S \times X \times (K \setminus \{0\}) : \\ y^* \in \partial(v^*f)(u+y) \cap (-\partial((z^*h) + \delta_S))(u)\Big\}$$

and

$$h_{W}^{FL}(v^*, y^*, z^*, u, y, r) = f(u+y) + \frac{(z^*h)(u) - \langle y^*, y\rangle}{\langle v^*, r\rangle} r,$$

further called the *vector Wolfe dual of Fenchel-Lagrange type*,

$$(DVC_{M}^{FL}) \qquad \operatorname*{Max}_{(v^*,z^*,u)\in\mathscr{B}_{M}^{FL}} h_{M}^{FL}(v^*, z^*, u),$$

where

$$\mathscr{B}_{M}^{FL} = \Big\{(v^*, z^*, u) \in K^{*0} \times C^* \times S : (z^*h)(u) \geq 0, h(u) \in -C, \\ 0 \in \partial(v^*f)(u) + \partial((z^*h) + \delta_S)(u)\Big\}$$

and

$$h_M^{FL}(v^*, z^*, u) = f(u),$$

and, for each $r \in K \setminus \{0\}$,

$$(DVC_{W^r}^{FL}) \qquad \operatorname*{Max}_{(v^*,y^*,z^*,u,y)\in\mathscr{B}_{W^r}^{FL}} h_{W^r}^{FL}(v^*, y^*, z^*, u, y),$$

where

$$\mathscr{B}_{W^r}^{FL} = \Big\{(v^*, y^*, z^*, u, y) \in K^{*0} \times X^* \times C^* \times S \times X : \langle v^*, r\rangle = 1,$$
$$y^* \in \partial(v^* f)(u + y) \cap (-\partial((z^*h) + \delta_S))(u)\Big\}$$

and

$$h_{W^r}^{FL}(v^*, y^*, z^*, u, y) = f(u + y) + \big((z^*h)(u) - \langle y^*, y\rangle\big)r.$$

Note that in the constraints of (DVC_M^{FL}) one can replace $(z^*h)(u) \geq 0$ with $(z^*h)(u) = 0$ without altering anything. Removing the constraint $h(u) \in -C$ from $\mathscr{B}_M^{FL}$, one obtains from (DVC_M^{FL}) the *vector Mond-Weir dual of Fenchel-Lagrange type* to *(PVC)*

$$(DVC_{MW}^{FL}) \qquad \operatorname*{Max}_{(v^*,z^*,u)\in\mathscr{B}_{MW}^{FL}} h_{MW}^{FL}(v^*, z^*, u),$$

where

$$\mathscr{B}_{MW}^{FL} = \Big\{(v^*, z^*, u) \in K^{*0} \times C^* \times S : (z^*h)(u) \geq 0, 0 \in \partial(v^* f)(u) + \partial((z^*h) + \delta_S)(u)\Big\}$$

and

$$h_{MW}^{FL}(v^*, z^*, u) = f(u).$$

Remark 5.33. Due to the way the vector duals we assigned to *(DVC)* are constructed it is clear that $h_M^{FL}(\mathscr{B}_M^{FL}) \subseteq h_{MW}^{FL}(\mathscr{B}_{MW}^{FL})$ and, via Proposition 5.4, it holds $h_M^{FL}(\mathscr{B}_M^{FL}) \subseteq \cup_{r\in K\setminus\{0\}} h_{W^r}^{FL}(\mathscr{B}_{W^r}^{FL}) = h_W^{FL}(\mathscr{B}_W^{FL})$. These inclusions are strict in general, for instance in the situations presented in Example 5.14 we have $h_M^{FL}(\mathscr{B}_M^{FL}) = \emptyset$, $h_{MW}^{FL}(\mathscr{B}_{MW}^{FL}) = \{(0,0)^\top\}$ and $(u, -u)^\top \in h_W^{FL}(\mathscr{B}_W^{FL})$ for all $u \geq 0$, respectively. Moreover, one can see that in the mentioned situation $h_W^{FL}(\mathscr{B}_W^{FL})$ is strictly larger than $h_{MW}^{FL}(\mathscr{B}_{MW}^{FL})$, but it is not known whether the latter is in general a subset of the former image set.

Remark 5.34. Results analogous to the one from Remark 5.6 can be given for the Fenchel-Lagrange type vector duals to *(PVC)*, too. Assuming that S is a convex set and h is a C-convex vector function, the satisfaction of any of the following conditions (cf. [48, Theorem 3.5.6])

(i) h is continuous at a point in $S \cap \operatorname{dom} h$;
(ii) $\operatorname{int} S \cap \operatorname{dom} h \neq \emptyset$;
(iii) X is a Fréchet space, S is closed, h is star C-lower semicontinuous and $0 \in \operatorname{sqri}(S - \operatorname{dom} h)$;
(iv) $\dim \operatorname{lin}(S - \operatorname{dom} h) < +\infty$ and $\operatorname{ri} S \cap \operatorname{ri} \operatorname{dom} h \neq \emptyset$;

ensures the fulfillment of the formula

$$\partial((z^*h) + \delta_S)(u) = \partial(z^*h)(u) + N_S(u) \ \forall u \in X \ \forall z^* \in C^*,$$

in which case the constraint involving $\partial((z^*h) + \delta_S)(u)$ in (DVC_W^{FL}), $(DVC_{W^r}^{FL})$, for $r \in K \setminus \{0\}$, (DVC_M^{FL}) and (DVC_{MW}^{FL}) can be correspondingly modified.

The results involving (PVG) and its vector duals can be particularized for the Fenchel-Lagrange type vector duals, too, but here we give only the weak and strong duality statements involving (PVC) and these vector duals.

Theorem 5.21. *There are no $x \in \mathscr{A}$ and $(v^*, y^*, z^*, u, y, r) \in \mathscr{B}_W^{FL}$ such that $f(x) \leq_K h_W^{FL}(v^*, y^*, z^*, u, y, r)$.*

Theorem 5.22. *There are no $x \in \mathscr{A}$ and $(v^*, z^*, u) \in \mathscr{B}_M^{FL}$ such that $f(x) \leq_K h_M^{FL}(v^*, z^*, u)$.*

Theorem 5.23. *Let $r \in K \setminus \{0\}$. Then there are no $x \in \mathscr{A}$ and $(v^*, y^*, z^*, u, y) \in \mathscr{B}_{W^r}^{FL}$ such that $f(x) \leq_K h_{W^r}^{FL}(v^*, y^*, z^*, u, y)$.*

Analogously, one can prove also the following weak duality statement involving (PVC) and (DVC_{MW}^{FL}).

Theorem 5.24. *There are no $x \in \mathscr{A}$ and $(v^*, z^*, u) \in \mathscr{B}_{MW}^{FL}$ such that $f(x) \leq_K h_{MW}^{FL}(v^*, z^*, u)$.*

For strong duality, which follows directly from Theorem 5.15, besides convexity assumptions which guarantee the K-convexity of the vector perturbation function Φ_v^{FL} we use regularity conditions, too, obtained by particularizing (RCV_i^G), $i \in \{1, 2, 3, 4\}$, namely (cf. [21, 48])

$$(RCV_1^{FL}) \Big| \exists x' \in \operatorname{dom} f \cap S \text{ such that } f \text{ is continuous at } x' \text{ and } h(x') \in -\operatorname{int} C,$$

$$(RCV_2^{FL}) \left| \begin{array}{l} X \text{ and } Y \text{ are Fréchet spaces, } S \text{ is closed, } f \text{ is } K\text{-lower} \\ \text{semicontinuous, } h \text{ is } C\text{-epi-closed and} \\ 0 \in \operatorname{sqri}\big(\operatorname{dom} f \times C - \operatorname{epi}_{-C}(-h) \cap (S \times Y)\big), \end{array} \right.$$

$$(RCV_3^{FL}) \left| \begin{array}{l} \dim \operatorname{lin}\left(\operatorname{dom} f \times C - \operatorname{epi}_{-C}(-h) \cap (S \times Z)\right) < +\infty \text{ and} \\ 0 \in \operatorname{ri}\left(\operatorname{dom} f \times C - \operatorname{epi}_{-C}(-h) \cap (S \times Z)\right). \end{array} \right.$$

and, respectively,

$$(RCV_4^{FL}) \left| \begin{array}{l} S \text{ is closed, } f \text{ is } K\text{-lower semicontinuous, } h \text{ is } C\text{-epi-closed and} \\ \text{for any } v^* \in K^{*0} \ \operatorname{epi}(v^* f)^* + \bigcup\limits_{z^* \in C^*} \operatorname{epi}((z^* h) + \delta_S)^* \text{ is} \\ \text{closed in the topology } \omega(X^*, X) \times \mathscr{R}. \end{array} \right. \tag{5.3.1}$$

The strong duality assertions concerning (DVC_W^{FL}) and (DVC_M^{FL}), respectively, follow via Theorem 5.15, while their counterparts for (DVC_{MW}^{FL}) and $(DVC_{W\bar{r}}^{FL})$, where $\bar{r} \in K \setminus \{0\}$, can be proven analogously.

Theorem 5.25. *Let $\bar{r} \in K \setminus \{0\}$. Assume that S is a convex set, f is a K-convex vector function, h is a C-convex vector function and one of the regularity conditions (RCV_i^{FL}), $i \in \{1,2,3,4\}$, is fulfilled. If $\bar{x} \in \mathscr{PE}_{LS}(PVC)$, then there exist $\bar{v}^* \in K^{*0}$, $\bar{y}^* \in X^*$ and $\bar{z}^* \in C^*$ such that $(\bar{v}^*, \bar{y}^*, \bar{z}^*, \bar{x}, 0, \bar{r}) \in \mathscr{E}(DVC_W^{FL})$, $(\bar{v}^*, \bar{y}^*, \bar{z}^*, \bar{x}, 0) \in \mathscr{E}(DVC_{W\bar{r}}^{FL})$, $(\bar{v}^*, \bar{z}^*, \bar{x}) \in \mathscr{E}(DVC_M^{FL}) \cap \mathscr{E}(DVC_{MW}^{FL})$ and $f(\bar{x}) = h_W^{FL}(\bar{v}^*, \bar{y}^*, \bar{z}^*, \bar{x}, 0, \bar{r}) = h_{W\bar{r}}^{FL}(\bar{v}^*, \bar{y}^*, \bar{z}^*, \bar{x}, 0) = h_M^{FL}(\bar{v}^*, \bar{z}^*, \bar{x}) = h_{MW}^{FL}(\bar{v}^*, \bar{z}^*, \bar{x})$.*

Remark 5.35. Like in the general case (see Remark 5.24), if $V = \mathbb{R}$ and $K = \mathbb{R}_+$, taking the functions $f : X \to \overline{\mathbb{R}}$ and $h : X \to Y^\bullet$ proper we rediscover the Wolfe and Mond-Weir duality schemes for constrained scalar optimization problems from Sect. 5.2.2, respectively. More precisely the problem *(PVC)* becomes then the constrained scalar optimization problem *(PC)* and the vector duals considered in this section turn out to be to the corresponding dual problems considered there to it.

5.3.3 *Wolfe and Mond-Weir Vector Duals of Classical Type for Unconstrained Vector Optimization Problems*

Consider again the framework of Sect. 5.3.1. Let $f : X \to V^\bullet$ and $g : Y \to V^\bullet$ be given proper vector functions and $A : X \to Y$ a linear continuous mapping such that the feasibility condition $\operatorname{dom} f \cap A^{-1}(\operatorname{dom} g) \neq \emptyset$ is fulfilled.

The primal unconstrained vector optimization problem we deal with is

$$(PVU) \qquad \operatorname*{Min}_{x \in X}[f(x) + g(Ax)].$$

We work with properly efficient solutions to *(PVU)*, while for the vector dual we assign to it in this section we consider efficient solutions. Since *(PVU)* is a special case of *(PVG)* obtained by taking $F = f + g \circ A$, we use the approach developed in Sect. 5.3.1 in order to deal with it via duality. More precisely, for a convenient choice of the vector perturbation function Φ we obtain vector duals to *(PVU)* which are special cases of (DVG_M) and (DVG_W).

In order to attach vector dual problems to (PVU), consider like in Sect. 4.4.2 the vector perturbation function

$$\Phi_v^U : X \times Y \to V^\bullet,\ \Phi_v^U(x,y) = f(x) + g(Ax+y).$$

For $v^* \in K^{*0}$, $u \in X$, $y \in Y$ and $y^* \in Y^*$ one has $(0,y^*) \in \partial(v^*\Phi_v^U)(u,y)$ if and only if $(v^*\Phi_v^U)^*(0,y^*) + (v^*\Phi_v^U)(u,y) = \langle y^*, y\rangle$. This is further equivalent to $(v^*f)^*(-A^*y^*) + (v^*g)^*(y^*) + f(u) + g(Au+y) = \langle y^*, y\rangle$. Using the Young-Fenchel inequality, the last equality yields that $(0,y^*) \in \partial(v^*\Phi_v^U)(u,y)$ if and only if $y^* \in \partial(v^*g)(Au+y)$ and $-A^*y^* \in \partial(v^*f)(u)$. Now we are ready to formulate the vector duals to (PVU) that are special cases of (DVG_M) and (DVG_W), namely

$$(DVU_W) \qquad \operatorname*{Max}_{(v^*,y^*,u,y,r)\in\mathcal{B}_W^U} h_W^U(v^*,y^*,u,y,r),$$

where

$$\mathcal{B}_W^U = \Big\{(v^*,y^*,u,y,r) \in K^{*0} \times Y^* \times X \times Y \times (K\setminus\{0\}) : \\ y^* \in (A^*)^{-1}(-\partial(v^*f)(u)) \cap \partial(v^*g)(Au+y)\Big\}$$

and

$$h_W^U(v^*,y^*,u,y,r) = f(u) + g(Au+y) - \frac{\langle y^*, y\rangle}{\langle v^*, r\rangle} r,$$

and, respectively,

$$(DVU_M) \qquad \operatorname*{Max}_{(v^*,u)\in\mathcal{B}_M^U} h_M^U(v^*,u),$$

where

$$\mathcal{B}_M^U = \Big\{(v^*,u) \in K^{*0} \times X : (A^*)^{-1}(-\partial(v^*f)(u)) \cap \partial(v^*g)(Au) \neq \emptyset\Big\}$$

and

$$h_W^U(v^*,u) = f(u) + g(Au).$$

One can also consider the particularizations of the family of vector duals introduced in Remark 5.20. For each $r \in K\setminus\{0\}$ we have the vector dual to (PVU)

$$(DVU_{W^r}) \qquad \operatorname*{Max}_{(v^*,y^*,u,y)\in\mathcal{B}_{W^r}^U} h_{W^r}^U(v^*,y^*,u,y),$$

where

$$\mathscr{B}^U_{Wr} = \Big\{(v^*, y^*, u, y) \in K^{*0} \times Y^* \times X \times Y : \langle v^*, r\rangle = 1,\\ y^* \in (A^*)^{-1}(-\partial(v^* f)(u)) \cap \partial(v^* g)(Au + y)\Big\}$$

and

$$h^U_W(v^*, y^*, u, y) = f(u) + g(Au + y) - \langle y^*, y\rangle r.$$

Observations similar to Remarks 5.11 and 5.15 can be made in the vector case, too. Note also that via Proposition 5.4, it holds $h^U_M(\mathscr{B}^U_M) \subseteq \cup_{r\in K\setminus\{0\}} h^U_{Wr}(\mathscr{B}^U_{Wr}) = h^U_W(\mathscr{B}^U_W)$.

Let us give now the weak and strong duality statements for these duals.

Theorem 5.26. *There are no $x \in X$ and $(v^*, y^*, u, y, r) \in \mathscr{B}^U_W$ such that $f(x) + g(Ax) \leq_K h^U_W(v^*, y^*, u, y, r)$.*

Theorem 5.27. *There are no $x \in X$ and $(v^*, u) \in \mathscr{B}^U_M$ such that $f(x) + g(Ax) \leq_K h^U_M(v^*, u)$.*

Theorem 5.28. *Let $r \in K\setminus\{0\}$. Then there are no $x \in X$ and $(v^*, y^*, u, y) \in \mathscr{B}^U_{Wr}$ such that $f(x) + g(Ax) \leq_K h^U_{Wr}(v^*, y^*, u, y)$.*

For strong duality, which follows directly from Theorem 5.15, besides convexity assumptions which guarantee the K-convexity of the vector perturbation function Φ^U_v we use regularity conditions, too, obtained by particularizing (RCV^G_i), $i \in \{1, 2, 3, 4\}$, namely (cf. [21, 48])

(RCV^U_1) | $\exists x' \in \operatorname{dom} f \cap A^{-1}(\operatorname{dom} g)$ such that g is continuous at Ax',

(RCV^U_2) | X and Y are Fréchet spaces, f and g are K-lower semicontinuous and $0 \in \operatorname{sqri}(\operatorname{dom} g - A(\operatorname{dom} f))$,

(RCV^U_3) | $\dim \operatorname{lin}(\operatorname{dom} g - A(\operatorname{dom} f)) < +\infty$ and $\operatorname{ri} A(\operatorname{dom} f) \cap \operatorname{ri} \operatorname{dom} g \neq \emptyset$,

and, respectively,

(RCV^U_4) | f and g are K-lower semicontinuous and for any $v^* \in K^{*0}$ $\operatorname{epi}(v^* f)^* + (A^* \times \operatorname{id}_{\mathbb{R}})(\operatorname{epi}(v^* g)^*)$ is closed in the topology $\omega(X^*, X) \times \mathscr{R}$. (5.3.2)

Theorem 5.29. *Let $\bar{r} \in K \setminus \{0\}$. Assume that f and g are K-convex vector functions and one of the regularity conditions (RCV^U_i), $i \in \{1, 2, 3, 4\}$, is fulfilled. If $\bar{x} \in \mathscr{PE}_{LS}(PVU)$, then there exist $\bar{v}^* \in K^{*0}$ and $\bar{y}^* \in Y^*$ such that $(\bar{v}^*, \bar{y}^*, \bar{x}, 0, \bar{r}) \in \mathscr{E}(DVU_W)$, $(\bar{v}^*, \bar{y}^*, \bar{x}, 0) \in \mathscr{E}(DVU_{W\bar{r}})$, $(\bar{v}^*, \bar{x}) \in \mathscr{E}(DVU_M)$ and $f(\bar{x}) + g(A\bar{x}) = h^U_W(\bar{v}^*, \bar{y}^*, \bar{x}, 0, \bar{r}) = h^U_{W\bar{r}}(\bar{v}^*, \bar{y}^*, \bar{x}, 0) = h^U_M(\bar{v}^*, \bar{x})$.*

Remark 5.36. In case $V = \mathbb{R}$ and $K = \mathbb{R}_+$, taking the functions $f : X \to \overline{\mathbb{R}}$ and $g : Y \to \overline{\mathbb{R}}$ proper we rediscover the Wolfe and Mond-Weir duality schemes for unconstrained scalar optimization problems from Sect. 5.2.3. More precisely the problem (PVU) becomes then the unconstrained scalar optimization problem (PU), the duals (DVU_W) and (DVU_{W^r}), $r > 0$, turn out to coincide with the scalar Wolfe type dual to (PU) denoted (DU_W) and (DVU_M) is nothing but its Mond-Weir type dual (DU_M).

One can see (PVC) as an unconstrained vector optimization problem, namely

$$(PVC) \qquad \operatorname*{Min}_{x \in X} \left[f(x) + \delta^v_{\mathscr{A}}(x) \right],$$

where the notations are consistent with the ones in Sect. 5.3.2. Then, taking $A := \mathrm{id}_X$, $f := f$ and $g := \delta^v_{\mathscr{A}}$, (DVU_W), (DVU_M) and (DVU^r_W) (where $r \in K \setminus \{0\}$) turn into

$$(DVC^F_W) \qquad \operatorname*{Max}_{(v^*, y^*, u, y, r) \in \mathscr{B}^F_W} h^F_W(v^*, y^*, u, y, r),$$

where

$$\mathscr{B}^F_W = \Big\{ (v^*, y^*, u, y, r) \in K^{*0} \times C^* \times X \times Y \times (K \setminus \{0\}) : \\ y^* \in \partial(v^* f)(u + y) \cap (-N_{\mathscr{A}}(u)) \Big\}$$

and

$$h^F_W(v^*, y^*, u, y, r) = f(u + y) - \frac{\langle y^*, y \rangle}{\langle v^*, r \rangle} r,$$

further referred to as the *vector Wolfe dual of Fenchel type*,

$$(DVC^F_M) \qquad \operatorname*{Max}_{(v^*, u) \in \mathscr{B}^F_M} h^F_M(v^*, u),$$

where

$$\mathscr{B}^F_M = \Big\{ (v^*, u) \in K^{*0} \times X : 0 \in \partial(v^* f)(u) + N_{\mathscr{A}}(u) \Big\}$$

and

$$h^F_M(v^*, u) = f(u),$$

and, respectively,

$$(DVC^F_{Wr}) \qquad \operatorname*{Max}_{(v^*,y^*,u,y)\in\mathscr{B}^F_{Wr}} h^F_{Wr}(v^*,y^*,u,y),$$

where

$$\mathscr{B}^F_{Wr} = \Big\{(v^*,y^*,u,y) \in K^{*0} \times C^* \times X \times Y : \langle v^*, r\rangle = 1, \\ y^* \in \partial(v^* f)(u+y) \cap (-N_{\mathscr{A}}(u))\Big\}$$

and

$$h^F_{Wr}(v^*,y^*,u,y) = f(u+y) - \langle y^*, y\rangle r.$$

Note that Proposition 5.4 yields $h^F_M(\mathscr{B}^F_M) \subseteq \cup_{r\in K\setminus\{0\}} h^F_{Wr}(\mathscr{B}^F_{Wr}) = h^F_W(\mathscr{B}^F_W)$.

Remark 5.37. These vector dual problems to (PVC) can be obtained directly from (DVG_W), (DVG_M) and (DVG^r_W) (where $r \in K \setminus \{0\}$), respectively, too, by using the vector perturbation function Φ^F_v introduced in Sect. 4.4.2.

Let us give now the weak and strong duality statements for these duals.

Theorem 5.30. *There are no $x \in \mathscr{A}$ and $(v^*,y^*,u,y,r) \in \mathscr{B}^F_W$ such that $f(x) \leq_K h^F_W(v^*,y^*,u,y,r)$.*

Theorem 5.31. *There are no $x \in \mathscr{A}$ and $(v^*,u) \in \mathscr{B}^F_M$ such that $f(x) \leq_K h^F_M(v^*,u)$.*

Theorem 5.32. *Let $r \in K\setminus\{0\}$. Then there are no $x \in \mathscr{A}$ and $(v^*,y^*,u,y) \in \mathscr{B}^F_{Wr}$ such that $f(x) \leq_K h^F_{Wr}(v^*,y^*,u,y)$.*

For strong duality, which follows directly from either Theorems 5.15 or 5.29, besides convexity assumptions which guarantee the K-convexity of the corresponding vector perturbation function (see also Remark 5.12) we use regularity conditions, too, obtained by particularizing (RCV^G_i) or (RCV^U_i), $i \in \{1,2,3,4\}$, respectively, namely (cf. [21, 48])

$$(RCV^F_1)\big|\ \exists x' \in \operatorname{dom} f \cap \mathscr{A} \text{ such that } f \text{ is continuous at } x',$$

$$(RCV^F_2)\left|\begin{array}{l} X \text{ is a Fréchet space}, \mathscr{A} \text{ is closed}, f \text{ is } K\text{-lower semicontinuous} \\ \text{and } 0 \in \operatorname{sqri}(\operatorname{dom} f - \mathscr{A}), \end{array}\right.$$

$$(RCV^F_3)\big|\ \dim\operatorname{lin}(\operatorname{dom} f - \mathscr{A}) < +\infty \text{ and } \operatorname{ri}\operatorname{dom} f \cap \operatorname{ri}\mathscr{A} \neq \emptyset,$$

and

$$(RCV^F_4)\left|\begin{array}{l} \mathscr{A} \text{ is closed}, f \text{ is } K\text{-lower semicontinuous and for any } v^* \in K^{*0} \\ \operatorname{epi}(v^* f)^* + \operatorname{epi}\sigma_{\mathscr{A}} \text{ is closed in the topology } \omega(X^*, X) \times \mathscr{R}. \end{array}\right.$$

Theorem 5.33. *Let $\bar{r} \in K \setminus \{0\}$. Assume that $\mathscr{A}$ is a convex set, f is a K-convex vector function and one of the regularity conditions (RCV_i^F), $i \in \{1,2,3,4\}$, is fulfilled. If $\bar{x} \in \mathscr{P}\mathscr{E}_{LS}(PVC)$, then $\bar{x} \in \mathscr{A}$ and there exist $\bar{v}^* \in K^{*0}$ and $\bar{y}^* \in Y^*$ such that $(\bar{v}^*, \bar{y}^*, \bar{x}, 0, \bar{r}) \in \mathscr{E}(DVC_W)$, $(\bar{v}^*, \bar{y}^*, \bar{x}, 0) \in \mathscr{E}(DVC_{W\bar{r}})$, $(\bar{v}^*, \bar{x}) \in \mathscr{E}(DVC_M)$ and $f(\bar{x}) = h_W^F(\bar{v}^*, \bar{y}^*, \bar{x}, 0, \bar{r}) = h_{W\bar{r}}^F(\bar{v}^*, \bar{y}^*, \bar{x}, 0) = h_M^F(\bar{v}^*, \bar{x})$.*

Remark 5.38. Sufficient conditions that ensure the equivalence of the corresponding duals of Lagrange type and Fenchel-Lagrange type, respectively, can be obtained via [48, Theorem 3.5.6], while for the equivalence of the corresponding duals of Fenchel type and Fenchel-Lagrange type, respectively, one can apply [48, Theorem 3.5.13].

5.4 Alternative General Wolfe and Mond-Weir Vector Duality

As mentioned in Sect. 5.1, there are two ways of assigning vector dual problems of Wolfe and Mond-Weir type to vector optimization problems. In this section we deal with the so-called *alternative* approach, where the objective vector function of the dual vector optimization problems consists of a vector $v \in V$, while its counterpart from the primal vector optimization problem appears only in the constraints. As we shall see later, the image sets of the alternative vector duals are larger than their counterparts of classical type and this can prove to be an advantage when trying to solve them numerically.

We begin our investigations with a general vector optimization problem to which alternative vector duals of both Wolfe and Mond-Weir types are assigned. Then we particularize the primal problem to be constrained and unconstrained, respectively, and the corresponding vector dual problems are derived, following the scheme from the previous sections.

5.4.1 General Vector Optimization Problems

Like in Sect. 5.3, let X, Y and V be Hausdorff locally convex vector spaces, with V partially ordered by the nontrivial pointed convex cone $K \subseteq V$. Let $F : X \to V^\bullet$ be a proper vector function and consider the general vector-minimization problem

$$(PVG) \qquad \operatorname*{Min}_{x \in X} F(x).$$

The solution concepts we consider for this vector optimization problem are the ones introduced in Sect. 5.3, too. Like in Sect. 5.3.1, we employ the proper vector perturbation function $\Phi : X \times Y \to V^\bullet$ which fulfills $\Phi(x, 0) = F(x)$ for all $x \in X$, thus also $0 \in \Pr_Y(\operatorname{dom} \Phi)$, in order to assign vector dual problems to (PVG).

Different to Sect. 5.3.1, we attach this time Wolfe and Mond-Weir type vector dual problems to (*PVG*) by employing an idea considered in [58, 59, 140] (see also [71]) and making use of the corresponding scalar dual problems introduced in Sect. 5.2.1.

The *alternative Wolfe type vector dual* to (*PVG*) we consider now is

$$(DVG^W) \qquad \operatorname*{Max}_{(v^*,y^*,v,u,y)\in \mathscr{B}_G^W} h_G^W(v^*,y^*,v,u,y)$$

where

$$\mathscr{B}_G^W = \Big\{(v^*,y^*,v,u,y) \in K^{*0} \times Y^* \times V \times X \times Y : (0,y^*) \in \partial(v^*\Phi)(u,y), \\ \langle v^*, v\rangle \le -(v^*\Phi)^*(0,y^*)\Big\}$$

and

$$h_G^W(v^*,y^*,v,u,y) = v,$$

while the *alternative Mond-Weir type vector dual* one is

$$(DVG^M) \qquad \operatorname*{Max}_{(v^*,y^*,v,u)\in \mathscr{B}_G^M} h_G^M(v^*,y^*,v,u)$$

where

$$\mathscr{B}_G^M = \Big\{(v^*,y^*,v,u) \in K^{*0} \times Y^* \times V \times X : (0,y^*) \in \partial(v^*\Phi)(u,0), \\ \langle v^*, v\rangle \le \langle v^*, \Phi(u,0)\rangle\Big\}$$

and

$$h_G^M(v^*,y^*,v,u) = v.$$

For these dual vector optimization problems we consider efficient solutions, defined analogously to the ones in Definition 5.1. Their image sets fulfill an inclusion similar to the one given in Proposition 5.4.

Proposition 5.7. *One has* $h_G^M(\mathscr{B}_G^M) \subseteq h_G^W(\mathscr{B}_G^W)$.

Proof. Whenever $(v^*,y^*,v,u) \in \mathscr{B}_G^M$, it is easy to see that $(v^*,y^*,v,u,0) \in \mathscr{B}_G^W$ and $h_G^M(v^*,y^*,v,u) = h_G^W(v^*,y^*,v,u,0)$. Therefore all the values taken by the objective function of (DVG^M) over its feasible set can be found also in $h_G^W(\mathscr{B}_G^W)$. □

Remark 5.39. The sets $h_G^M(\mathscr{B}_G^M)$ and $h_G^W(\mathscr{B}_G^W)$ do not coincide in general. Situations where the inclusion in Proposition 5.7 is strictly fulfilled can be found in Examples 5.16 and 5.17.

Remark 5.40. One can consider other vector dual problems to (PVG) by replacing in (DVG^M) and (DVG^W) the inequalities involving $\langle v^*, v\rangle$ by the corresponding equalities. However, we will not consider further these vector dual problems since one can easily show that $(v^*, y^*, v, u, y) \in \mathscr{E}(DVG^W)$ yields $\langle v^*, v\rangle = -(v^*\Phi)^*(0, y^*)$, while $(v^*, y^*, v, u) \in \mathscr{E}(DVG^M)$ implies $\langle v^*, v\rangle = \langle v^*, \Phi(u, 0)\rangle$. Of course this observation can be extended for all the special instances of these vector duals considered later in this chapter.

For the newly introduced dual problems one can easily show that the weak duality holds.

Theorem 5.34. *There are no $x \in X$ and $(v^*, y^*, v, u, y) \in \mathscr{B}_G^W$ such that $F(x) \leq_K h_G^W(v^*, y^*, v, u, y)$.*

Proof. Assume to the contrary that there are some $x \in X$ and $(v^*, y^*, v, u, y) \in \mathscr{B}_G^W$ fulfilling $F(x) \leq_K h_G^W(v^*, y^*, v, u, y)$. Then $x \in \operatorname{dom} F$ and it follows $\langle v^*, v - \Phi(x, 0)\rangle > 0$. On the other hand, from the feasibility of (v^*, y^*, v, u, y) to (DVG^W), it follows $\langle v^*, v\rangle \leq -(v^*\Phi)^*(0, y^*)$ and since $-(v^*\Phi)^*(0, y^*) \leq (v^*\Phi)(x, 0)$, one gets $\langle v^*, v - \Phi(x, 0)\rangle \leq 0$, which contradicts the strict inequality obtained above. □

Using Proposition 5.7 and Theorem 5.34, one can easily prove the following weak duality statement, too.

Theorem 5.35. *There are no $x \in X$ and $(v^*, y^*, v, u) \in \mathscr{B}_G^M$ such that $F(x) \leq_K h_G^M(v^*, y^*, v, u)$.*

Remark 5.41. If $\mathscr{B}_G^M \neq \emptyset$, i.e. there exists a feasible element $(v^*, y^*, v, u) \in \mathscr{B}_G^M$, one can notice that $(v^*, y^*, \Phi(u, 0), u) \in \mathscr{B}_G^M$ and $(\alpha v^*, \alpha y^*, v, u) \in \mathscr{B}_G^M$ whenever $\alpha > 0$, too, and it also follows immediately that $(v^*, y^*, \Phi(u, 0), u, 0) \in \mathscr{B}_G^W$. Moreover, employing Theorem 5.35 one obtains that $(v^*, y^*, \Phi(u, 0), u) \in \mathscr{E}(DVG^M)$, while Theorem 5.34 yields $(v^*, y^*, \Phi(u, 0), u, 0) \in \mathscr{E}(DVG^W)$. On the other hand, if $(v^*, y^*, v, u, y) \in \mathscr{B}_G^W$, then $(v^*, y^*, \Phi(u, y), u, y) \in \mathscr{B}_G^W$ and $(\alpha v^*, \alpha y^*, v, u, y) \in \mathscr{B}_G^W$ for all $\alpha > 0$, too.

A statement analogous to Corollary 5.6 can be given for the alternative vector duals to (PVG), too.

Remark 5.42. If $(v^*, y^*, v, u, 0) \in \mathscr{B}_G^W$, then $(v^*, y^*, \Phi(u, 0), u) \in \mathscr{E}(DVG^M)$, $(v^*, y^*, \Phi(u, 0), u, 0) \in \mathscr{E}(DVG^W)$, $u \in \mathscr{PE}_{LS}(PVG)$ and $F(\bar{u}) = h_G^M(v^*, y^*, \Phi(u, 0), u) = h_G^W(v^*, y^*, \Phi(u, 0), u, 0)$.

For the strong duality statements concerning the vector optimization problem (PVG) and its two newly introduced vector dual problems we employ the ones considered in Sect. 5.3.1, as follows.

Theorem 5.36. *Assume that Φ is a K-convex function and one of the regularity conditions (RCV_i^G), $i \in \{1, 2, 3, 4\}$, is fulfilled. If $\bar{x} \in \mathscr{PE}_{LS}(PVG)$, then there* *exist* $\bar{v}^* \in K^{*0}$, $\bar{y}^* \in Y^*$ *and* $\bar{v} \in V$ *such that* $(\bar{v}^*, \bar{y}^*, \bar{v}, \bar{u}, 0) \in \mathcal{E}(DVG^W)$, $(\bar{v}^*, \bar{y}^*, \bar{v}, \bar{u}) \in \mathcal{E}(DVG^M)$ *and* $F(\bar{x}) = h_G^W(\bar{v}^*, \bar{y}^*, \bar{v}, \bar{u}, 0) = h_G^M(\bar{v}^*, \bar{y}^*, \bar{v}, \bar{u})$.

Proof. Since $\bar{x} \in \mathcal{PE}_{LS}(PVG)$, there exists a $\bar{v}^* \in K^{*0}$ such that $\langle \bar{v}^*, F(\bar{x})\rangle \leq \langle \bar{v}^*, F(x)\rangle$ for all $x \in X$. As each of the regularity conditions (RCV_i^G), $i \in \{1, 2, 3, 4\}$, ensures (cf. Corollary 2.6 and Remark 2.5) the stability of the scalar optimization problem

$$\inf_{x \in X} (\bar{v}^* F)(x),$$

with respect to the perturbation function Φ. Then, via Theorem 5.2 (see also Remark 5.4), there is strong duality for it and its Wolfe type dual

$$\sup_{\substack{u \in X, y \in Y, y^* \in Y^*, \\ (0, y^*) \in \partial(\bar{v}^* \Phi)(u, y)}} \big\{ -(\bar{v}^* \Phi)^*(0, y^*) \big\},$$

i.e. there exists a $\bar{y}^* \in Y^*$ such that

$$-(\bar{v}^* \Phi)^*(0, \bar{y}^*) = \sup_{y^* \in Y^*} \{-(\bar{v}^* \Phi)^*(0, y^*)\} = \inf_{x \in X} \langle \bar{v}^*, F(x)\rangle = \langle \bar{v}^*, F(\bar{x})\rangle,$$

and $(0, \bar{y}^*) \in \partial(\bar{v}^* \Phi)(\bar{x}, 0)$. Taking $\bar{v} = F(\bar{x})$, one sees that $(\bar{v}^*, \bar{y}^*, \bar{v}, \bar{x}, 0) \in \mathcal{B}_G^W$.

Moreover, $(\bar{v}^*, \bar{y}^*, \bar{v}, \bar{u}, 0) \in \mathcal{E}(DVG^W)$ via Theorem 5.34. Indeed, if $(\bar{v}^*, \bar{y}^*, \bar{v}, \bar{u}, \bar{y})$ were not an efficient solution to problem (DVG^W) there would exist an element $(v^*, y^*, v, u, y) \in \mathcal{B}_G^W$ such that $h_G^W(v^*, y^*, v, u, y) = v \geq_K \bar{v} = F(\bar{x})$. But this contradicts the weak duality statement.

In order to deal with problem (DVG^M) we consider the Mond-Weir type dual to $\inf_{x \in X}(\bar{v}^* \Phi)(x, 0)$, namely

$$\sup_{\substack{u \in X, y^* \in Y^*, \\ (0, y^*) \in \partial(\bar{v}^* \Phi)(u, 0)}} \langle \bar{v}^*, \Phi(x, 0)\rangle.$$

The conclusion follows analogously. □

Remark 5.43. In case $V = \mathbb{R}$ and $K = \mathbb{R}_+$, identifying $V^\bullet$ with $\mathbb{R} \cup \{+\infty\}$ and $\infty_{\mathbb{R}_+}$ with $+\infty$, and taking the function $F : X \to \overline{\mathbb{R}}$ proper we rediscover the Wolfe and Mond-Weir type scalar duality schemes from the scalar case presented in Sect. 5.2.1. More precisely the problem (PVG) becomes then the general scalar optimization problem (PG), while the duals (DVG^W) and (DVG^M) turn out to coincide with the scalar Wolfe and Mond-Weir type duals to (PG) introduced there, namely (DG_W) and (DG_M), respectively.

Remark 5.44. Similar observations to the ones formulated in Remarks 5.25–5.28 can be given within this section, too.

In the next subsections we consider as special instances of (PVG) the two main classes of vector optimization problems, namely we work with a constrained and an unconstrained vector optimization problem, respectively. To these problems we attach vector duals that are special cases of (DVG^M) and (DVG^W), respectively, obtained for different choices of the vector perturbation function Φ.

5.4.2 *Alternative Wolfe and Mond-Weir Type Vector Duals for Constrained Vector Optimization Problems*

Besides the standing framework we let the space Y be partially ordered by the nonempty convex cone $C \subseteq Y$, like in Sect. 5.3.2 and the notations we use are consistent with the ones considered there, namely $S \subseteq X$ is a nonempty set and $f : X \to V^\bullet$ and $h : X \to Y^\bullet$ are proper vector functions fulfilling the feasibility condition $\operatorname{dom} f \cap S \cap h^{-1}(C) \neq \emptyset$. Using the same vector perturbation functions as in the mentioned subsection, we attach to the primal constrained vector optimization problem

$$(PVC) \qquad \operatorname*{Min}_{x \in \mathscr{A}} f(x),$$

where

$$\mathscr{A} = \{x \in S : h(x) \in -C\},$$

vector dual problems obtained as special cases of (DVG^W) and (DVG^M), respectively. Making use of Φ_v^L, one gets the *alternative Wolfe vector dual of Lagrange type*

$$(DVC_L^W) \qquad \operatorname*{Max}_{(v^*,z^*,v,u,z)\in\mathscr{B}_L^{\tilde{W}}} h_L^{\tilde{W}}(v^*,z^*,v,u,z)$$

where

$$\mathscr{B}_L^{\tilde{W}} = \Big\{(v^*,z^*,v,u,z) \in K^{*0} \times (-C^*) \times V \times S \times Y : \langle v^*, v - f(u)\rangle$$
$$\leq -(z^*h)(u),\ 0 \in \partial((v^*f) - (z^*h) + \delta_S)(u), \delta_{-C}(h(u)-z) - \langle z^*, h(u)-z\rangle = 0\Big\}$$

and

$$h_L^{\tilde{W}}(v^*,z^*,v,u,z) = v,$$

which can be equivalently rewritten as

$$(DVC_L^W) \qquad \underset{(v^*,z^*,v,u)\in\mathscr{B}_L^W}{\text{Max}} h_L^W(v^*,z^*,v,u)$$

where

$$\mathscr{B}_L^W = \Big\{(v^*,z^*,v,u) \in K^{*0} \times C^* \times V \times S : \langle v^*, v - f(u)\rangle \leq (z^*h)(u), \\ 0 \in \partial((v^*f) + (z^*h) + \delta_S)(u)\Big\}$$

and

$$h_L^W(v^*,z^*,v,u) = v,$$

and, respectively,

$$(DVC_L^M) \qquad \underset{(v^*,z^*,v,u)\in\mathscr{B}_L^M}{\text{Max}} h_L^M(v^*,z^*,v,u)$$

where

$$\mathscr{B}_L^M = \Big\{(v^*,z^*,v,u) \in K^{*0} \times C^* \times V \times S : (z^*h)(u) \geq 0, h(u) \in -C, \\ \langle v^*, v\rangle \leq (v^*f)(u), 0 \in \partial((v^*f) + (z^*h) + \delta_S)(u)\Big\}$$

and

$$h_L^M(v^*,z^*,v,u) = v.$$

Note that in the constraints of (DVC_L^M) one can replace $(z^*h)(u) \geq 0$ by $(z^*h)(u) = 0$ without altering anything since $h(u) \in -C$ and $z^* \in C^*$. Like in Sect. 5.3.2, removing from this vector dual the constraint $h(u) \in -C$, we obtain a new vector dual to (*PVC*), namely the *alternative Mond-Weir vector dual of Lagrange type* to it

$$(DVC_L^{MW}) \qquad \underset{(v^*,z^*,v,u)\in\mathscr{B}_L^{MW}}{\text{Max}} h_L^{MW}(v^*,z^*,v,u)$$

where

$$\mathscr{B}_L^{MW} = \Big\{(v^*,z^*,v,u) \in K^{*0} \times C^* \times V \times S : (z^*h)(u) \geq 0, \\ \langle v^*, v\rangle \leq (v^*f)(u), 0 \in \partial((v^*f) + (z^*h) + \delta_S)(u)\Big\}$$

and

$$h_L^{MW}(v^*,z^*,v,u) = v.$$

Remark 5.45. Due to the way the vector duals we assigned above to (*PVC*) are constructed it is clear that $h_L^M(\mathscr{B}_L^M) \subseteq h_L^{MW}(\mathscr{B}_L^{MW})$. Moreover, since for all $(v^*, z^*, v, u) \in \mathscr{B}_L^{MW}$ the double inequality $\langle v^*, v - (v^* f)(u)\rangle \leq 0 \leq (z^* h)(u)$ yields $\langle v^*, v - (v^* f)\,(u)\rangle - (z^* h)(u) \leq 0$, it follows that $h_L^{MW}(\mathscr{B}_L^{MW}) \subseteq h_L^W(\mathscr{B}_L^W)$. Consequently, even without resorting to Proposition 5.7 one obtains $h_L^M(\mathscr{B}_L^M) \subseteq h_L^W(\mathscr{B}_L^W)$, too. Situations where these inclusion are strictly fulfilled can be found below.

Example 5.16. Consider the situation from Example 5.14. One can easily show that $\mathscr{B}_L^M = \emptyset$. On the other hand, as $0 \in \partial((v^* f) + (0h) + \delta_S)(0) = (-\infty, 1]$ for all $v^* \in \operatorname{int}\mathbb{R}_+^2$, $f(0) = (0,0)^\top$ and $(0h)(0) = 0$, it follows that whenever $v \in \mathbb{R}^2$ fulfills $v^{*\top} v \leq 0$, one has $(v^*, 0, v, 0) \in \mathscr{B}_L^{MW}$. Consequently, $\{v \in \mathbb{R}^2 : v^{*\top} v \leq 0, v^* \in \operatorname{int}\mathbb{R}_+^2\} \subseteq h_L^{MW}(\mathscr{B}_L^{MW})$.

Therefore, $h_L^M(\mathscr{B}_L^M) \subsetneq h_L^{MW}(\mathscr{B}_L^{MW})$ in this case. Taking into consideration Remark 5.45, it follows that the inclusion given in Proposition 5.7 can in general be strictly fulfilled.

Example 5.17. Consider the situation from Example 5.15. One can easily show that $h_L^M(\mathscr{B}_L^M) = h_L^{MW}(\mathscr{B}_L^{MW}) = \emptyset$, while $(-2,-2)^\top \in h_L^W(\mathscr{B}_L^W)$.

Therefore, $h_L^M(\mathscr{B}_L^M) = h_L^{MW}(\mathscr{B}_L^{MW}) \subsetneq h_L^W(\mathscr{B}_L^W)$ in this case.

Remark 5.46. A statement similar to the one given in Remark 5.30 can be given for the vector duals of Lagrange type assigned to (*PVC*) within this subsection, too, one being able to split the subdifferential $\partial((v^* f) + (z^* h) + \delta_S)(u)$ under the same hypotheses.

Remark 5.47. A vector dual similar to (DVC_L^W), but with respect to weakly efficient solutions, was introduced in [71], under quasidifferentiability hypotheses for the functions involved. Later, it was mentioned also in [212], where the functions were taken differentiable.

Like in the previous section, the results involving (*PVG*) and its vector duals can be particularized for the problems introduced above, however we give here only the weak and strong duality statements involving (*PVC*) and its vector duals of Lagrange type.

Theorem 5.37. *There are no $x \in \mathscr{A}$ and $(v^*, z^*, v, u) \in \mathscr{B}_L^W$ such that $f(x) \leq_K h_L^W(v^*, z^*, v, u)$.*

Theorem 5.38. *There are no $x \in \mathscr{A}$ and $(v^*, z^*, v, u) \in \mathscr{B}_L^M$ such that $f(x) \leq_K h_L^M(v^*, z^*, v, u)$.*

Analogously, one can prove also the following weak duality statement involving (*PVC*) and (DVC_L^{MW}).

Theorem 5.39. *There are no $x \in \mathscr{A}$ and $(v^*, z^*, v, u) \in \mathscr{B}_L^{MW}$ such that $f(x) \leq_K h_L^{MW}(v^*, z^*, v, u)$.*

In order to achieve strong duality for the vector duals of Lagrange type we just assigned to (PVC), we need, besides convexity assumptions which guarantee the K-convexity of the vector perturbation function Φ_v^L, the fulfillment of some sufficient conditions. To this end, we employ the regularity conditions used in Sect. 5.3.2. The strong duality assertions concerning (DVC_L^W) and (DVC_L^M), respectively, follow via Theorem 5.36, while their counterpart for (DVC_L^{MW}) can be proven analogously.

Theorem 5.40. *Assume that S is a convex set, f is a K-convex vector function, h is a C-convex vector function and one of the regularity conditions (RCV_i^L), $i \in \{1,2,3,4\}$, is fulfilled. If $\bar{x} \in \mathcal{PE}_{LS}(PVC)$, then there exist $\bar{v}^* \in K^{*0}$ and $\bar{z}^* \in C^*$ such that $(\bar{v}^*, \bar{z}^*, f(\bar{x}), \bar{x}) \in \mathcal{E}(DVC_L^W) \cap \mathcal{E}(DVC_L^M) \cap \mathcal{E}(DVC_L^{MW})$ and $f(\bar{x}) = h_L^W(\bar{v}^*, \bar{z}^*, f(\bar{x}), \bar{x}) = h_L^M(\bar{v}^*, \bar{z}^*, f(\bar{x}), \bar{x}) = h_L^{MW}(\bar{v}^*, \bar{z}^*, f(\bar{x}), \bar{x})$.*

Another vector perturbation function considered in Sect. 5.4.2 in order to assign a vector dual problem to (PVC) is the Fenchel-Lagrange type vector perturbation function Φ_v^{FL}. It particularizes (DVG^W) to the following *Wolfe vector dual of Fenchel-Lagrange type* to (PVC)

$$(DVC_{FL}^W) \qquad \operatorname*{Max}_{(v^*,y^*,z^*,v,u,y)\in\mathcal{B}_{FL}^W} h_{FL}^W(v^*,y^*,z^*,v,u,y)$$

where

$$\mathcal{B}_{FL}^W = \Big\{(v^*,y^*,z^*,v,u,y) \in K^{*0} \times X^* \times C^* \times V \times S \times X : \langle v^*, v\rangle \leq \langle y^*, u\rangle - (v^*f)^*(y^*) + (z^*h)(u),\ y^* \in \partial(v^*f)(u+y) \cap (-\partial((z^*h)+\delta_S)(u))\Big\}$$

and

$$h_{FL}^W(v^*,y^*,z^*,v,u,y) = v,$$

and (DVG^M), respectively, into

$$(DVC_{FL}^M) \qquad \operatorname*{Max}_{(v^*,z^*,v,u)\in\mathcal{B}_{FL}^M} h_{FL}^M(v^*,z^*,v,u)$$

where

$$\mathcal{B}_{FL}^M = \Big\{(v^*,z^*,v,u) \in K^{*0} \times C^* \times V \times S : (z^*h)(u) \geq 0, h(u) \in -C,\ \langle v^*, v\rangle \leq (v^*f)(u), 0 \in \partial(v^*f)(u) + \partial((z^*h)+\delta_S)(u)\Big\}$$

and

$$h_{FL}^M(v^*,z^*,v,u) = v.$$

Note that in its constraints one can replace $(z^*h)(u) \geq 0$ by $(z^*h)(u) = 0$ without altering anything since $h(u) \in -C$ and $z^* \in C^*$. Like in the Lagrange case, removing from (DVC^M_{FL}) the constraint $h(u) \in -C$, one obtains another vector dual to (PVC), namely its *Mond-Weir vector dual of Fenchel-Lagrange type*

$$(DVC^{MW}_{FL}) \qquad \underset{(v^*,z^*,v,u)\in\mathscr{B}^{MW}_{FL}}{\text{Max}} h^{MW}_{FL}(v^*,z^*,v,u)$$

where

$$\mathscr{B}^{MW}_{FL} = \Big\{(v^*,z^*,v,u) \in K^{*0} \times C^* \times V \times S : (z^*h)(u) \geq 0,$$
$$\langle v^*,v\rangle \leq (v^*f)(u), 0 \in \partial(v^*f)(u) + \partial((z^*h) + \delta_S)(u)\Big\}$$

and

$$h^{MW}_{FL}(v^*,z^*,v,u) = v.$$

Remark 5.48. Due to the way the vector duals we assigned above to (PVC) are constructed it is clear that $h^M_{FL}(\mathscr{B}^M_{FL}) \subseteq h^{MW}_{FL}(\mathscr{B}^{MW}_{FL})$. Moreover, for all $(v^*,z^*,v,u) \in \mathscr{B}^{MW}_{FL}$ the double inequality $\langle v^*, v - (v^*f)(u)\rangle \leq 0 \leq (z^*h)(u)$ yields $\langle v^*, v - (v^*f)\ (u)\rangle - (z^*h)(u) \leq 0$, while the constraint involving the subdifferentials ensures the existence of a $y^* \in \partial(v^*f)(u) \cap (-\partial((z^*h) + \delta_S)(u))$. Thus, $\langle v^*,v\rangle - \langle y^*,u\rangle + (v^*f)^*(y^*) - (z^*h)(u) = \langle v^*, v - (v^*f)(u)\rangle - (z^*h)(u) \leq 0$, i.e. $(v^*,y^*,z^*,v,u,0) \in \mathscr{B}^W_{FL}$. As $h^W_{FL}(v^*,\ y^*,z^*,v,u,0) = v = h^{MW}_{FL}(v^*,z^*,v,u)$, it follows that $h^{MW}_{FL}(\mathscr{B}^{MW}_{FL}) \subseteq h^W_{FL}(\mathscr{B}^W_{FL})$. Consequently, even without resorting to Proposition 5.7 one obtains $h^M_{FL}(\mathscr{B}^M_{FL}) \subseteq h^W_{FL}(\mathscr{B}^W_{FL})$, too. Situations where these inclusion are strictly fulfilled can be found below.

Example 5.18. Consider again the situation from Examples 5.14 and 5.16. One can easily show that $\mathscr{B}^M_{FL} = \emptyset$, while $\{v \in \mathbb{R}^2 : v^{*\top}v \leq 0, v^* \in \operatorname{int}\mathbb{R}^2_+\} \subseteq h^{MW}_{FL}(\mathscr{B}^{MW}_{FL})$.

Therefore, $h^M_{FL}(\mathscr{B}^M_{FL}) \subsetneq h^{MW}_{FL}(\mathscr{B}^{MW}_{FL})$ in this case.

Example 5.19. Consider the situation from Examples 5.15 and 5.17. One can easily show that $h^M_{FL}(\mathscr{B}^M_{FL}) = h^{MW}_{FL}(\mathscr{B}^{MW}_{FL}) = \emptyset$, while $(-2,-2)^\top \in h^W_{FL}(\mathscr{B}^W_{FL})$.

Therefore, $h^M_{FL}(\mathscr{B}^M_{FL}) = h^{MW}_{FL}(\mathscr{B}^{MW}_{FL}) \subsetneq h^W_{FL}(\mathscr{B}^W_{FL})$ in this case.

Remark 5.49. A statement similar to the one given in Remark 5.34 can be given for the vector duals of Fenchel-Lagrange type assigned to (PVC) within this subsection, too, one being able to split the subdifferential $\partial((z^*h) + \delta_S)(u)$ under the same hypotheses.

Like in the previous section, the results involving (PVG) and its vector duals can be particularized for the problems introduced above, however we give here only the weak and strong duality statements involving (PVC) and its vector duals of Lagrange type.

Theorem 5.41. *There are no $x \in \mathscr{A}$ and $(v^*, y^*, z^*, v, u, y) \in \mathscr{B}^W_{FL}$ such that $f(x) \leq_K h^W_{FL}(v^*, y^*, z^*, v, u, y)$.*

Theorem 5.42. *There are no $x \in \mathscr{A}$ and $(v^*, z^*, v, u) \in \mathscr{B}^M_{FL}$ such that $f(x) \leq_K h^M_{FL}(v^*, z^*, v, u)$.*

Analogously, one can prove also the following weak duality statement involving (PVC) and (DVC^{MW}_{FL}).

Theorem 5.43. *There are no $x \in \mathscr{A}$ and $(v^*, z^*, v, u) \in \mathscr{B}^{MW}_{FL}$ such that $f(x) \leq_K h^{MW}_{FL}(v^*, z^*, v, u)$.*

In order to achieve strong duality for the vector duals of Lagrange type we just assigned to (PVC), we need, besides convexity assumptions which guarantee the K-convexity of the vector perturbation function Φ^{FL}_v, the fulfillment of some sufficient conditions. To this end, we employ the regularity conditions used in Sect. 5.3.2. The strong duality assertions concerning (DVC^W_{FL}) and (DVC^M_{FL}), respectively, follow via Theorem 5.36, while their counterpart for (DVC^{MW}_{FL}) can be proven analogously.

Theorem 5.44. *Assume that S is a convex set, f is a K-convex vector function, h is a C-convex vector function and one of the regularity conditions (RCV^{FL}_i), $i \in \{1, 2, 3, 4\}$, is fulfilled. If $\bar{x} \in \mathscr{P}\mathscr{E}_{LS}(PVC)$, then there exist $\bar{v}^* \in K^{*0}$, $\bar{y}^* \in X^*$ and $\bar{z}^* \in C^*$ such that $(\bar{v}^*, \bar{y}^*, \bar{z}^*, f(\bar{x}), \bar{x}, 0) \in \mathscr{E}(DVC^W_{FL})$, $(\bar{v}^*, \bar{z}^*, f(\bar{x}), \bar{x}) \in \mathscr{E}(DVC^M_{FL}) \cap \mathscr{E}(DVC^{MW}_{FL})$ and $f(\bar{x}) = h^W_{FL}(\bar{v}^*, \bar{z}^*, f(\bar{x}), \bar{x}, 0) = h^M_{FL}(\bar{v}^*, \bar{z}^*, f(\bar{x}), \bar{x}) = h^{MW}_{FL}(\bar{v}^*, \bar{z}^*, f(\bar{x}), \bar{x})$.*

Remark 5.50. Like in the general case (see Remark 5.24), if $V = \mathbb{R}$ and $K = \mathbb{R}_+$, taking the functions $f : X \to \overline{\mathbb{R}}$ and $h : X \to Y^\bullet$ proper we rediscover the Wolfe and Mond-Weir duality schemes for constrained scalar optimization problems from Sect. 5.2.2, respectively. More precisely the problem (PVC) becomes then the constrained scalar optimization problem (PC) and the vector duals considered in this section turn out to be to the corresponding dual problems considered there to it.

5.4.3 *Alternative Wolfe and Mond-Weir Type Vector Duals for Unconstrained Vector Optimization Problems*

Consider again the framework of Sect. 5.3.3 and the notations used there, namely $f : X \to V^\bullet$ and $g : Y \to V^\bullet$ are proper vector functions and $A : X \to Y$ a linear continuous mapping such that the feasibility condition $\operatorname{dom} f \cap A^{-1}(\operatorname{dom} g) \neq \emptyset$ is fulfilled. To the primal unconstrained vector optimization problem

$$(PVU) \qquad \operatorname*{Min}_{x \in X} [f(x) + g(Ax)],$$

we assign vector dual problems obtained as special cases of (DVG^W) and (DVG^M), respectively, by making use of the the vector perturbation function Φ_v^U, namely

$$(DVU^W) \qquad \operatorname*{Max}_{(v^*,y^*,v,u,y)\in\mathscr{B}_U^W} h_U^W(v^*,y^*,v,u,y)$$

where

$$\mathscr{B}_U^W = \Big\{(v^*,y^*,v,u,y) \in K^{*0} \times Y^* \times V \times X \times Y : y^* \in (A^*)^{-1}(-\partial(v^* f)(u)) \cap \partial(v^* g)^*(Au+y) \text{ and } \langle v^*, v\rangle \leq -(v^* f)^*(-A^* y^*) + (v^* g)^*(y^*)\Big\}$$

and

$$h_U^W(v^*,y^*,v,u,y) = v,$$

and, respectively,

$$(DVU^M) \qquad \operatorname*{Max}_{(v^*,v,u)\in\mathscr{B}_U^M} h_U^M(v^*,v,u)$$

where

$$\mathscr{B}_U^M = \Big\{(v^*,v,u) \in K^{*0} \times V \times X : 0 \in (A^*)^{-1}(-\partial(v^* f)(u)) - \partial(v^* g)(Au) \text{ and } \langle v^*, v\rangle \leq \langle v^*, f(u) + g(Au)\rangle\Big\}$$

and

$$h_U^M(v^*,v,u) = v.$$

Observations similar to Remarks 5.11 and 5.15 can be made in the vector case, too. Note also that via Proposition 5.7, it holds $h_U^M(\mathscr{B}_U^M) \subseteq h_U^W(\mathscr{B}_U^W)$. For the primal vector problem (PVU) and its Wolfe type and Mond-Weir type vector duals (DVU^W) and (DVU^M), respectively, the weak and strong duality statements follow from the general case.

Theorem 5.45. *There are no $x \in X$ and $(v^*,y^*,v,u,y) \in \mathscr{B}_U^W$ such that $f(x) + g(Ax) \leq_K h_U^W(v^*,y^*,v,u,y)$.*

Theorem 5.46. *There are no $x \in X$ and $(v^*,v,u) \in \mathscr{B}_U^M$ such that $f(x) + g(Ax) \leq_K h_U^M(v^*,v,u)$.*

In order to achieve strong duality, besides convexity assumptions which guarantee the K-convexity of the vector perturbation function Φ_v^U, one needs the fulfillment of some sufficient conditions. To this end, we employ the regularity conditions used in Sect. 5.3.3. The strong duality assertions concerning (DVU^W) and (DVU^M), respectively, follow then directly from Theorem 5.36.

Theorem 5.47. *Assume that f and g are K-convex vector functions and one of the regularity conditions (RCV_i^U), $i \in \{1,2,3,4\}$, is fulfilled. If $\bar{x} \in \mathscr{PE}_{LS}(PVU)$, then there exist $\bar{v}^* \in K^{*0}$ and $\bar{y}^* \in Y^*$ such that $(\bar{v}^*, \bar{y}^*, f(\bar{x}) + g(A\bar{x}), \bar{x}, 0) \in \mathscr{E}(DVU^W)$, $(\bar{v}^*, f(\bar{x}) + g(A\bar{x}), \bar{x}) \in \mathscr{E}(DVU^M)$ and $f(\bar{x}) + g(A\bar{x}) = h_U^W(\bar{v}^*, \bar{y}^*, f(\bar{x}) + g(A\bar{x}), \bar{x}, 0) = h_U^M(\bar{v}^*, f(\bar{x}) + g(A\bar{x}), \bar{x})$.*

Remark 5.51. In case $V = \mathbb{R}$ and $K = \mathbb{R}_+$, taking the functions $f : X \to \overline{\mathbb{R}}$ and $g : Y \to \overline{\mathbb{R}}$ proper we rediscover the Wolfe and Mond-Weir duality schemes for unconstrained scalar optimization problems from Sect. 5.2.3. More precisely the problem (PVU) becomes then the unconstrained scalar optimization problem (PU), the dual (DVU^W) turns out to coincide with the scalar Wolfe type dual to (PU) denoted (DU_W) and (DVU^M) is nothing but its Mond-Weir type dual (DU_M).

As mentioned in Sect. 5.3.3, one can see (PVC) as an unconstrained vector optimization problem, namely

$$(PVC) \qquad \operatorname*{Min}_{x \in X} \left[f(x) + \delta^v_{\mathscr{A}}(x) \right],$$

where the notations are consistent with the ones in Sect. 5.4.2. The vector dual problem assigned to (PVC) in this case as special cases of (DVU^W) and (DVU^M) are

$$(DV_F^W) \qquad \operatorname*{Max}_{(v^*, y^*, v, u, y) \in \mathscr{B}_F^W} h_F^W(v^*, y^*, v, u, y)$$

where

$$\mathscr{B}_F^W = \Big\{ (v^*, y^*, v, u, y) \in K^{*0} \times Y^* \times V \times X \times X : \langle v^*, v \rangle \leq \langle y^*, u \rangle \\ -(v^* f)^*(y^*), y^* \in \partial(v^* f)(u + y) \cap (-N_{\mathscr{A}}(u)) \Big\}$$

and

$$h_F^W(v^*, y^*, v, u, y) = v,$$

and, respectively,

$$(DV_F^M) \qquad \operatorname*{Max}_{(v^*, v, u) \in \mathscr{B}_F^M} h_F^M(v^*, v, u)$$

where

$$\mathscr{B}_F^M = \left\{(v^*, v, u) \in K^{*0} \times V \times X : \langle v^*, v\rangle \leq (v^* f)(u),\ 0 \in \partial(v^* f)(u) + N_{\mathscr{A}}(u)\right\}$$

and

$$h_F^M(v^*, v, u) = v.$$

Note that Proposition 5.7 yields $h_F^M(\mathscr{B}_F^M) \subseteq h_F^W(\mathscr{B}_F^W)$.

Remark 5.52. These vector dual problems to (*PVC*) can be obtained directly from (DVG^W) and (DVG^M), respectively, too, by using the vector perturbation function Φ_v^F considered in Remark 5.37.

Let us give now the weak and strong duality statements for these duals.

Theorem 5.48. *There are no $x \in \mathscr{A}$ and $(v^*, y^*, v, u, y) \in \mathscr{B}_F^W$ such that $f(x) \leq_K h_F^W(v^*, y^*, v, u, y)$.*

Theorem 5.49. *There are no $x \in \mathscr{A}$ and $(v^*, v, u) \in \mathscr{B}_F^M$ such that $f(x) \leq_K h_F^M(v^*, v, u)$.*

For strong duality, besides the usual convexity assumptions which guarantee the K-convexity of the corresponding vector perturbation function, the fulfillment of some sufficient conditions is required. To this end, we employ the ones used in Sect. 5.3.3. The strong duality assertions concerning (DV_F^W) and (DV_F^M), respectively, follow then directly from Theorems 5.36 or 5.47.

Theorem 5.50. *Assume that $\mathscr{A}$ is a convex set, f is a K-convex vector function and one of the regularity conditions (RCV_i^F), $i \in \{1, 2, 3, 4\}$, is fulfilled. If $\bar{x} \in \mathscr{PE}_{LS}(PVC)$, then $\bar{x} \in \mathscr{A}$ and there exist $\bar{v}^* \in K^{*0}$ and $\bar{y}^* \in Y^*$ such that $(\bar{v}^*, \bar{y}^*, f(\bar{x}), \bar{x}, 0) \in \mathscr{E}(DVC^W)$, $(\bar{v}^*, f(\bar{x}), \bar{x}) \in \mathscr{E}(DVC^M)$ and $f(\bar{x}) = h_F^W(\bar{v}^*, \bar{y}^*, f(\bar{x}), \bar{x}, 0) = h_F^M(\bar{v}^*, f(\bar{x}), \bar{x})$.*

Remark 5.53. Sufficient conditions that ensure the equivalence of the corresponding duals of Lagrange type and Fenchel-Lagrange type, respectively, to (*PVC*) can be obtained via [48, Theorem 3.5.6], while for the equivalence of the corresponding duals of Fenchel type and Fenchel-Lagrange type, respectively, one can apply [48, Theorem 3.5.13].

5.5 Comparisons Between the Vector Duals

After assigning several different vector dual problems to the same primal vector optimization problem it is a legitimate task to try to compare their image sets. The notations considered within this section are consistent with the ones used in Sects. 5.3 and 5.4.

5.5.1 Duals to General Vector Optimization Problems

First let us compare the image sets of the corresponding dual within the two classes of vector dual problems we considered to the general vector optimization problem (*PVG*), namely the ones of classical type from Sect. 5.3 and the alternative ones from Sect. 5.4.

Theorem 5.51. *One has* $h_W^G(\mathcal{B}_W^G) \subseteq h_G^W(\mathcal{B}_G^W)$ *and* $h_M^G(\mathcal{B}_M^G) \subseteq h_G^M(\mathcal{B}_G^M)$.

Proof. Whenever $(v^*, y^*, u, y, r) \in \mathcal{B}_W^G$, one has $(0, y^*) \in \partial(v^*\Phi)(u, y)$, which yields $(v^*\Phi)(u, y) + (v^*\Phi)^*(0, y^*) = \langle y^*, y\rangle$, $\Phi(u, y) \in V$ and

$$\langle v^*, h_W^G(v^*, y^*, u, y, r) - \Phi(u, y)\rangle = \left\langle v^*, -\frac{\langle y^*, y\rangle}{\langle v^*, r\rangle} r\right\rangle = -\langle y^*, y\rangle,$$

thus $\langle v^*, h_W^G(v^*, y^*, u, y, r)\rangle = \langle v^*, \Phi(u, y)\rangle - \langle y^*, y\rangle = -(v^*\Phi)^*(0, y^*)$. Then, it follows that $(v^*, y^*, h_W^G(v^*, y^*, u, y, r), u, y) \in \mathcal{B}_G^W$ and $h_G^W(v^*, y^*, h_W^G(v^*, y^*, u, y, r), u, y) = h_W^G(v^*, y^*, u, y, r)$, therefore $h_W^G(\mathcal{B}_W^G) \subseteq h_G^W(\mathcal{B}_G^W)$.

The inclusion $h_M^G(\mathcal{B}_M^G) \subseteq h_G^M(\mathcal{B}_G^M)$ can be proven analogously. □

Remark 5.54. The inclusions proven in Theorem 5.51 are in general strict, as the situation depicted in Example 5.20 shows.

Example 5.20. Let $X = \mathbb{R}^2$, $Y = \mathbb{R}$, $C = \mathbb{R}_+$, $V = \mathbb{R}^2$, $K = \mathbb{R}_+^2$, $V^\bullet = \mathbb{R}^2 \cup \{\infty_{\mathbb{R}_+^2}\}$,

$$S = \left\{(x_1, x_2)^\top \in \mathbb{R}^2 : 0 \le x_1 \le 2, \begin{array}{l} 3 \le x_2 \le 4, \text{ if } x_1 = 0, \\ 1 \le x_2 \le 4, \text{ if } x_1 \in (0, 2] \end{array}\right\},$$

$$f : \mathbb{R}^2 \to (\mathbb{R}^2)^\bullet,\ f(x_1, x_2) = \begin{cases} \begin{pmatrix} 1 \\ 1 \end{pmatrix} x_2, & \text{if } x_1 \le 0, \\ \infty_{\mathbb{R}_+^2}, & \text{otherwise,} \end{cases}$$

and $h : \mathbb{R}^2 \to \mathbb{R}$, $h(x_1, x_2) = 0$ for all $(x_1, x_2)^\top \in \mathbb{R}^2$. Thus, $h(u) \in C$ whenever $u \in \mathbb{R}^2$. Then, for $v^* = (1/2, 1/2)^\top$, $u = (0, 3)^\top$ and any $z^* \in \mathbb{R}_+$ we get $(0, 0) \in \partial((v^* f) + (z^* h) + \delta_S)(0, 3)$ and $(z^* h)(u) = 0$, thus $(3, 3)^\top \in h_M^L(\mathcal{B}_M^L) \cap h_W^L(\mathcal{B}_W^L)$. Moreover, since the values taken by f consist of vectors with equal entries, when $(a, b)^\top \in h_M^L(\mathcal{B}_M^L)$ it is binding to have $a = b$, and, since h is everywhere equal to 0, one can conclude that whenever $(a, b)^\top \in h_W^L(\mathcal{B}_W^L)$ it must hold $a = b$, too.

On the other hand, taking $v = (2, 3)^\top$, it holds $v^{*\top}(v - f(u)) = -1/2 < 0 = (z^* h)(u)$ whenever $z^* \in \mathbb{R}_+$, thus $(v^*, z^*, v, u) \in \mathcal{B}_L^M \cap \mathcal{B}_L^W$ for any $z^* \in \mathbb{R}_+$, and consequently $(2, 3)^\top \in h_L^M(\mathcal{B}_L^M) \cap h_L^W(\mathcal{B}_L^W)$. But, as noted above, $(2, 3)^\top \notin h_M^L(\mathcal{B}_M^L) \cup h_W^L(\mathcal{B}_W^L)$.

Therefore, $h_M^L(\mathscr{B}_M^L) \subsetneq h_L^M(\mathscr{B}_L^M)$ and, respectively, $h_W^L(\mathscr{B}_W^L) \subsetneq h_L^W(\mathscr{B}_L^W)$ in this case. This shows that in general one has $h_M^G(\mathscr{B}_M^G) \subsetneq h_G^M(\mathscr{B}_G^M)$ and, respectively, $h_W^G(\mathscr{B}_W^G) \subsetneq h_G^W(\mathscr{B}_G^W)$.

Of interest would be to compare the maximal sets of the vector duals we assigned to *(PVG)*. Besides the results already given in Remarks 5.23, 5.42, Propositions 5.5 and 5.6, Corollaries 5.6 and 5.7, we were able to provide the following statement.

Theorem 5.52. *One has*

$$\mathrm{Max}(h_M^G(\mathscr{B}_M^G), K) \subseteq \mathrm{Max}(h_G^M(\mathscr{B}_G^M), K).$$

Proof. Combining Proposition 5.5 and Theorem 5.51, one gets $\mathrm{Max}(h_M^G(\mathscr{B}_M^G), K) = h_M^G(\mathscr{B}_M^G) \subseteq h_G^M(\mathscr{B}_G^M)$. Thus, if there is some $(v^*, y^*, u) \in \mathscr{B}_M^G$, then $(v^*, y^*, \Phi(u,0), u) \in \mathscr{B}_G^M$, while by Proposition 5.5 it follows that $u \in \mathscr{P}\mathscr{E}_{LS}(PVG)$. Assuming that $(v^*, y^*, \Phi(u,0), u) \notin \mathscr{E}(DVG^M)$, one would obtain then a contradiction to Theorem 5.35, therefore $(v^*, y^*, \Phi(u,0), u) \in \mathscr{E}(DVG^M)$. The conclusion follows. □

Whether the inclusion proven in Theorem 5.52 is in general strict or not it is not known at the moment. As direct consequences of Theorem 5.52 one has the following statement.

Corollary 5.8. *If* $(v^*, y^*, v, u) \in \mathscr{B}_G^M$, *then* $(v^*, y^*, u, 0, r) \in \mathscr{B}_W^G \cap \mathscr{E}(DVG_W)$ *for all* $r \in K \setminus \{0\}$ *and, respectively,* $(v^*, y^*, \Phi(u,0), u, 0) \in \mathscr{B}_G^W \cap \mathscr{E}(DVG^W)$.

Regarding the Wolfe vector dual problems we assigned to *(PVG)*, it is not known whether a statement similar to Theorem 5.52 can be proven for them. However, for some of the efficient solutions to the mentioned duals we have the following result (recall also Corollary 5.6 and Remark 5.42).

Theorem 5.53. *If* $(v^*, y^*, u, 0, r) \in \mathscr{B}_W^G$, *then* $(v^*, y^*, \Phi(u,0), u, 0) \in \mathscr{E}(DVG^W)$, *while* $(v^*, y^*, v, u, 0) \in \mathscr{B}_G^W$ *yields* $(v^*, y^*, u, 0, r) \in \mathscr{E}(DVG_W)$ *for all* $r \in K \setminus \{0\}$.

Proof. Under any of the hypotheses $\Phi(u,0)$ belongs to the image set of both vector duals (DVG_W) and (DVG^W). The conclusions follow after employing the corresponding weak duality statement, namely Theorems 5.34 and 5.12, respectively. □

Finally, let us notice an interesting connection between the infeasibility of the two classes of vector dual problems we assigned to *(PVG)*.

Theorem 5.54. *One has* $h_W^G(\mathscr{B}_W^G) = \emptyset$ *if and only if* $h_G^W(\mathscr{B}_G^W) = \emptyset$ *and, respectively,* $h_M^G(\mathscr{B}_M^G) = \emptyset$ *if and only if* $h_G^M(\mathscr{B}_G^M) = \emptyset$.

Proof. The sufficiency in both equivalences is a direct consequence of Theorem 5.51. To prove the necessity it is enough to note that the alternative vector duals contain all the constraints of the vector duals of classical type and moreover an inequality that controls the values taken by the vector v. Consequently, when the vector duals of classical type are infeasible, so are their alternative counterparts, too. □

In the next subsection we present different relations of inclusion between the image sets of the vector duals assigned to the constrained vector optimization problem *(PVC)*, extending the investigations performed in the scalar case in Sect. 5.2.4.

5.5.2 Duals to Constrained Vector Optimization Problems

Besides the inclusion relations that can be obtained as particularizations of Propositions 5.4, 5.7, Theorems 5.51, 5.52 and the ones given in Remarks 5.29, 5.33, 5.45 and 5.48, there are other inclusions between the images of the feasible sets of the vector duals to *(PVC)* introduced in Sects. 5.3.2 and 5.4.2 through their objective functions. In the following we prove some of them. First we deal with the vector duals obtained from (DVG_M). The notations in this subsection are consistent with the ones in Sects. 5.3.2 and 5.4.2.

Proposition 5.8. *It holds*

$$h_M^{FL}(\mathscr{B}_M^{FL}) \subseteq h_M^{C_F}(\mathscr{B}_M^{C_F}) \text{ and } h_M^{FL}(\mathscr{B}_M^{FL}) \subseteq h_M^{L}(\mathscr{B}_M^{L}).$$

Proof. Let $(v^*, z^*, u) \in \mathscr{B}_M^{FL}$. This means that $(v^*, z^*, u) \in K^{*0} \times C^* \times S$, $(z^*h)(u) \geq 0$, $h(u) \in -C$ and $0 \in \partial(v^* f)(u) + \partial((z^*h) + \delta_S)(u)$. But $0 \in \partial(v^* f)(u) + \partial((z^*h) + \delta_S)(u) \subseteq \partial\big((v^* f)(u) + (z^*h) + \delta_S\big)(u)$, consequently $(v^*, z^*, u) \in \mathscr{B}_M^{L}$. As $h_M^{FL}(v^*, z^*, u) = f(u) = h_M^{L}(v^*, z^*, u)$, the conclusion follows.

On the other hand, $(v^*, z^*, u) \in \mathscr{B}_M^{FL}$ yields $u \in \mathscr{A}$. As $\partial((z^*h) + \delta_S)(u) \subseteq N_{\mathscr{A}}(u)$, one gets $0 \in \partial(v^* f)(u) + N_{\mathscr{A}}(u)$. Consequently, $(v^*, u) \in \mathscr{B}_M^{F}$ and, since $h_M^{FL}(v^*, z^*, u) = f(u) = h_M^{F}(v^*, u)$, we are done. □

One can similarly prove its counterpart regarding the alternative vector duals to *(PVC)*.

Proposition 5.9. *It holds*

$$h_{FL}^{M}(\mathscr{B}_{FL}^{M}) \subseteq h_{L}^{M}(\mathscr{B}_{L}^{M}) \text{ and } h_{FL}^{M}(\mathscr{B}_{FL}^{M}) \subseteq h_{F}^{M}(\mathscr{B}_{F}^{M}).$$

Proof. Let $(v^*, z^*, v, u) \in \mathscr{B}_{FL}^{M}$. Then $0 \in \partial(v^* f)(u) + \partial((z^*h) + \delta_S)(u)$, $\langle v^*, v\rangle \leq (v^* f)(u)$, $(z^*h)(u) \geq 0$ and $h(u) \in -C$. As $\partial(v^* f)(u) + \partial((z^*h) + \delta_S)(u) \subseteq \partial\big((v^* f)\ (u) + (z^*h) + \delta_S\big)(u)$, it follows immediately that $(v^*, z^*, v, u) \in \mathscr{B}_{L}^{M}$. Since $h_{FL}^{M}(v^*, z^*, v, u) = v = h_{L}^{M}(v^*, z^*, v, u)$, the conclusion follows.

On the other hand, $(v^*,z^*,v,u) \in \mathscr{B}_{FL}^M$ yields $u \in \mathscr{A}$, so it holds $\partial((z^*h) + \delta_S)(u) \subseteq N_{\mathscr{A}}(u)$. Consequently, $(v^*,v,u) \in \mathscr{B}_F^M$ and, since $h_{FL}^M(v^*,z^*,v,u) = v = h_F^M(v^*,v,u)$, we are done. □

Analogously one can prove the following inclusion concerning the Mond-Weir vector duals to *(PVC)* of Lagrange and Fenchel-Lagrange type, respectively.

Proposition 5.10. *It holds*

$$h_{MW}^{FL}(\mathscr{B}_{MW}^{FL}) \subseteq h_{MW}^{L}(\mathscr{B}_{MW}^{L}) \text{ and } h_{FL}^{MW}(\mathscr{B}_{FL}^{MW}) \subseteq h_{L}^{MW}(\mathscr{B}_{L}^{MW}).$$

Moreover, one can provide an extension of Proposition 5.3 to the vector case, which at the moment is known to hold only for the alternative vector duals to *(PVC)*.

Proposition 5.11. *It holds*

$$h_{FL}^{MW}(\mathscr{B}_{FL}^{MW}) \subseteq h_{L}^{W}(\mathscr{B}_{L}^{W}).$$

Proof. Let $(v^*,z^*,v,u) \in \mathscr{B}_{FL}^{MW}$. Then $0 \in \partial(v^*f)(u) + \partial((z^*h) + \delta_S)(u)$, $\langle v^*, v - (v^*f)(u)\rangle \leq 0$ and $(z^*h)(u) \geq 0$. But $\partial(v^*f)(u) + \partial((z^*h) + \delta_S)(u) \subseteq \partial\big((v^*f)(u) + (z^*h) + \delta_S\big)(u)$ and $\langle v^*, v - (v^*f)(u)\rangle \leq 0 \leq (z^*h)(u)$ yield $0 \in \partial\big((v^*f)(u) + (z^*h) + \delta_S\big)(u)$ and $\langle v^*, v - (v^*f)(u)\rangle \leq (z^*h)(u)$, respectively, consequently $(v^*,z^*,v,u) \in \mathscr{B}_L^W$. As $h_{FL}^{MW}(v^*,z^*,v,u) = v = h_L^W(v^*,z^*,v,u)$, the conclusion follows. □

Situations where the inclusions in Propositions 5.8–5.11 are strictly fulfilled can be found below.

Example 5.21. Consider again the situation from Example 5.20. Then $(3,3)^\top \in h_M^L(\mathscr{B}_M^L) \cap h_L^M(\mathscr{B}_L^M)$. Moreover, by Remarks 5.29 and 5.45, one gets $(3,3)^\top \in h_{MW}^L(\mathscr{B}_{MW}^L) \cap h_L^{MW}(\mathscr{B}_L^{MW})$, too.

On the other hand, taking without loss of generality $z^* = 1$, to have, for some $v^* = (v_1^*,v_2^*)^\top \in \operatorname{int}\mathbb{R}_+^2$ and $u = (u_1,u_2)^\top \in S$, that $0 \in \partial(v^*f)(u) + \partial((z^*h) + \delta_S)(u)$ means actually that there exists a $y^* \in \partial(v^*f)(u) \cap (-N_S(u))$. From $y^* \in \partial(v^*f)(u)$ we obtain that $y^* = (y_1^*,y_2^*)^\top \in \mathbb{R}_+ \times \{v_1^* + v_2^*\}$ and $u_1 = 0$. Consequently, $y_2^* = v_1^* + v_2^*$. Let us see now for what $y_1^* \in \mathbb{R}_+$ does one obtain $(-y_1^*, -v_1^* - v_2^*)) \in N_S(0,u_2)$. We have $(-y_1^*, -v_1^* - v_2^*) \in N_S(0,u_2)$ if and only if $\sigma_S(-y_1^*, -v_1^* - v_2^*) = -(v_1^* + v_2^*)u_2$. This yields $u_2 = 1$, but $(0,1) \notin S$, consequently (DVC_M^{FL}) and (DVC_{MW}^{FL}) are infeasible, and, via Theorem 5.54, so is (DVC_{FL}^M). Analogously it follows that (DVC_{FL}^{MW}) is infeasible, too. As $\mathscr{A} = S$ and since $(z^*h)(u) = 0$ for all $u \in S$, it follows that (DCV_M^F) is equivalent to (DCV_M^{FL}) and (DCV_F^M) to (DCV_{FL}^M), so these vector dual problems are infeasible, too.

Moreover, as $\mathscr{A} = S$ and since $(z^*h)(u) = 0$ for all $u \in S$ one immediately notes that (DCV_W^F) is equivalent to (DCV_W^{FL}) and, respectively, (DCV_F^W) to (DCV_{FL}^W). Taking again without loss of generality $z^* = 1$, for some $v^* = (v_1^*,v_2^*)^\top \in \operatorname{int}\mathbb{R}_+^2$ and $u = (u_1,u_2)^\top \in S$ one finds a $y^* \in \partial(v^*f)(u + y) \cap (-N_S(u))$ for some $y \in \mathbb{R}^2$. From $y^* \in \partial(v^*f)(u + y)$ we obtain that

$y^* = (y_1^*, y_2^*)^\top \in \mathbb{R}_+ \times \{v_1^* + v_2^*\}$. Consequently, $y_2^* = v_1^* + v_2^*$. Let us see now for what $y_1^* \in \mathbb{R}_+$ does one obtain $(-y_1^*, -v_1^* - v_2^*) \in N_S(u_1, u_2)$. We have $(-y_1^*, -v_1^* - v_2^*) \in N_S(u_1, u_2)$ if and only if $\sigma_S(-y_1^*, -v_1^* - v_2^*) = -y_1^* u_1 - (v_1^* + v_2^*)u_2$. Since this can take place only if $y_1^* = 0$, it follows that $u_1 \in (0, 2]$, $u_2 = 1$ and $(0, v_1^* + v_2^*)^\top$ is the only possible value for y^*. Trying to find an $r = (r_1, r_2)^\top \in \mathbb{R}_+^2 \setminus \{0\}$ such that $h_W^F(v^*, y^*, u, y, r)$ has equal entries yields $r_1 = r_2$, but then $h_W^F(v^*, y^*, u, y, r) = (u_2, u_2)^\top = (1, 1)^\top$, consequently $(3, 3)^\top \notin h_W^F(\mathscr{B}_W^F)$ and $(3, 3)^\top \notin h_W^{FL}(\mathscr{B}_W^{FL})$. On the other hand, assuming that there exists a $v \in \mathbb{R}^2$ such that $h_F^W(v^*, y^*, v, u, y) = (3, 3)^\top$ yields $v = (3, 3)^\top$, but in this case $(v^*, y^*, v, u, y) \notin \mathscr{B}_F^W$ for any $y \in \mathbb{R}^2$. Thus, $(3, 3)^\top \notin h_F^W(\mathscr{B}_F^W) \cup h_{FL}^W(\mathscr{B}_{FL}^W)$, too.

Therefore, $h_M^{FL}(\mathscr{B}_M^{FL}) \subsetneq h_M^L(\mathscr{B}_M^L)$, $h_{FL}^M(\mathscr{B}_{FL}^M) \subsetneq h_L^M(\mathscr{B}_L^M)$, $h_{MW}^{FL}(\mathscr{B}_{MW}^{FL}) \subsetneq h_{MW}^L(\mathscr{B}_{MW}^L)$, $h_{FL}^{MW}(\mathscr{B}_{FL}^{MW}) \subsetneq h_L^{MW}(\mathscr{B}_L^{MW})$ and $h_{FL}^{MW}(\mathscr{B}_{FL}^{MW}) \subsetneq h_L^W(\mathscr{B}_L^W)$ in this case. Moreover, neither of $h_M^L(\mathscr{B}_M^L)$, $h_L^M(\mathscr{B}_L^M)$, $h_{MW}^L(\mathscr{B}_{MW}^L)$, $h_L^{MW}(\mathscr{B}_L^{MW})$, $h_W^L(\mathscr{B}_W^L)$ and $h_L^W(\mathscr{B}_L^W)$ is in general a subset of either $h_M^F(\mathscr{B}_M^F)$, $h_F^M(\mathscr{B}_F^M)$, $h_W^F(\mathscr{B}_W^F)$, $h_F^W(\mathscr{B}_F^W)$, $h_W^{FL}(\mathscr{B}_W^{FL})$ or $h_{FL}^W(\mathscr{B}_{FL}^W)$.

The question if similar inclusions to the ones in Propositions 5.8 or 5.9 are valid for the Wolfe vector duals to (PVC) comes very natural, but, even if (PVC) is a convex vector optimization problem, has a negative answer, like its scalar counterpart. In Example 5.21 we have already seen that the image sets of the Wolfe vector duals of Lagrange type (DVC_W^L) and (DVC_L^W) are in general not included in the ones of their counterparts of Fenchel-Lagrange or Fenchel type. Now let us show that the other possible inclusions do not hold in general.

Example 5.22. Let $X = \mathbb{R}$, $Y = \mathbb{R}$, $C = \mathbb{R}_+$, $V = \mathbb{R}^2$, $K = \mathbb{R}_+^2$, $V^\bullet = \mathbb{R}^2 \cup \{\infty_{\mathbb{R}_+^2}\}$, $S = \mathbb{R}$,

$$f : \mathbb{R} \to (\mathbb{R}^2)^\bullet,\ f(x) = \begin{cases} \begin{pmatrix} 1 \\ 1 \end{pmatrix} x, & \text{if } x > 0, \\ \infty_{\mathbb{R}_+^2}, & \text{otherwise}, \end{cases}$$

and

$$h : \mathbb{R} \to \mathbb{R},\ h(x) = \begin{cases} -x, & \text{if } x \leq 0, \\ 0, & \text{otherwise}. \end{cases}$$

Note that $\mathscr{A} = \mathbb{R}_+$ and for all $v^* = (v_1^*, v_2^*)^\top \in \operatorname{int} \mathbb{R}_+^2$ and $z^* \geq 0$ one has

$$\partial((v^* f) + (z^* h) + \delta_S)(u) = \partial(v^* f)(u) = \begin{cases} \{v_1^* + v_2^*\}, & \text{if } u > 0, \\ \emptyset, & \text{otherwise}. \end{cases}$$

Consequently, $\mathscr{B}_W^L = \mathscr{B}_L^W = \emptyset$.

On the other hand it can be shown that $\left((1/2,1/2)^\top,1,1,0,1,(1,1)^\top\right) \in \mathscr{B}_W^{FL}$, thus $(0,0)^\top \in h_W^{FL}(\mathscr{B}_W^{FL})$, while $((1/2,1/2)^\top,1,1,(0,0)^\top,0,1) \in \mathscr{B}_{FL}^W$, thus $(0,0)^\top \in h_{FL}^W(\mathscr{B}_{FL}^W)$, too. Indeed, for $v^* = (1/2,1/2)^\top$, $t^* = 1$, $y^* = 1$, $v = (0,0)^\top$, $u = 0$ and $t = 1$, the validity of the subdifferential constraint was proven in Example 5.11, while the inequality constraint that appears in (DVC_{FL}^W) means $\langle v^*,v\rangle - \langle y^*,u\rangle + (v^*f)^*(y^*) + (z^*h)(u) = \langle (1/2,1/2)^\top,(0,0)^\top\rangle - \langle 1,0\rangle + ((1/2,1/2)^\top f)^*(1) + (1h)\,(0) = 0$, which is true. Moreover, $N_{\mathscr{A}}(0) = (-\infty,0]$, so $-1 \in N_{\mathscr{A}}$, consequently, $\left((1/2,1/2)^\top,1,0,1,(1,1)^\top\right) \in \mathscr{B}_W^F$, while $((1/2,1/2)^\top,1,(0,0)^\top,0,1) \in \mathscr{B}_F^W$, thus $(0,0)^\top \in h_W^F(\mathscr{B}_W^F) \cap h_F^W(\mathscr{B}_F^W)$.

Therefore, neither of $h_W^F(\mathscr{B}_W^F)$, $h_F^W(\mathscr{B}_F^W)$, $h_W^{FL}(\mathscr{B}_W^{FL})$ and $h_{FL}^W(\mathscr{B}_{FL}^W)$ is in general a subset of either $h_W^L(\mathscr{B}_W^L)$ or $h_L^W(\mathscr{B}_L^W)$.

Example 5.23. Let $X = \mathbb{R}^2$, $Y = \mathbb{R}$, $C = \mathbb{R}_+$, $V = \mathbb{R}^2$, $K = \mathbb{R}_+^2$,

$$S = \left\{(x_1,x_2)^\top \in \mathbb{R}^2 : 0 \le x_1 \le 2, \begin{array}{l} 3 \le x_2 \le 4, \text{ if } x_1 = 0, \\ 1 \le x_2 \le 4, \text{ if } x_1 \in (0,2] \end{array}\right\},$$

$f : \mathbb{R}^2 \to \mathbb{R}^2$, $f(x_1,x_2) = (x_2,x_2)^\top$ and $h : \mathbb{R}^2 \to \mathbb{R}$, $h(x_1,x_2) = x_1$. Then $\mathscr{A} = \{0\} \times [3,4]$.

Since, for $\bar{v}^* = (1/2,1/2)^\top$ it holds $(0,1)^\top \in \partial(\bar{v}^*f)(0,3) \cap (-N_{\mathscr{A}}(0,3))$, it follows that $(3,3)^\top \in h_W^F(\mathscr{B}_W^F)$ and, via Theorem 5.51, $(3,3)^\top \in h_F^W(\mathscr{B}_F^W)$, too.

On the other hand, for $v^* \in \operatorname{int}\mathbb{R}_+^2$, $u \in S$, $y^* \in \mathbb{R}^2$ and $z^* \ge 0$, $0 \in \partial(v^*f)(u) + \partial((z^*h) + \delta_S)(u)$ if and only if one concomitantly has $(0,1)^\top \in \partial(v^*f)(u) \cap (-\partial((z^*h) + \delta_S)(u))$, $z^* = 0$ and $u \in (0,2] \times \{1\}$. But then $h(u) > 0$, so (DCV_M^{FL}) is infeasible.

As f and h are continuous, the condition (ii) in Remark 5.30 is fulfilled and it follows that for the vector optimization problem we are dealing with the vector dual problems (DCV_W^{FL}) and (DCV_W^L) are equivalent. The same conclusion can be drawn for the pairs of vector problems (DCV_{FL}^W) and (DCV_L^W), (DCV_{MW}^{FL}) and (DCV_{MW}^L), and (DCV_{FL}^{MW}) and (DCV_L^{MW}), respectively, too.

For some $v^* \in \operatorname{int}\mathbb{R}_+^2$, $u \in S$, $z^* \ge 0$ and $r \in \mathbb{R}^2 \setminus \{0\}$, one has $0 \in \partial\big((v^*f) + (z^*h) + \delta_S\big)(u)$ if and only if $z^* = 0$ and $u \in (0,2] \times \{1\}$. Then $h_W^L(v^*,z^*,u,r) = h_{MW}^L(v^*,z^*,u) = f(u) = (u_2,u_2)^\top = (1,1)^\top$, consequently $h_W^L(\mathscr{B}_W^L) = h_W^{FL}(\mathscr{B}_W^{FL}) = h_{MW}^L(\mathscr{B}_{MW}^L) = h_{MW}^{FL}(\mathscr{B}_{MW}^{FL}) = \{(1,1)^\top\}$.

Moreover, assuming that $(3,3)^\top \in h_L^W(\mathscr{B}_L^W)$, it follows that for some $v^* = (v_1^*,v_2^*)^\top \in \operatorname{int}\mathbb{R}_+^2$ one has $(v_1^* + v_2^*)(3-1) \le 0$, which cannot happen and analogously one can prove that $(3,3)^\top \notin h_L^{MW}(\mathscr{B}_L^{MW})$. Actually, it holds

$$h_L^W(\mathscr{B}_L^W) = h_{FL}^W(\mathscr{B}_{FL}^W) = h_L^{MW}(\mathscr{B}_L^{MW}) = h_{FL}^{MW}(\mathscr{B}_{FL}^{MW})$$

$$= \left\{\begin{pmatrix} v_1 \\ v_2 \end{pmatrix} \in \mathbb{R}^2 : v_1^*(v_1 - 1) + v_2^*(v_2 - 1) \le 0,\ v_1^*, v_2^* > 0\right\}.$$

Therefore, neither of $h_W^F(\mathscr{B}_W^F)$ and $h_F^W(\mathscr{B}_F^W)$ is in general a subset of either $h_{MW}^{FL}(\mathscr{B}_{MW}^{FL})$, $h_{FL}^{MW}(\mathscr{B}_{FL}^{MW})$, $h_{MW}^{L}(\mathscr{B}_{MW}^{L})$, $h_L^{MW}(\mathscr{B}_L^{MW})$, $h_W^{FL}(\mathscr{B}_W^{FL})$, $h_{FL}^{W}(\mathscr{B}_{FL}^{W})$, $h_W^L(\mathscr{B}_W^L)$ or $h_L^W(\mathscr{B}_L^W)$. Moreover one can note that, as a byproduct, we have identified another situation where the image set of a vector dual of classical type is strictly included in its alternative counterpart.

Example 5.24. Consider again the situation from Examples 5.14 and 5.16. We have $\mathscr{A} = (0, +\infty)$, $N_{\mathscr{A}}(u) = \{0\}$ for all $u \in \mathscr{A}$, $\partial(v^* f)(u) = \{v_1^* + v_2^*\}$ for all $v^* = (v_1^*, v_2^*)^\top \in \operatorname{int} \mathbb{R}_+^2$ and $u \in \mathbb{R}$, thus $\partial(v^* f)(u + y) \cap (-N_{\mathscr{A}}(u)) = \emptyset$ for all $u \in S$ and all $y \in \mathbb{R}$. Consequently, (DVC_W^F) and (DVC_F^W) are infeasible. Moreover, via Propositions 5.4 and 5.7, respectively, (DVC_M^F) and (DVC_F^M) are infeasible, too.

On the other hand, we have seen in Examples 5.14 and 5.16 that $(0,0)^\top \in h_{MW}^{FL}(\mathscr{B}_{MW}^{FL}) \cap h_{FL}^{MW}(\mathscr{B}_{FL}^{MW})$ and it can be easily shown that $\big((1/2,1/2)^\top, 0, 1, 0, 1, (1,1)^\top\big) \in \mathscr{B}_W^{FL}$ and $\big((1/2,1/2)^\top, 0, 0, (0,0)^\top, 0, 0\big) \in \mathscr{B}_{FL}^W$, thus $(0,0)^\top \in h_W^{FL}(\mathscr{B}_W^{FL}) \cap h_{FL}^W(\mathscr{B}_{FL}^W)$, too.

Therefore, neither of $h_{MW}^{FL}(\mathscr{B}_{MW}^{FL})$, $h_{FL}^{MW}(\mathscr{B}_{FL}^{MW})$, $h_W^{FL}(\mathscr{B}_W^{FL})$ and $h_{FL}^W(\mathscr{B}_{FL}^W)$ is in general a subset of either $h_M^F(\mathscr{B}_M^F)$, $h_F^M(\mathscr{B}_F^M)$, $h_W^F(\mathscr{B}_W^F)$ or $h_F^W(\mathscr{B}_F^W)$.

Remark 5.55. Fixing $r \in K \setminus \{0\}$, one can show similar facts for the vector dual (DVG_{W^r}) and its special cases. Note also that the situations presented in Examples 5.23 and 5.24 prove that a counterpart of Proposition 5.11 with the Fenchel type vector dual instead of the Lagrange one considered there is not valid in general.

Remark 5.56. Investigations on vector duality similar to the ones performed in Sects. 5.3 and 5.4 can be made with respect to weakly efficient solutions, too. The differences consist in the fact that the vector duals are reformulated by taking the variable v^* to belong to $K^* \setminus \{0\}$ instead of K^{*0}, while for the Wolfe vector duals of classical type one shall also take $r \in \operatorname{int} K$ or $r \in \operatorname{qi} K$ with K closed, and in the fact that instead of efficient and properly efficient solutions we deal then only with weakly efficient solutions.

Chapter 6
Vector Duality for Linear and Semidefinite Vector Optimization Problems

6.1 Historical Overview and Motivation

While the scalar linear optimization problems were intensively studied, inclusive via duality, and the things regarding them are settled down, the investigations on their vector counterparts are far from being complete. The first papers on linear vector duality were due to Gale, Kuhn and Tucker (cf. [96]), Kornbluth (cf. [150]), Schönefeld (cf. [187]) and Rödder (cf. [181]), while Isermann was the one who introduced in [132, 133] the *classical vector dual problem* to a primal linear vector optimization problem in finitely dimensional spaces. Moreover, he compared his results to the previously mentioned ones, pointing which of them could be recovered as special cases of his approach. He has also proven in [131] that the proper efficient solutions in the sense of linear scalarization of a linear vector optimization problem in finitely dimensional spaces coincide with its efficient ones, result which was shown, independently, also by Focke in [94] and Hartley in [118]. Another vector dual problem to a linear vector optimization one is the so-called *abstract vector dual* mentioned in [138] where its equivalence to Isermann's one was proven. Moreover, Jahn has noted in [138, 140] that both these vector duals have as major drawback the fact that when the constraint consists of a homogenous linear inequality system the vector strong duality fails. This issue solved by the Lagrange type vector dual proposed by Jahn in [138] (see also the vector dual due to Kolumbán from [149]) and also in [117], where a new vector dual for the classical linear vector optimization problem was introduced by considering as the objective function a set-valued mapping. Specializing the Fenchel-Lagrange type vector dual introduced in [54, 55, 200] for a linear vector optimization problem delivers for the latter a vector dual lacking the mentioned weak point, too. Let us also mention the so-called *geometric vector dual* to a linear vector optimization problem due to Nakayama (cf. [170] – not to be confused with the geometric duality due to Peterson, shown in [36] to be a special case of the Fenchel-Lagrange duality) and the *parametric vector dual* proposed by Luc in [156], that encompasses as special case another

S.-M. Grad, *Vector Optimization and Monotone Operators via Convex Duality*, Vector Optimization, DOI 10.1007/978-3-319-08900-3_6

so-called *geometric vector dual* to the primal classical linear vector optimization problem introduced in [121]. However, most of the mentioned vector duals to a linear vector optimization problem were given only in finitely dimensional spaces, with the image space of the primal problem partially ordered by the corresponding nonnegative orthant.

Motivated by this situation, we assign in Sect. 6.2 a new vector dual with respect to efficient solutions to the classical vector optimization problem defined in finitely dimensional spaces, but with the image space of the primal problem partially ordered by an arbitrary convex cone, following our paper [51]. This vector dual represents an extension to the case when the image space of the primal problem is partially ordered by an arbitrary convex cone of the Fenchel-Lagrange type vector dual introduced in [54, 55, 200], originally considered only with the mentioned image space partially ordered by the corresponding nonnegative orthant. Our investigations show that the vector dual we propose, unlike the mentioned Lagrange type one from [138], retains the form of the classical vector dual due to Isermann, "curing" moreover its vulnerability, and, unlike its counterpart from [117], is defined without resorting to set-valued optimization constructions and do not involve the determination of the sets of efficient solutions of two vector optimization problems. Moreover, we deal with some vector duals to the classical linear vector optimization problem with respect to weakly efficient solutions, too. As one can see in Sect. 6.3, the vector dual with respect to efficient solutions we introduced can be extended to infinitely dimensional spaces (cf. [34]). In the latter setting not all the nice properties of the finitely dimensional linear vector duality are preserved, for instance in the strong and converse duality statements one needs now the fulfillment of a regularity condition. Another vector dual we extend to this framework is the mentioned one from [117] and we show that the inclusions of the image sets of the vector duals we consider can be extended from finitely to infinitely dimensional spaces, too. Similar investigations are done with respect to weakly efficient solutions in the latter framework, too.

Matrix functions play an important role in optimization, too, especially in connection to the cone of symmetric positive semidefinite matrices which induces the Löwner partial ordering on the corresponding space of symmetric matrices. Besides numerous papers on scalar optimization, one can find contributions to vector optimization where such functions are involved. For instance in [201] we dealt with vector duality for convex vector optimization problems subject to semidefinite constraints, while in [111, 112] there were considered vector optimization problems consisting in vector minimizing matrix functions with respect to the cone of the symmetric positive semidefinite matrices under semidefinite constraints. Motivated by them and also by the discussions with Y. Ledyaev and L.M. Graña Drummond, respectively, at some conferences, regarding how can the results from Sect. 6.2 be extended for semidefinite vector problems, we propose a vector duality approach inspired by the one considered for the linear vector optimization problems for dealing via vector duality with vector optimization problem similar to the ones from [111, 112] mentioned above.

6.2 Linear Vector Duality in Finitely Dimensional Spaces

In this section we will revisit the vector duality for the classical linear vector optimization problem in finitely dimensional spaces, providing a new vector dual problem to this problem.

6.2.1 The Classical Linear Vector Optimization Problem

Let the space $\mathbb{R}^k$ be partially ordered by a nontrivial pointed closed convex cone $K \subseteq \mathbb{R}^k$, $L \in \mathbb{R}^{k\times n}$, $A \in \mathbb{R}^{m\times n}$ and $b \in \mathbb{R}^m$. Note that since K is a closed convex cone in a finitely dimensional space, it holds $K^{*0} = \operatorname{qi} K^* = \operatorname{int} K^*$.

The *classical linear vector optimization problem* is

$$(PLF) \qquad \operatorname*{Min}_{x\in\mathscr{A}} Lx,$$

where

$$\mathscr{A} = \{x \in \mathbb{R}^n_+ : Ax = b\}.$$

Recall that an element $\bar{x} \in \mathscr{A}$ is said to be a *properly efficient solution* to (PLF) in the sense of linear scalarization if $L\bar{x} \in \mathrm{PMin}_{LS}(L(\mathscr{A}), K)$, i.e. there exists a $\lambda \in \operatorname{int} K^*$ such that $\lambda^\top(L\bar{x}) \leq \lambda^\top(Lx)$ for all $x \in \mathscr{A}$, and the set of all the properly efficient solutions to (PLF) in the sense of linear scalarization is denoted by $\mathscr{P}\mathscr{E}_{LS}(PLF)$. An element $\bar{x} \in \mathscr{A}$ is said to be an *efficient solution* to (PLF) if $L\bar{x} \in \mathrm{Min}(L(\mathscr{A}), K)$, i.e. there exists no $x \in \mathscr{A}$ such that $Lx \leq_K L\bar{x}$, and the set of all the efficient solutions to (PLF) is denoted by $\mathscr{E}(PLF)$. Of course each properly efficient solution $\bar{x}$ to (PLF) in the sense of linear scalarization is also efficient to it. Let us show that for (PLF) these two classes of solutions actually coincide.

Theorem 6.1. *One has* $\mathscr{P}\mathscr{E}_{LS}(PLF) = \mathscr{E}(PLF)$.

Proof. Since $\mathscr{P}\mathscr{E}_{LS}(PLF) \subseteq \mathscr{E}(PLF)$, we have to prove only the opposite inclusion. Let $\bar{x} \in \mathscr{A}$ be an efficient solution to (PLF). As $\mathscr{A}$ is a polyhedral set, [178, Theorem 19.3] yields that $L(\mathscr{A})$ is polyhedral, too. Consequently, also $L(\mathscr{A}) - L\bar{x}$ is a polyhedral set. The efficiency of $\bar{x}$ to (PLF) yields $(L(\mathscr{A})-L\bar{x})\cap(-K) = \{0\}$, thus we are allowed to apply Lemma 1.3, which guarantees the existence of a $\gamma \in \mathbb{R}^k \setminus \{0\}$ for which

$$\gamma^\top(-k) < 0 \leq \gamma^\top(Lx - L\bar{x}) \ \forall k \in K \setminus \{0\} \ \forall x \in \mathscr{A}. \tag{6.2.1}$$

Since $\gamma^\top k > 0$ for all $k \in K \setminus \{0\}$, it follows that $\gamma \in \operatorname{int} K^*$. From (6.2.1) we obtain $\gamma^\top(L\bar{x}) \leq \gamma^\top(Lx)$ for all $x \in \mathscr{A}$, which, taking into account that $\gamma \in \operatorname{int} K^*$, means actually that $\bar{x} \in \mathscr{P}\mathscr{E}_{LS}(PLF)$ and the conclusion follows. □

Remark 6.1. In Theorem 6.1 we extend the classical result proven in [94, 131] for the special case $K = \mathbb{R}^k_+$. The statement remains valid when the feasible set of (*PLF*) is replaced by a set $\widetilde{\mathscr{A}}$ for which $L(\widetilde{\mathscr{A}})$ is polyhedral. Actually, the assertion of Theorem 6.1 can be extended for an arbitrary polyhedral set $M \subseteq \mathbb{R}^k$, any minimal point of which being properly minimal to it in the sense of linear scalarization, too, as mentioned in Remark 3.24. The same conclusion can be extracted from [118, Theorem 5.4], via [48, Proposition 2.1.1].

Remark 6.2. In the literature there were proposed several concepts of properly efficient solutions to a vector optimization problem, that can be derived from the proper minimality notions mentioned in Sect. 3.2. Taking into account (3.2.1) and [48, Proposition 2.4.7], Theorem 6.1 yields that for (*PLF*) the properly efficient solutions in the sense of linear scalarization coincide also with the properly efficient solutions to (*PLF*) in the senses of Geoffrion, Hurwicz, Borwein, Benson, Henig and Lampe and generalized Borwein, respectively. It is obvious then that it is enough to deal only with the efficient solutions to (*PLF*), since they coincide with most of the types of properly efficient solutions considered in the literature.

6.2.2 *Vector Duals to the Classical Linear Vector Optimization Problem*

The first relevant contributions to the study of vector duality for (*PLF*) were brought by Isermann in [132, 133] for the case $K = \mathbb{R}^k_+$. The vector dual he assigned to it, extended to the present framework to

$$(DLF^I) \qquad \operatorname*{Max}_{U \in \mathscr{B}^I_F} h^I_F(U),$$

where

$$\mathscr{B}^I_F = \left\{U \in \mathbb{R}^{k \times m} : (L - UA)(\mathbb{R}^n_+) \cap (-K) = \{0\}\right\}$$

and

$$h^I_F(U) = Ub,$$

turned out to work well only when $b \neq 0$ (see [138]), otherwise its image set containing only the element 0 when the dual is feasible. The same drawback was noticed in [138, 140] also for the so-called *dual abstract optimization problem* to (*PLF*)

$$(DLF^J) \qquad \operatorname*{Max}_{(\lambda,U) \in \mathscr{B}^J_F} h^J_F(\lambda, U),$$

where

$$\mathscr{B}_F^J = \left\{(\lambda, U) \in \operatorname{int} K^* \times \mathbb{R}^{k\times m} : (L - UA)^\top \lambda \in \mathbb{R}_+^n\right\}$$

and

$$h_F^J(\lambda, U) = Ub.$$

This issue was solved first by particularizing the general vector Lagrange type dual introduced in [138] (see also [149]), a vector dual to (PLF) for which vector duality statements can be given for every choice of $b \in \mathbb{R}^m$ being obtained, namely

$$(DLF^L) \qquad \underset{(\lambda,z,v)\in\mathscr{B}_F^L}{\operatorname{Max}} \; h_F^L(\lambda, z, v),$$

where

$$\mathscr{B}_F^L = \left\{(\lambda, z, v) \in \operatorname{int} K^* \times \mathbb{R}^m \times \mathbb{R}^k : \lambda^\top v - z^\top b \leq 0,\; L^\top\lambda - A^\top z \in \mathbb{R}_+^n\right\}$$

and

$$h_F^L(\lambda, z, v) = v.$$

But this vector dual has a different construction than the previous ones, so, recently, in [117] another vector dual to (PLF) was proposed, namely

$$(DLF^H) \qquad \underset{U\in\mathscr{B}_F^H}{\operatorname{Max}} \; h_F^H(U),$$

where

$$\mathscr{B}_F^H = \left\{U \in \mathbb{R}^{k\times m} : (L - UA)(\mathbb{R}_+^n) \cap (-K) = \{0\}\right\}$$

and

$$h_F^H(U) = Ub + \operatorname{Min}\big((L - UA)(\mathbb{R}_+^n), K\big).$$

The objective function of this vector dual extends the ones of the classical vector duals to (PLF) mentioned above, but it contains itself a linear vector optimization problem, too. Moreover, the vector duality assertions for this vector dual problem were shown via quite complicated set-valued optimization techniques. Other multiobjective dual problems with set-valued objective functions to a linear vector optimization problem were proposed in [122, 123]. As one can notice, the dual given in the first mentioned paper reduces to the one of Lagrange type as given in [138], while the objective of the second one is expressed by means of the efficient set of a polyhedral set, as happens in (DLF^H). Several primal-dual pairs of linear vector problems and some relations between them were treated in [137, 156], too. For primal problems of type (PLF) one can rediscover (DLF^I), (DLF^L), two vector duals of Wolfe type, which, as shown in [137], reduce to the Lagrange type one, and the mentioned parametric vector dual from [156]. In [54, 200] a vector duality theory with dual vector-valued function objectives for general convex primal vector optimization problems in finitely dimensional spaces with the image space partially ordered by the corresponding nonnegative orthant and with convex geometric and inequality constraints was introduced. The linear vector optimization case is covered, also in the case of $b = 0$, but not explicitly handled until it was revisited in [48].

Starting from the latter vector dual, we proposed in [51] a vector dual problem to (PLF) for the framework considered within this section, i.e. with the image space of the primal problem partially ordered by an arbitrary nontrivial pointed closed convex cone, namely

$$(DLF) \qquad \underset{(\lambda,U,v)\in\mathscr{B}_F}{\operatorname{Max}} \; h_F(\lambda, U, v),$$

where

$$\mathscr{B}_F = \left\{(\lambda, U, v) \in \operatorname{int} K^* \times \mathbb{R}^{k\times m} \times \mathbb{R}^k : \lambda^\top v = 0,\ (L - UA)^\top \lambda \in \mathbb{R}^n_+\right\}$$

and

$$h_F(\lambda, U, v) = Ub + v.$$

Remark 6.3. As suggested in [48, Remark 5.2.5.], one can modify the constraint $\lambda^\top v = 0$ in (DLF) into $\lambda^\top v \leq 0$, obtaining another vector dual problem to (PLF), namely

$$(DLF^D) \qquad \underset{(\lambda,U,v)\in\mathscr{B}_F^D}{\operatorname{Max}} \; h_F^D(\lambda, U, v),$$

where

$$\mathscr{B}_F^D = \left\{(\lambda, U, v) \in \operatorname{int} K^* \times \mathbb{R}^{k\times m} \times \mathbb{R}^k : \lambda^\top v \leq 0,\ (L - UA)^\top \lambda \in \mathbb{R}^n_+\right\}$$

and

$$h_F^D(\lambda, U, v) = Ub + v,$$

afterwards considered also in [156] for the case $K = \mathbb{R}^k_+$.

Remark 6.4. If $(\lambda, U, v) \in \mathscr{B}_F$, one can easily note that $v \notin (K \cup (-K)) \setminus \{0\}$, while when $(\lambda, U, v) \in \mathscr{B}_F^D$ it follows that $v \notin K \setminus \{0\}$.

We delivered a complete analysis of the inclusion relations between the image sets of the vector dual problems to (PLF) introduced above in the case $K = \mathbb{R}^k_+$ in [48, Section 5.5]. Let us show that the inclusion schemes given in [48, Remark 5.5.3] remain valid in the more general framework considered here, too.

We begin with a Farkas type result which allows us to formulate the feasible sets of the vector dual problems to (PLF) in a different manner, yielding moreover the coincidence of the image sets of (DLF^I) and (DLF^J).

Proposition 6.1. *Let $U \in \mathbb{R}^{k\times m}$. Then $(L - UA)(\mathbb{R}^n_+) \cap (-K) = \{0\}$ if and only if there exists a $\lambda \in \operatorname{int} K^*$ such that $(L - UA)^\top \lambda \in \mathbb{R}^n_+$.*

Proof. "$\Rightarrow$" The set $(L-UA)(\mathbb{R}^n_+)$ is polyhedral and has with the nontrivial pointed closed convex cone $-K$ only the origin as a common element. Applying Lemma 1.3 we obtain a $\lambda \in \mathbb{R}^k \setminus \{0\}$ for which

$$\lambda^\top(-k) < 0 \le \lambda^\top((L - UA)x) \ \forall x \in \mathbb{R}^n_+ \ \forall k \in K \setminus \{0\}. \tag{6.2.2}$$

Like in the proof of Theorem 6.1 we obtain that $\lambda \in \operatorname{int} K^*$ and by (6.2.2) it follows immediately that $(L - UA)^\top \lambda \in \mathbb{R}^n_+$.

"$\Leftarrow$" Assuming the existence of an $x \in \mathbb{R}^n_+$ for which $(L - UA)x \in (-K) \setminus \{0\}$, it follows $\lambda^\top((L - UA)x) < 0$, but $\lambda^\top((L - UA)x) = ((L - UA)^\top\lambda)^\top x \ge 0$ since $(L - UA)^\top\lambda \in \mathbb{R}^n_+$ and $x \in \mathbb{R}^n_+$. The so-obtained contradiction yields $(L - UA)(\mathbb{R}^n_+) \cap (-K) = \{0\}$. □

A direct consequence of this statement is the following assertion, already proven in case $K = \mathbb{R}^k_+$ in [138].

Proposition 6.2. *One has $h_F^I(\mathscr{B}_F^I) = h_F^J(\mathscr{B}_F^J)$.*

Remark 6.5. In [117] it is mentioned that $h_F^J(\mathscr{B}_F^J) \subseteq h_F^H(\mathscr{B}_F^H)$, an example when these sets do not coincide being also provided.

For the image sets of (DLF) and (DLF^H) we have the following assertion.

Proposition 6.3. *One has $h_F^H(\mathscr{B}_F^H) \subseteq h_F(\mathscr{B}_F)$.*

Proof. Let $d \in h_F^H(\mathscr{B}_F^H)$. Thus, there exist $\bar{U} \in \mathscr{B}_F^H$ and an efficient solution $\bar{x} \in \mathbb{R}^n_+$ to the vector optimization problem

$$\operatorname*{Min}_{x\in\mathbb{R}^n_+} (L - \bar{U}A)x, \tag{6.2.3}$$

such that $d = h_F^H(\bar{U}) = \bar{U}b + (L - \bar{U}A)\bar{x}$.

The efficiency of $\bar{x}$ to the problem (6.2.3) yields, via Theorem 6.1, that $\bar{x}$ is a properly efficient solution to this problem, too. Consequently, there exists a $\gamma \in$

int K^* such that

$$\gamma^\top((L - \bar{U}A)\bar{x}) \leq \gamma^\top((L - \bar{U}A)x) \ \forall x \in \mathbb{R}^n_+. \tag{6.2.4}$$

This yields $\gamma^\top((L - \bar{U}A)\bar{x}) \leq 0$. On the other hand, taking in (6.2.4) $x := x + \bar{x} \in \mathbb{R}^n_+$ it follows immediately $\gamma^\top((L - \bar{U}A)x) \geq 0$ for all $x \in \mathbb{R}^n_+$. Therefore $\gamma^\top((L - \bar{U}A)\bar{x}) \geq 0$, consequently $\gamma^\top((L - \bar{U}A)\bar{x}) = 0$. Taking $\bar{v} = (L - \bar{U}A)\bar{x}$, it follows $\gamma^\top\bar{v} = 0$ and, since $\gamma^\top(L - \bar{U}A) \in \mathbb{R}^n_+$, also $(\gamma, \bar{U}, \bar{v}) \in \mathscr{B}_F$. As $d = h^H(\bar{U}) = \bar{U}b + (L - \bar{U}A)\bar{x} = \bar{U}b + \bar{v} = h(\gamma, \bar{U}, \bar{v}) \in h(\mathscr{B}_F)$, we obtain $h^H_F(\mathscr{B}^H_F) \subseteq h_F(\mathscr{B}_F)$. □

Remark 6.6. The inclusion given in Proposition 6.3 is in general strict, as the situation presented in Example 6.1 shows.

Example 6.1 (cf. [61], see also [48,117]). Let $L = (1, -1)^\top, n = 1, k = 2, A = 0$ and $b = 0$. The classical linear vector optimization primal problem is now

$$(PLF) \ \operatorname*{Min}_{x \in \mathbb{R}_+} \begin{pmatrix} x \\ -x \end{pmatrix}.$$

It is not difficult to note that (DLF^H) actually coincides with (PLF), therefore $h^H_F(\mathscr{B}^H_F) = \{(x, -x) : x \in \mathbb{R}_+\}$. On the other hand, (DLF) turns into

$$(DLF) \qquad \operatorname*{Max}_{\substack{\lambda_1 \geq \lambda_2 > 0, \\ v \in \mathbb{R}}} \begin{pmatrix} -\frac{\lambda_2}{\lambda_1} \\ 1 \end{pmatrix} v,$$

It is clear that, for instance $(-1/2, 1)^\top \in h_F(\mathscr{B}_F) \setminus h^H_F(\mathscr{B}^H_F)$. Therefore, $h^H_F(\mathscr{B}^H_F) \subsetneq h_F(\mathscr{B}_F)$.

Remark 6.7. By construction it is obvious that one has $h_F(\mathscr{B}_F) \subseteq h^D_F(\mathscr{B}^D_F)$.

For the image sets of (DLF^D) and (DLF^L) we have the following assertion.

Proposition 6.4. *One has* $h^D_F(\mathscr{B}^D_F) = h^L_F(\mathscr{B}^L_F)$.

Proof. "$\subseteq$" Let $d \in h^D_F(\mathscr{B}^D_F)$. Thus, there exist $(\lambda, U, v) \in \mathscr{B}^D_F$ such that $d = h^D_F(\lambda, U, v) = Ub + v$. Take $z := U^\top\lambda$. Then $\lambda^\top d = \lambda^\top(Ub + v) = (\lambda^\top(Ub + v))^\top = b^\top(U^\top\lambda) + v^\top\lambda \leq b^\top z$, while $L^\top\lambda - A^\top z = L^\top\lambda - A^\top(U^\top\lambda) = (L - UA)^\top\lambda \in \mathbb{R}^n_+$. Consequently, $(\lambda, z, d) \in \mathscr{B}^L_F$ and, since $d = h^D_F(\lambda, U, v) = h^L_F(\lambda, z, d) \in h^L_F(\mathscr{B}^L_F)$, it follows that $h^D_F(\mathscr{B}^D_F) \subseteq h^L_F(\mathscr{B}^L_F)$.

"$\supseteq$" Let now $d \in h^L_F(\mathscr{B}^L_F)$. Thus, there exist $(\lambda, z, d) \in \mathscr{B}^L_F$ such that $d = h^L_F(\lambda, z, d)$. As $\lambda \in \operatorname{int} K^*$, there exists a $\tilde{\lambda} \in K$ such that $\lambda^\top\tilde{\lambda} = 1$. Take $U := \tilde{\lambda}z^\top$ and $v := d - Ub$. Then $\lambda^\top v = \lambda^\top d - \lambda^\top(\tilde{\lambda}z^\top)b = \lambda^\top d - z^\top b \leq 0$ and $(L - UA)^\top\lambda = L^\top\lambda - A^\top(z\tilde{\lambda}^\top)\lambda = L^\top\lambda - A^\top z \in \mathbb{R}^n_+$. Consequently, $(\lambda, U, v) \in \mathscr{B}^D_F$ and, since $d = h^L_F(\lambda, z, d) = Ub + v = h^D_F(\lambda, U, v) \in h^D_F(\mathscr{B}^D_F)$, it follows that $h^D_F(\mathscr{B}^D_F) \supseteq h^L_F(\mathscr{B}^L_F)$. □

Remark 6.8. The inclusion given in Remark 6.7 is in general strict, as the situation presented in Example 6.2 shows, via Proposition 6.4.

Example 6.2 (cf. [51], see also [48]). Let $n = 1$, $k = 2$, $m = 2$, $L = (0,0)^\top$, $A = (1,1)^\top$, $b = (-1,-1)^\top$ and $K = \mathbb{R}^2_+$. Then, for $v = (-1,-1)^\top$, $\lambda = (1,1)^\top$ and $z = (0,0)^\top$ we have $(\lambda, z, v) \in \mathscr{B}^L$ since $\lambda^\top v = -2 \leq 0 = z^\top b$ and $L^\top \lambda - A^\top z = (0,0)^\top \in \mathbb{R}^2_+$. Consequently, $h^L_F(\lambda, z, v) = (-1,-1)^\top \in h^L_F(\mathscr{B}^L_F)$.

On the other hand, assuming that $(-1,-1)^\top \in h_F(\mathscr{B}_F)$, there must exist $(\bar{\lambda}, \bar{U}, \bar{v}) \in \mathscr{B}_F$ such that $h_F(\bar{\lambda}, \bar{U}, \bar{v}) = \bar{U}b + \bar{v} = (-1,-1)^\top$. Then $\bar{\lambda}^\top(\bar{U}b + \bar{v}) = -\bar{\lambda}_1 - \bar{\lambda}_2 < 0$, where $\bar{\lambda} = (\bar{\lambda}_1, \bar{\lambda}_2)^\top \in \operatorname{int} \mathbb{R}^2_+$. But $\bar{\lambda}^\top(\bar{U}b + \bar{v}) = \bar{\lambda}^\top \bar{U}(-1,-1)^\top = -(\bar{U}A)^\top \bar{\lambda} = (L - \bar{U}A)^\top \bar{\lambda} \geq 0$, which contradicts what we obtained above as a consequence of the assumption we made, hence $(-1,-1)^\top \notin h_F(\mathscr{B}_F)$.

Therefore, $h_F(\mathscr{B}_F) \subsetneq h^L_F(\mathscr{B}^L_F) = h^D_F(\mathscr{B}^D_F)$.

Taking into consideration what we have proven in this subsection, one can conclude that the images of the feasible sets through their objective functions of the vector duals to (*PLF*) we dealt with satisfy the following inclusions chain

$$h^I_F(\mathscr{B}^I_F) = h^J_F(\mathscr{B}^J_F) \subsetneq h^H_F(\mathscr{B}^H_F) \subsetneq h_F(\mathscr{B}_F) \subsetneq h^D_F(\mathscr{B}^D_F) = h^L_F(\mathscr{B}^L_F), \qquad (6.2.5)$$

extending thus the scheme from [48, Remark 5.5.3] to the situation when the image space of the considered vector problems is partially ordered by a nontrivial pointed closed convex cone $K \subseteq \mathbb{R}^k$.

6.2.3 Duality Results for the Classical Linear Vector Optimization Problem and Its Vector Duals

Now let us deliver for the primal-dual pair of vector optimization problems (*PLF*)–(*DLF*) weak, strong and converse duality statements.

Theorem 6.2. *There exist no $x \in \mathscr{A}$ and $(\lambda, U, v) \in \mathscr{B}_F$ such that $Lx \leq_K Ub + v$.*

Proof. Assume the existence of $x \in \mathscr{A}$ and $(\lambda, U, v) \in \mathscr{B}_F$ such that $Lx \leq_K Ub+v$. Then $0 < \lambda^\top(Ub + v - Lx) = \lambda^\top(U(Ax) - Lx) = -((L - UA)^\top \lambda)^\top x \leq 0$, since $(L - UA)^\top \lambda \in \mathbb{R}^n_+$ and $x \in \mathbb{R}^n_+$. But this cannot happen, therefore the assumption we made is false. □

Like in the scalar linear case, for strong and converse vector duality no regularity conditions have to be satisfied.

Theorem 6.3. *If $\bar{x} \in \mathscr{E}(PLF)$, there exists $(\bar{\lambda}, \bar{U}, \bar{v}) \in \mathscr{E}(DLF)$ such that $L\bar{x} = \bar{U}b + \bar{v}$.*

Proof. The efficiency of $\bar{x}$ to (*PLF*) yields via Theorem 6.1 that $\bar{x} \in \mathscr{P}\mathscr{E}_{LS}(PLF)$. Thus there exists a $\bar{\lambda} \in \operatorname{int} K^*$ such that $\bar{\lambda}^\top(L\bar{x}) \leq \bar{\lambda}^\top(Lx)$ for all $x \in \mathscr{A}$. Consequently, one has strong duality for the scalar optimization problem

$$\inf_{x \in \mathscr{A}} \bar{\lambda}^\top (Lx)$$

and its dual

$$\sup_{\substack{\eta \in \mathbb{R}^m, \\ L^\top \bar{\lambda} + A^\top \eta \in \mathbb{R}^n_+}} \Big\{ -\eta^\top b \Big\},$$

i.e. their optimal objective values coincide and the dual has an optimal solution, say $\bar{\eta}$. Consequently, $\bar{\lambda}^\top (L\bar{x}) + \bar{\eta}^\top b = 0$ and $L^\top \bar{\lambda} + A^\top \bar{\eta} \in \mathbb{R}^n_+$.

As $\bar{\lambda} \in \operatorname{int} K^*$, there exists a $\tilde{\lambda} \in K \setminus \{0\}$ such that $\bar{\lambda}^\top \tilde{\lambda} = 1$. Let $\bar{U} := -\tilde{\lambda}\bar{\eta}^\top$ and $\bar{v} := L\bar{x} - \bar{U}b$. It is obvious that $\bar{U} \in \mathbb{R}^{k \times m}$ and $\bar{v} \in \mathbb{R}^k$. Moreover, $\bar{\lambda}^\top \bar{v} = \bar{\lambda}^\top (L\bar{x} - \bar{U}b) = \bar{\lambda}^\top (L\bar{x}) + \bar{\eta}^\top b = 0$ and $(L - \bar{U}A)^\top \bar{\lambda} = L^\top \bar{\lambda} + A^\top \bar{\eta} \in \mathbb{R}^n_+$. Consequently, $(\bar{\lambda}, \bar{U}, \bar{v}) \in \mathscr{B}_F$ and $\bar{U}b + \bar{v} = \bar{U}b + L\bar{x} - \bar{U}b = L\bar{x}$. Assuming that $(\bar{\lambda}, \bar{U}, \bar{v}) \notin \mathscr{E}(DLF)$, i.e. the existence of another feasible solution $(\lambda, U, v) \in \mathscr{B}_F$ satisfying $\bar{U}b + \bar{v} \leq_K Ub + v$, it follows $L\bar{x} \leq_K Ub + v$, which contradicts Theorem 6.2. Consequently, $(\bar{\lambda}, \bar{U}, \bar{v}) \in \mathscr{E}(DLF)$ and $L\bar{x} = \bar{U}b + \bar{v}$. □

Theorem 6.4. *If $(\bar{\lambda}, \bar{U}, \bar{v}) \in \mathscr{E}(DLF)$, there exists an $\bar{x} \in \mathscr{E}(PLF)$ such that $L\bar{x} = \bar{U}b + \bar{v}$.*

Proof. Let $\bar{d} := \bar{U}b + \bar{v} \in \operatorname{Max}(h_F(\mathscr{B}_F), K)$. Assume that $\mathscr{A} = \emptyset$. Then $b \neq 0$ and, by the Farkas Lemma there exists a $\bar{z} \in \mathbb{R}^m$ such that $b^\top \bar{z} > 0$ and $A^\top \bar{z} \in -\mathbb{R}^n_+$. As $\bar{\lambda} \in \operatorname{int} K^*$, there exists a $\tilde{\lambda} \in K \setminus \{0\}$ such that $\bar{\lambda}^\top \tilde{\lambda} = 1$. Let $\tilde{U} := \tilde{\lambda}\bar{z}^\top + \bar{U} \in \mathbb{R}^{k \times m}$. We have $(L - \tilde{U}A)^\top \bar{\lambda} = (L - \bar{U}A)^\top \bar{\lambda} - A^\top \bar{z} \in \mathbb{R}^n_+$, thus $(\bar{\lambda}, \tilde{U}, \bar{v}) \in \mathscr{B}_F$. But $h(\bar{\lambda}, \tilde{U}, \bar{v}) = \tilde{U}b + \bar{v} = \tilde{\lambda}\bar{z}^\top b + \bar{U}b + \bar{v} = \tilde{\lambda}\bar{z}^\top b + \bar{d} \geq_K \bar{d}$, which contradicts the efficiency of $(\bar{\lambda}, \bar{U}, \bar{v})$ to (DLF). Consequently, $\mathscr{A} \neq \emptyset$.

Suppose now that $\bar{d} \notin L(\mathscr{A})$. Using Theorem 6.2, it follows easily that $\bar{d} \notin L(\mathscr{A}) + K$, too. Since $\mathscr{A} = A^{-1}(b) \cap \mathbb{R}^n_+$, we have $0^+\mathscr{A} = 0^+(A^{-1}(b)) \cap 0^+\mathbb{R}^n_+$. As $0^+(A^{-1}(b)) = A^{-1}(0^+\{b\}) = A^{-1}(0)$ and $0^+\mathbb{R}^n_+ = \mathbb{R}^n_+$, it follows $0^+\mathscr{A} = A^{-1}(0) \cap \mathbb{R}^n_+$. Then $0^+L(\mathscr{A}) = L(0^+\mathscr{A}) = L(A^{-1}(0) \cap \mathbb{R}^n_+) = \{Lx : x \in \mathbb{R}^n_+, Ax = 0\} \subseteq (L - \bar{U}A)(\mathbb{R}^n_+)$ and, obviously, $0 \in 0^+L(\mathscr{A})$.

Using Proposition 6.1, we obtain $(L - \bar{U}A)(\mathbb{R}^n_+) \cap (-K) = \{0\}$, thus, taking into account the inclusions from above, we obtain $0^+L(\mathscr{A}) \cap (-K) = \{0\} \subseteq K = 0^+K$. This assertion and the fact that $L(\mathscr{A})$ is polyhedral and K is closed convex yield, via [178, Theorem 20.3], that $L(\mathscr{A}) + K$ is a closed convex set. Applying [178, Corollary 11.4.2] we obtain a $\gamma \in \mathbb{R}^k \setminus \{0\}$ and an $\alpha \in \mathbb{R}$ such that

$$\gamma^\top \bar{d} < \alpha < \gamma^\top (Lx + k) \ \forall x \in \mathscr{A} \ \forall k \in K. \tag{6.2.6}$$

As K is a cone, $\gamma \in K^*$. Taking $k = 0$ in (6.2.6) it follows

$$\gamma^\top \bar{d} < \alpha < \gamma^\top (Lx) \ \forall x \in \mathscr{A}. \tag{6.2.7}$$

On the other hand, for all $x \in \mathscr{A}$ one has $0 \leq \bar{\lambda}^\top((L - \bar{U}A)x) = \bar{\lambda}^\top(Lx - \bar{U}b) = \bar{\lambda}^\top(Lx - \bar{U}b) - \bar{\lambda}^\top \bar{v} = \bar{\lambda}^\top(Lx - \bar{d})$, therefore

$$\bar{\lambda}^\top \bar{d} \leq \bar{\lambda}^\top(Lx) \ \forall x \in \mathscr{A}. \tag{6.2.8}$$

Now, taking $\delta := \alpha - \gamma^\top \bar{d} > 0$ it follows $\bar{d}^\top(s\bar{\lambda} + (1-s)\gamma) = \alpha - \delta + s(\bar{\lambda}^\top \bar{d} - \alpha + \delta)$ for all $s \in \mathbb{R}$. Note that there exists an $\bar{s} \in (0, 1)$ such that $\bar{s}(\bar{\lambda}^\top \bar{d} - \alpha + \delta) < \delta/2$ and $\bar{s}(\bar{\lambda}^\top \bar{d} - \alpha) > -\delta/2$, and let $\lambda := \bar{s}\bar{\lambda} + (1 - \bar{s})\gamma$. It is clear that $\lambda \in \operatorname{int} K^*$.

By (6.2.7) and (6.2.8) it follows $s\bar{\lambda}^\top \bar{d} + (1 - s)\alpha < (s\bar{\lambda} + (1 - s)\gamma)^\top(Lx)$ for all $x \in \mathscr{A}$ and all $s \in (0, 1)$, consequently

$$\begin{aligned} \lambda^\top \bar{d} &= \bar{s}\bar{\lambda}^\top \bar{d} + (1 - \bar{s})\gamma^\top \bar{d} = \bar{s}\bar{\lambda}^\top \bar{d} + (1 - \bar{s})(\alpha - \delta) \\ &< \frac{\delta}{2} + \bar{s}(\alpha - \delta) + (1 - \bar{s})(\alpha - \delta) = \alpha - \frac{\delta}{2} < \lambda^\top(Lx) \ \forall x \in \mathscr{A}. \end{aligned} \tag{6.2.9}$$

Since there is strong duality for the scalar linear optimization problem

$$\inf_{x \in \mathscr{A}} \lambda^\top(Lx)$$

and its Lagrange dual

$$\sup_{\substack{\eta \in \mathbb{R}^m, \\ L^\top \lambda + A^\top \eta \in \mathbb{R}^n_+}} \Big\{ -\eta^\top b \Big\},$$

the latter has an optimal solution, say $\bar{\eta}$, and $\inf_{x \in \mathscr{A}} \lambda^\top(Lx) + \bar{\eta}^\top b = 0$ and $L^\top \lambda + A^\top \bar{\eta} \in \mathbb{R}^n_+$. As $\lambda \in \operatorname{int} K^*$, there exists a $\tilde{\lambda} \in K \setminus \{0\}$ such that $\lambda^\top \tilde{\lambda} = 1$. Let $U := -\tilde{\lambda}\bar{\eta}^\top$. It follows that $(L - UA)^\top \lambda \in \mathbb{R}^n_+$ and $\inf_{x \in \mathscr{A}} \lambda^\top(Lx) = \lambda^\top(Ub)$.

Consider now the hyperplane $\mathscr{H} := \{Ub + v : \lambda^\top v = 0\}$, which is nothing but the set $\{w \in \mathbb{R}^k : \lambda^\top w = \lambda^\top(Ub)\}$. Consequently, $\mathscr{H} \subseteq h_F(\mathscr{B}_F)$. On the other hand, (6.2.9) yields $\lambda^\top \bar{d} < \lambda^\top(Ub)$. Then there exists a $\bar{k} \in K \setminus \{0\}$ such that $\lambda^\top(\bar{d} + \bar{k}) = \lambda^\top(Ub)$, which has as consequence that $\bar{d} + \bar{k} \in \mathscr{H} \subseteq h_F(\mathscr{B}_F)$. Noting that $\bar{d} \leq_K \bar{d} + \bar{k}$, we have just arrived to a contradiction to the maximality of $\bar{d}$ to the set $h_F(\mathscr{B}_F)$. Therefore our initial supposition is false, consequently $\bar{d} \in L(\mathscr{A})$. Then there exists an $\bar{x} \in \mathscr{A}$ such that $L\bar{x} = \bar{d} = \bar{U}b + \bar{v}$. Employing Theorem 6.2, it follows $\bar{x} \in \mathscr{E}(PLF)$. □

Regarding necessary and optimality conditions for the primal-dual pair of vector optimization problems (*PLF*) – (*DLF*) we make the following observation.

Remark 6.9. If $\bar{x} \in \mathscr{A}$ and $(\bar{\lambda}, \bar{U}, \bar{v}) \in \mathscr{B}_F$ are, like in Theorems 6.3 or 6.4, such that $L\bar{x} = \bar{U}b + \bar{v}$, then the *complementarity condition* $\bar{x}^\top(L - \bar{U}A)^\top \bar{\lambda} = 0$ is fulfilled.

Analogously to [117, Theorem 3.14], we summarize the duality results proven above in a general duality statement for (*PLF*) and (*DLF*).

Corollary 6.1. *One has* $\mathrm{Min}(L(\mathscr{A}), K) = \mathrm{Max}(h_F(\mathscr{B}_F), K)$.

To complete the investigation on the primal-dual pair of vector optimization problems (PLF) – (DLF) we give also the following assertions concerning the infeasibility cases.

Theorem 6.5. *If* $\mathscr{A} \neq \emptyset$, *one has* $\mathscr{E}(PLF) = \emptyset$ *if and only if* $\mathscr{B}_F = \emptyset$.

Proof. "⇒" By [117, Lemma 2.1], the lack of efficient solutions to (PLF) yields $0^+L(\mathscr{A}) \cap (-K) \setminus \{0\} \neq \emptyset$. Then $(L - UA)(\mathbb{R}^n_+) \cap (-K) \setminus \{0\} \neq \emptyset$ for all $U \in \mathbb{R}^{k \times m}$ and employing Proposition 6.1 we see that $\mathscr{B}_F$ cannot contain in this situation any element.

"⇐" Assuming that (PLF) has efficient solutions, Theorem 6.3 yields that also (DLF) has an efficient solution. But this cannot happen since the dual has no feasible elements, consequently (PLF) has no efficient solutions. □

Theorem 6.6. *If* $\mathscr{B}_F \neq \emptyset$, *one has* $\mathscr{E}(DLF) = \emptyset$ *if and only if* $\mathscr{A} = \emptyset$.

Proof. "⇒" Assume that $\mathscr{A} \neq \emptyset$. If (PLF) has no efficient solutions, Theorem 6.5 would yield $\mathscr{B}_F = \emptyset$, but this is false, therefore (PLF) must have at least an efficient solution. Employing Theorem 6.3 it follows that (DLF) has an efficient solution, too, contradicting the assumption we made. Therefore $\mathscr{A} = \emptyset$.

"⇐" Assuming that (DLF) has an efficient solution, Theorem 6.4 yields that (PLF) has an efficient solution, too. But this cannot happen since this problem has no feasible elements, consequently (DLF) has no efficient solutions. □

For the classical vector dual problems to (PLF), (DLF^I) and (DLF^J) weak duality holds in general, but for strong and converse duality one needs to impose additionally the condition $b \neq 0$ to the hypotheses of the corresponding theorems given for (DLF) (see also [48, Section 4.5 and Section 5.5]). For (DLF^H) all three duality statements hold under the hypotheses of the corresponding theorems regarding (DLF), as proven in [117]. Concerning (DLF^L), weak and strong duality were shown (for instance in [48, 140]), but the converse duality statement does not follow directly from a more general case because there a regularity condition is needed, so we prove it below.

Theorem 6.7. *If* $(\bar{\lambda}, \bar{z}, \bar{v}) \in \mathscr{E}(DLF^L)$, *there exists an* $\bar{x} \in \mathscr{E}(PLF)$ *such that* $L\bar{x} = \bar{v}$.

Proof. Analogously to the proof of Theorem 6.4 it can be easily shown that $\mathscr{A} \neq \emptyset$. Since $\bar{\lambda} \in \operatorname{int} K^*$, there exists a $\tilde{\lambda} \in K \setminus \{0\}$ such that $\bar{\lambda}^\top \tilde{\lambda} = 1$. Let $U := (\bar{z}\tilde{\lambda}^\top)^\top$. Then $U^\top \bar{\lambda} = \bar{z}\tilde{\lambda}^\top \bar{\lambda} = \bar{z}$. Thus, $(L - UA)^\top \bar{\lambda} = L^\top \bar{\lambda} - A^\top U^\top \bar{\lambda} = L^\top \bar{\lambda} - A^\top \bar{z} \in \mathbb{R}^n_+$. Assuming the existence of an $x \in \mathbb{R}^n_+$ for which $(L - UA)x \in -K \setminus \{0\}$, it follows $\bar{\lambda}^\top((L - UA)x) < 0$. But $\bar{\lambda}^\top((L - UA)x) = x^\top((L - UA)^\top \bar{\lambda}) \geq 0$, since $x \in \mathbb{R}^n_+$ and $(L - UA)^\top \bar{\lambda} \in \mathbb{R}^n_+$. This contradiction yields $(L - UA)(\mathbb{R}^n_+) \cap (-K) = \{0\}$. Like in the proof of Theorem 6.4, this result, together with the facts that $L(\mathscr{A})$ is polyhedral and K is closed convex implies, via [178, Theorem 20.3], that $L(\mathscr{A}) + K$ is a closed convex set. The existence of an $\bar{x} \in \mathscr{A}$ that is a properly efficient, thus also efficient, solution to (*PLF*) fulfilling $L\bar{x} = \bar{v}$ follows in the lines of [48, Theorem 4.3.4] (see also [48, Section 4.5.1]). $\square$

The results given within this subsection can be summarized in the following chain of inclusions and equalities involving the sets of maximal elements of the sets mentioned in (6.2.5), namely

$$\begin{aligned}\mathrm{Max}(h_F^I(\mathscr{B}_F^I), K) &= \mathrm{Max}(h_F^J(\mathscr{B}_F^J), K) \subsetneq \mathrm{Min}(L(\mathscr{A}), K) = \mathrm{Max}(h_F^H(\mathscr{B}_F^H), K)\\ &= \mathrm{Max}(h_F(\mathscr{B}_F), K) = \mathrm{Max}(h_F^D(\mathscr{B}_F^D), K) = \mathrm{Max}(h_F^L(\mathscr{B}_F^L), K).\end{aligned}$$

Note that the inclusion turns into equality whenever $b \neq 0$. Thus the second scheme from [48, Remark 5.5.3] remains valid when the image space of the considered vector problems is partially ordered by a nontrivial pointed closed convex cone $K \subseteq \mathbb{R}^k$.

Finally, let us provide as a byproduct a generalization of the classical alternative statement due to Gale (see, for instance, [159, p. 35]), that was used in the literature in investigations closely related to the ones from this section, as done for instance in [48, Proposition 5.5.7].

Proposition 6.5. *Let $K \subseteq \mathbb{R}^k$ be a nontrivial closed convex pointed cone, $S \subseteq \mathbb{R}^n$ a nontrivial convex cone, $W \in \mathbb{R}^{k\times n}$ and $v \in \mathbb{R}^k$, such that $W(S) \cap (-K) = \{0\}$ and $W(S)$ is closed. Then the system*

$$\begin{cases}\lambda^\top v < 0,\\ \lambda^\top W \in S^*,\end{cases}$$

has a solution $\lambda \in \operatorname{int} K^*$ if and only if there is no $x \in S$ such that $Wx - v \in -K$.

Proof. "$\Rightarrow$" Assuming the existence of an $x \in S$ such that $Wx - v \in -K$, it follows $0 > \lambda^\top(Wx - v) > 0$, which is a contradiction. Thus, no such $x \in S$ exists.

"$\Leftarrow$" The set $W(S) - v$ is convex and closed, too, and due to the hypothesis has no common elements with $-K$. Moreover, $0^+(W(S) - v) = 0^+(W(S)) = W(S)$ and $W(S) \cap (-K) = \{0\}$. Applying [117, Lemma 2.2(ii)], one obtains the existence of a $\lambda \in \mathbb{R}^k \setminus \{0\}$ such that $\lambda^\top k < 0 < \lambda^\top w$ for all $k \in -K \setminus \{0\}$ and all $w \in W(S) - v$. Because of the left inequality it follows that $\lambda \in \operatorname{int} K^*$. Moreover, $0 \in W(S)$, and the second inequality yields then $0 < \lambda^\top(0 - v)$, therefore $\lambda^\top v < 0$. On the other hand, assuming that there is some $x \in S$ such that $\lambda^\top Wx < 0$ yields, because S is a cone, that there exists an $\alpha > 0$ such that $\lambda^\top(\alpha Wx - v) < 0$, contradicting the second inequality from above. Consequently, $\lambda^\top Wx \geq 0$ for all $x \in S$, which yields $\lambda^\top W \in S^*$. $\square$

Remark 6.10. A situation where the hypotheses of Proposition 6.5 are fulfilled happens for instance when S is a polyhedral cone, in which case $W(S)$ is closed. Taking $S = \mathbb{R}^n$ and $K = \mathbb{R}^k_+$ Proposition 6.5 turns into the celebrated Gale's alternative theorem.

6.2.4 Duality with Respect to Weakly Efficient Solutions

In this subsection we deliver vector duality statements for the classical linear vector optimization problem in finitely dimensional spaces and its vector dual problems with respect to weakly efficient solutions. To the framework considered in the rest of this section we add the hypothesis that $\operatorname{int} K \neq \emptyset$. The *classical linear vector optimization problem* is in this case

$$(PLF_w) \qquad \operatorname*{WMin}_{x \in \mathscr{A}} Lx,$$

where

$$\mathscr{A} = \{x \in \mathbb{R}^n_+ : Ax = b\}.$$

Recall that an element $\bar{x} \in \mathscr{A}$ is said to be a *weakly efficient solution* to (PLF_w) if $L\bar{x} \in \operatorname{WMin}(L(\mathscr{A}), K)$, and the set of all the weakly efficient solutions to (PLF_w) is denoted by $\mathscr{W}\mathscr{E}(PLF_w)$. Since the hypotheses of Theorem 3.8 are satisfied, an $\bar{x} \in \mathscr{A}$ turns out to be a *weakly efficient solution* to (PLF_w) if and only if there exists a $\lambda \in K^* \setminus \{0\}$ such that $\lambda^\top (L\bar{x}) \leq \lambda^\top (Lx)$ for all $x \in \mathscr{A}$.

Remark 6.11. We have seen in Theorem 6.1 that the properly efficient solutions to (PLF) in the sense of linear scalarization coincide with its efficient ones. This rises the question whether the weakly efficient solutions to (PLF_w) and its efficient ones coincide as well. The situation presented in Example 6.3 shows that in general these classes of solutions to a linear vector optimization problem do not coincide.

Example 6.3. Let the classical vector optimization problem (PLF) (respectively (PLF_w)) for $n = 1$, $m = k = 2$, $A = (0,0)^\top$, $b = (0,0)^\top$, $L = (0,1)^\top$ and $K = \mathbb{R}^2_+$. Then $\mathscr{A} = \mathbb{R}_+$ and $L(\mathscr{A}) = \{0\} \times \mathbb{R}_+$. Whenever $\bar{x} \geq 0$ there exists a $\bar{\lambda} = (1,0) \in \mathbb{R}^2_+ \setminus \{0\}$ such that $\bar{\lambda}^\top L(\bar{x}) = 0 \leq \bar{\lambda}^\top L(x) = 0$ for all $x \in \mathscr{A}$. Consequently, $\mathscr{A} = \mathscr{W}\mathscr{E}(PLF_w)$. On the other hand, whenever $\bar{x} > 0$, one has $L(\bar{x}/2) = (0, \bar{x}/2)^\top \leq_{\mathbb{R}^2_+} (0, \bar{x})^\top = L(\bar{x})$, thus $\bar{x} \notin \mathscr{E}(PLF)$. Obviously, $\mathscr{E}(PLF) \neq \mathscr{W}\mathscr{E}(PLF_w)$ in this situation.

The vector dual problems we assign to (PLF_w) with respect to weakly efficient solutions are similar to the ones considered in Sect. 6.2.2. While for the vector duals with respect to efficient solutions which have as a variable $\lambda \in \operatorname{int} K^*$ one only has to change this constraint into $\lambda \in K^* \setminus \{0\}$, in order to formulate the counterparts with respect to weakly efficient solutions of the others we need a Farkas type result similar to the one delivered in Proposition 6.1. Note that its proof does not use the fact that the cone K is closed.

Proposition 6.6. *Let $U \in \mathbb{R}^{k \times m}$. Then $(L - UA)(\mathbb{R}^n_+) \cap (-\operatorname{int} K) = \emptyset$ if and only if there exists a $\lambda \in K^* \setminus \{0\}$ such that $(L - UA)^\top \lambda \in \mathbb{R}^n_+$.*

Proof. "⇒" The set $(L-UA)(\mathbb{R}^n_+)$ is polyhedral, thus convex, and has no common elements with the interior of the nontrivial pointed convex cone $-\operatorname{int} K \cup \{0\}$. Applying Eidelheit's separation statement for these sets we obtain a $\lambda \in \mathbb{R}^k \setminus \{0\}$ for which

$$\lambda^\top(-k) < 0 \le \lambda^\top((L-UA)x) \ \forall x \in \mathbb{R}^n_+ \ \forall k \in \operatorname{int} K. \tag{6.2.10}$$

Assuming that $\lambda \notin K^*$ yields the existence of a $k \in \operatorname{int} K$ such that $\lambda^\top(-k) \ge 0$, that contradicts (6.2.10). Consequently, $\lambda \in K^* \setminus \{0\}$ and by (6.2.10) it follows immediately that $(L-UA)^\top\lambda \in \mathbb{R}^n_+$.

"⇐" Assuming the existence of an $x \in \mathbb{R}^n_+$ for which $(L-UA)x \in -\operatorname{int} K$, it follows $\lambda^\top((L-UA)x) < 0$, but $\lambda^\top((L-UA)x) = ((L-UA)^\top\lambda)^\top x \ge 0$ since $(L-UA)^\top\lambda \in \mathbb{R}^n_+$ and $x \in \mathbb{R}^n_+$. The so-obtained contradiction yields $(L-UA)(\mathbb{R}^n_+) \cap (-\operatorname{int} K) = \{0\}$. □

Therefore, the Isermann type vector dual to (PLF_w) is

$$(DLF^I_w) \qquad \operatorname*{WMax}_{U \in \mathcal{B}^I_{F_w}} h^I_{F_w}(U),$$

where

$$\mathcal{B}^I_{F_w} = \left\{U \in \mathbb{R}^{k\times m} : (L-UA)(\mathbb{R}^n_+) \cap (-\operatorname{int} K) = \emptyset\right\}$$

and

$$h^I_{F_w}(U) = Ub,$$

while the *dual abstract optimization problem* to (PLF_w) with respect to weakly efficient solutions is

$$(DLF^J_w) \qquad \operatorname*{WMax}_{(\lambda,U) \in \mathcal{B}^J_{F_w}} h^J_{F_w}(\lambda, U),$$

where

$$\mathcal{B}^J_{F_w} = \left\{(\lambda, U) \in (K^* \setminus \{0\}) \times \mathbb{R}^{k\times m} : (L-UA)^\top\lambda \in \mathbb{R}^n_+\right\}$$

and

$$h^J_{F_w}(\lambda, U) = Ub.$$

Like their counterparts considered with respect to efficient solutions, these vector dual problems are not the best choices to work with in case $b = 0$ as, when feasible, their image sets coincide with the set $\{0\}$. However, the other vector duals we assign to (PLF_w) do not share the mentioned vulnerability. The vector Lagrange type dual (cf. [138], see also [149]) is

$$(DLF_w^L) \qquad \underset{(\lambda,z,v)\in\mathscr{B}_{F_w}^L}{\operatorname{WMax}} h_{F_w}^L(\lambda,z,v),$$

where

$$\mathscr{B}_{F_w}^L = \left\{(\lambda,z,v) \in (K^* \setminus \{0\}) \times \mathbb{R}^m \times \mathbb{R}^k : \lambda^\top v - z^\top b \leq 0,\ L^\top\lambda - A^\top z \in \mathbb{R}_+^n\right\}$$

and

$$h_{F_w}^L(\lambda,z,v) = v,$$

while the counterpart with respect to weakly efficient solutions of (DLF) turns out to be

$$(DLF_w) \qquad \underset{(\lambda,U,v)\in\mathscr{B}_{F_w}}{\operatorname{WMax}} h_{F_w}(\lambda,U,v),$$

where

$$\mathscr{B}_{F_w} = \left\{(\lambda,U,v) \in (K^* \setminus \{0\}) \times \mathbb{R}^{k\times m} \times \mathbb{R}^k : \lambda^\top v = 0,\ (L-UA)^\top\lambda \in \mathbb{R}_+^n\right\}$$

and

$$h_{F_w}(\lambda,U,v) = Ub + v.$$

One can consider the counterparts with respect to weakly efficient solutions of (DLF^D), too, namely (see also [48, 156])

$$(DLF_w^D) \qquad \underset{(\lambda,U,v)\in\mathscr{B}_{F_w}^D}{\operatorname{WMax}} h_{F_w}^D(\lambda,U,v),$$

where

$$\mathscr{B}_{F_w}^D = \left\{(\lambda,U,v) \in (K^* \setminus \{0\}) \times \mathbb{R}^{k\times m} \times \mathbb{R}^k : \lambda^\top v \leq 0,\ (L-UA)^\top\lambda \in \mathbb{R}_+^n\right\}$$

and

$$h_{F_w}^D(\lambda,U,v) = Ub + v,$$

and (DLF^H)

$$(DLF_w^H) \qquad \underset{U\in\mathscr{B}_{F_w}^H}{\operatorname{WMax}} h_{F_w}^H(U),$$

where

$$\mathscr{B}_{F_w}^H = \left\{U \in \mathbb{R}^{k \times m} : (L - UA)(\mathbb{R}_+^n) \cap (-\operatorname{int} K) = \emptyset\right\}$$

and

$$h_{F_w}^H(U) = Ub + \operatorname{WMin}\big((L - UA)(\mathbb{R}_+^n), K\big).$$

Remark 6.12. If $(\lambda, U, v) \in \mathscr{B}_{F_w}$, one can easily note that $v \notin \operatorname{int} K \cup (-\operatorname{int} K)$, while when $(\lambda, U, v) \in \mathscr{B}_{F_w}^D$ it follows that $v \notin \operatorname{int} K$.

An inclusion chain similar to the ones given for their counterparts with respect to efficient solutions in (6.2.5) holds for these vector duals to (PLF_w), too, extending thus the one given in [48, Section 5.5] for only some of them in case $K = \mathbb{R}_+^k$. Therefore, it holds

$$h_{F_w}^I(\mathscr{B}_{F_w}^I) = h_{F_w}^J(\mathscr{B}_{F_w}^J) \subsetneq h_{F_w}^H(\mathscr{B}_{F_w}^H) \subsetneq h_{F_w}(\mathscr{B}_{F_w}) \subsetneq h_{F_w}^D(\mathscr{B}_{F_w}^D) = h_{F_w}^L(\mathscr{B}_{F_w}^L), \tag{6.2.11}$$

where the proofs and counterexamples can be directly adapted from the ones provided or mentioned in Sect. 6.2.2.

For the primal-dual pair of vector optimization problems $(PLF_w) - (DLF_w)$ weak, strong and converse duality statements can be proven analogously to their counterparts provided in Sect. 6.2.3 for the primal-dual pair $(PLF) - (DLF)$.

Theorem 6.8. *There exist no $x \in \mathscr{A}$ and $(\lambda, U, v) \in \mathscr{B}_{F_w}$ such that $Lx <_K Ub + v$.*

Like in the scalar linear case and in Sect. 6.2.3, for strong and converse vector duality no regularity conditions need to be satisfied.

Theorem 6.9. *If $\bar{x} \in \mathscr{W}\mathscr{E}(PLF_w)$, there exists $(\bar{\lambda}, \bar{U}, \bar{v}) \in \mathscr{W}\mathscr{E}(DLF_w)$ such that $L\bar{x} = \bar{U}b + \bar{v}$.*

Proof. As $\bar{x} \in \mathscr{W}\mathscr{E}(PLF_w)$, there exists a $\bar{\lambda} \in K^*\setminus\{0\}$ such that $\bar{\lambda}^\top(L\bar{x}) \leq \bar{\lambda}^\top(Lx)$ for all $x \in \mathscr{A}$. Then there is strong duality for the scalar optimization problem $\inf_{x\in\mathscr{A}} \bar{\lambda}^\top(Lx)$ and dual problem which has thus an optimal solution, say $\bar{\eta} \in \mathbb{R}^m$. Consequently, $\bar{\lambda}^\top(L\bar{x}) + \bar{\eta}^\top b = 0$ and $L^\top\bar{\lambda} + A^\top\bar{\eta} \in \mathbb{R}_+^n$. The rest of the proof follows in the lines of the one of Theorem 6.3, with the necessary modifications, namely here $\bar{\lambda} \in K^* \setminus \{0\}$, $\tilde{\lambda} \in \operatorname{int} K$ and, finally, $(\bar{\lambda}, \bar{U}, \bar{v}) \in \mathscr{W}\mathscr{E}(DLF_w)$. □

Theorem 6.10. *If $(\bar{\lambda}, \bar{U}, \bar{v}) \in \mathscr{W}\mathscr{E}(DLF_w)$, then $\bar{U}b + \bar{v} \in \operatorname{WMin}(L(\mathscr{A}) + K, K)$.*

Proof. Let $\bar{d} := \bar{U}b + \bar{v} \in \operatorname{WMax}(h_{F_w}(\mathscr{B}_{F_w}), K)$. Analogously to the proof of Theorem 6.4 one can show that $\mathscr{A} \neq \emptyset$. Suppose now that $\bar{d} \notin L(\mathscr{A}) + K$. Using the steps from the proof of Theorem 6.4, it follows that $L(\mathscr{A}) + K$ is a closed convex set. Applying [178, Corollary 11.4.2] we obtain a $\gamma \in \mathbb{R}^k \setminus \{0\}$ and an

$\alpha \in \mathbb{R}$ such that

$$\gamma^\top \bar{d} < \alpha < \gamma^\top (Lx + k) \ \forall x \in \mathscr{A} \ \forall k \in K. \tag{6.2.12}$$

Furthermore, $\gamma \in K^* \setminus \{0\}$. In the lines of the proof of Theorem 6.4 we can construct a $\lambda \in K^* \setminus \{0\}$. There is strong duality for the scalar linear optimization problem

$$\inf_{x \in \mathscr{A}} \lambda^\top (Lx)$$

and its Lagrange dual

$$\sup_{\substack{\eta \in \mathbb{R}^m, \\ L^\top \lambda + A^\top \eta \in \mathbb{R}^n_+}} \left\{ -\eta^\top b \right\},$$

so the latter has an optimal solution, say $\bar{\eta}$, and $\inf_{x \in \mathscr{A}} \lambda^\top (Lx) + \bar{\eta}^\top b = 0$ and $L^\top \lambda + A^\top \bar{\eta} \in \mathbb{R}^n_+$. As $\bar{\lambda} \in K^* \setminus \{0\}$, there exists a $\tilde{\lambda} \in \operatorname{int} K$ such that $\bar{\lambda}^\top \tilde{\lambda} = 1$. Let $U := -\tilde{\lambda}\bar{\eta}^\top$. It follows that $(L - UA)^\top \lambda \in \mathbb{R}^n_+$ and $\inf_{x \in \mathscr{A}} \lambda^\top (Lx) = \lambda^\top (Ub)$ and, moreover, $\lambda^\top \bar{d} < \lambda^\top (Ub)$. Then there exists a $\bar{k} \in \operatorname{int} K$ such that $\lambda^\top (\bar{d} + \bar{k}) = \lambda^\top (Ub)$.

Considering the hyperplane $\mathscr{H} := \{Ub + v : \lambda^\top v = 0\} = \{w \in \mathbb{R}^k : \lambda^\top w = \lambda^\top (Ub)\} \subseteq h_{F_w}(\mathscr{B}_{F_w})$, $\bar{d} + \bar{k}$ lies in it, thus also in $h_{F_w}(\mathscr{B}_{F_w})$. But $\bar{d} <_K \bar{d} + \bar{k}$, so we have just arrived to a contradiction to the weak maximality of $\bar{d}$ to the image set of the vector dual problem (DFL_w). Therefore our initial supposition is false, consequently $\bar{d} \in L(\mathscr{A}) + K$.

Then there exist $\bar{x} \in \mathscr{A}$ such that $L\bar{x} - \bar{U}b - \bar{v} \in -K$ and the conclusion follows via Theorem 6.8. □

Regarding necessary and optimality conditions for the primal-dual pair of vector optimization problems $(PLF_w) - (DLF_w)$ we make the following observation.

Remark 6.13. If $\bar{x} \in \mathscr{A}$ and $(\bar{\lambda}, \bar{U}, \bar{v}) \in \mathscr{B}_{F_w}$ are, like in Theorem 6.9, such that $L\bar{x} = \bar{U}b + \bar{v}$, then the *complementarity condition* $\bar{x}^\top (L - \bar{U}A)^\top \bar{\lambda} = 0$ is fulfilled.

Analogously to Corollary 6.1, we summarize the duality results proven above in a general duality statement for (PLF) and (DLF).

Corollary 6.2. *One has*

$$\operatorname{WMin}(L(\mathscr{A}), K) \subseteq \operatorname{WMax}(h_{F_w}(\mathscr{B}_{F_w}), K) \subseteq \operatorname{WMin}(L(\mathscr{A}) + K, K).$$

To complete the investigation on the primal-dual pair of vector optimization problems $(PLF_w) - (DLF_w)$ we give also the following assertions.

Theorem 6.11. *If $\mathscr{A} \neq \emptyset$, one has $\mathscr{W}\mathscr{E}(PLF_w) = \emptyset$ if and only if $\mathscr{B}_{F_w} = \emptyset$.*

Proof. "$\Rightarrow$" The lack of weakly efficient solutions to (PLF_w) yields $0^+L(\mathscr{A}) \cap (-\operatorname{int} K) \neq \emptyset$. Indeed, assuming the contrary there must exist a $y \in L(\mathscr{A})$ such that $y - \operatorname{int} K \notin L(\mathscr{A})$. Applying Eidelheit's separation statement for the nonempty convex sets $y - \operatorname{int} K$ and $L(\mathscr{A})$, we obtain a $\lambda \in \mathbb{R}^k \setminus \{0\}$ for which

$$\lambda^\top(y-k) < 0 \leq \lambda^\top(Lx) \ \forall x \in \mathscr{A} \ \forall k \in \operatorname{int} K. \tag{6.2.13}$$

Assuming that $\lambda \notin K^*$ yields the existence of a $k \in \operatorname{int} K$ such that $\lambda^\top(-k) \geq 0$, that contradicts (6.2.13). Consequently, $\lambda \in K^* \setminus \{0\}$ and by (6.2.13) it follows immediately that $\lambda^\top y \leq \lambda^\top(Lx)$ for all $x \in \mathscr{A}$, i.e. $y \in \mathscr{W}\mathscr{E}(PLF_w)$. But $\mathscr{W}\mathscr{E}(PLF_w) = \emptyset$, so a contradiction has been reached, consequently $0^+L(\mathscr{A}) \cap (-\operatorname{int} K) \neq \emptyset$. Then $(L - UA)(\mathbb{R}^n_+) \cap (-\operatorname{int} K) \neq \emptyset$ for all $U \in \mathbb{R}^{k\times m}$ and employing Proposition 6.6 we see that $\mathscr{B}_{F_w}$ cannot contain in this situation any element.

"$\Leftarrow$" Assuming that (PLF_w) has weakly efficient solutions, Theorem 6.9 yields that (DLF_w) has a weakly efficient solution, too. But this cannot happen since the dual has no feasible elements, consequently $\mathscr{W}\mathscr{E}(PLF_w) = \emptyset$. □

Remark 6.14. In the proof of Theorem 6.11 we have shown that $\mathscr{W}\mathscr{E}(PLF_w) = \emptyset$ yields $0^+L(\mathscr{A}) \cap (-\operatorname{int} K) \neq \emptyset$. The reverse implication holds, too, and can be shown in the lines of [117, Lemma 2.1$(i) \Rightarrow (ii)$]. If $0^+L(\mathscr{A}) \cap (-\operatorname{int} K) \neq \emptyset$, there exists a $\bar{y}$ belonging to both these sets. Then, whenever $y \in L(\mathscr{A})$, one has $y + \bar{y} \in L(\mathscr{A})$. But $y - (y + \bar{y}) = -\bar{y} \in \operatorname{int} K$. Therefore, for all $y \in L(\mathscr{A})$ there exists a $y + \bar{y} \in L(\mathscr{A})$ such that $y + \bar{y} <_K y$, i.e. $\mathscr{W}\mathscr{E}(PLF_w) = \emptyset$. As in the above considerations $L(\mathscr{A})$ can be replaced by any nonempty convex subset $M \subseteq \mathbb{R}^k$, one obtains for the latter that $\operatorname{WMin}(M, K) \neq \emptyset$ if and only if $0^+M \cap (-\operatorname{int} K) = \emptyset$. In this way we extend the equivalence $(i) \Leftrightarrow (ii)$ of [117, Lemma 2.1] to weakly minimal elements, too.

Theorem 6.12. *If $\mathscr{B}_{F_w} \neq \emptyset$, one has $\mathscr{W}\mathscr{E}(DLF_w) = \emptyset$ if and only if $\mathscr{A} = \emptyset$.*

Proof. "$\Rightarrow$" Assume that $\mathscr{A} \neq \emptyset$. If $\mathscr{W}\mathscr{E}(PLF_w) = \emptyset$, Theorem 6.11 would yield the false assertion $\mathscr{B}_{F_w} = \emptyset$, therefore $\mathscr{W}\mathscr{E}(PLF_w) \neq \emptyset$. Employing Theorem 6.9 it follows that $\mathscr{W}\mathscr{E}(DLF_w) \neq \emptyset$, contradicting the assumption we made. Therefore $\mathscr{A} = \emptyset$.

"$\Leftarrow$" Assuming that $\mathscr{W}\mathscr{E}(DLF_w) \neq \emptyset$, Theorem 6.10 yields that $\operatorname{WMin}(L(\mathscr{A}) + K, K) \neq \emptyset$, thus $\mathscr{A} \neq \emptyset$. But $\mathscr{A} = \emptyset$, thus $\mathscr{W}\mathscr{E}(DLF_w) = \emptyset$. □

For (DLF_w^J) weak duality holds in general, but for strong and converse duality one needs to impose additionally the condition $b \neq 0$ to the hypotheses of the corresponding theorems given for (DLF_w) (see also [48, Sections 4.5 and 5.5]). Since (DLF_w^I) has the same image set, the duality assertions concerning it require the same hypotheses.

For (DLF_w^H) the weak and strong duality statements hold under the hypotheses of the corresponding theorems regarding (DLF), too.

Theorem 6.13. *There exist no* $x \in \mathscr{A}$ *and* $U \in \mathscr{B}^H_{F_w}$ *such that* $Lx <_K Ub + v$, *where* $v \in \mathrm{WMin}\big((L-UA)(\mathbb{R}^n_+), K\big)$.

Proof. Assume the existence of some $x \in \mathscr{A}$ and $U \in \mathscr{B}^H_{F_w}$ such that $Lx <_K Ub+v$, where $v \in \mathrm{WMin}\big((L-UA)(\mathbb{R}^n_+), K\big)$. Since $v \in \mathrm{WMin}\big((L-UA)(\mathbb{R}^n_+), K\big)$, there exists a $\lambda \in K^* \setminus \{0\}$ such that $\lambda^\top v \leq \lambda^\top((L-UA)y)$ for all $y \in \mathbb{R}^n_+$. Then, as $x \in \mathscr{A}$, one has $\lambda^\top(Lx-Ub-v) = \lambda^\top((L-UA)x-v) \geq 0$. But $\lambda^\top(Lx-Ub-v) < 0$, due to the assumption we made above. A contradiction is reached, consequently the assumption we made is false and, therefore, the assertion is proven. □

The proof of the strong duality statement is different and simpler than its counterpart for the vector dual problem to (PLF) with respect to efficient solutions given in [117].

Theorem 6.14. *If* $\bar{x} \in \mathscr{W}\mathscr{E}(PLF_w)$, *there exist* $\bar{U} \in \mathscr{W}\mathscr{E}(DLF^H_w)$ *and* $\bar{v} \in \mathrm{WMin}\big((L-UA)(\mathbb{R}^n_+), K\big)$ *such that* $L\bar{x} = \bar{U}b + \bar{v}$.

Proof. As $\bar{x} \in \mathscr{W}\mathscr{E}(PLF_w)$, there exists a $\bar{\lambda} \in K^* \setminus \{0\}$ such that $\bar{\lambda}^\top(L\bar{x}) \leq \bar{\lambda}^\top(Lx)$ for all $x \in \mathscr{A}$. Like in the proof of Theorem 6.9, a $\bar{U} \in \mathbb{R}^{k\times m}$ for which $(L-\bar{U}A)^\top\bar{\lambda} \in \mathbb{R}^n_+$ can be obtained. Then, via Proposition 6.6, $(L-UA)(\mathbb{R}^n_+) \cap (-\operatorname{int} K) = \emptyset$, consequently $\bar{U} \in \mathscr{B}^H_{F_w}$. Denoting $\bar{v} := L\bar{x} - \bar{U}b \in \mathbb{R}^k$, one can easily verify that $\bar{v} = (L-UA)\bar{x} \in (L-UA)(\mathbb{R}^n_+)$ and $\bar{\lambda}^\top\bar{v} = 0$. As $(L-\bar{U}A)^\top\bar{\lambda} \in \mathbb{R}^n_+$, it follows that $\bar{\lambda}^\top((L-\bar{U}A)x) \geq 0$ for all $x \in \mathbb{R}^n_+$. Consequently, $\bar{v} \in \mathrm{WMin}\big((L-UA)(\mathbb{R}^n_+), K\big)$. Assuming that $\bar{U} \notin \mathscr{W}\mathscr{E}(DLF^H_w)$ one gets immediately a contradiction to Theorem 6.13. Consequently, $\bar{U} \in \mathscr{W}\mathscr{E}(DLF^H_w)$. Moreover, $L\bar{x} = \bar{U}b + \bar{v}$. □

Concerning (DLF^L_w) and (DLF^D_w), which have the same image set, weak and strong duality were shown (for instance in [48, 140]), but the converse duality statement, proven in case $K = \mathbb{R}^k_+$ in [48, 156], does not follow directly in the more general framework considered here. However, it can be proven in the lines of Theorem 6.7, modified according to the proof of Theorem 6.10.

Theorem 6.15. *If* $(\bar{\lambda}, \bar{z}, \bar{v}) \in \mathscr{W}\mathscr{E}(DLF^L_w)$, *then* $\bar{v} \in \mathrm{WMin}(L(\mathscr{A}) + K, K)$.

The results given within this subsection can be summarized in the following chains of inclusions and equalities involving the sets of maximal elements of the sets mentioned in (6.2.11), namely, when $b \neq 0$

$$\mathrm{WMin}(L(\mathscr{A}), K) \subseteq \mathrm{WMax}(h^I_{F_w}(\mathscr{B}^I_{F_w}), K) = \mathrm{WMax}(h^J_{F_w}(\mathscr{B}^J_{F_w}), K)$$
$$\subseteq \mathrm{WMax}(h_{F_w}(\mathscr{B}_{F_w}), K) = \mathrm{WMax}(h^D_{F_w}(\mathscr{B}^D_{F_w}), K) = \mathrm{WMax}(h^L_{F_w}(\mathscr{B}^L_{F_w}), K)$$
$$\subseteq \mathrm{WMin}(L(\mathscr{A}) + K, K),$$

while in case $b = 0$ it holds

$$\begin{aligned}\mathrm{WMax}(h^I_{F_w}(\mathscr{B}^I_{F_w}), K) &= \mathrm{WMax}(h^J_{F_w}(\mathscr{B}^J_{F_w}), K) \subseteq \mathrm{WMin}(L(\mathscr{A}), K)\\ &\subseteq \mathrm{WMax}(h_{F_w}(\mathscr{B}_{F_w}), K) = \mathrm{WMax}(h^D_{F_w}(\mathscr{B}^D_{F_w}), K)\\ &= \mathrm{WMax}(h^L_{F_w}(\mathscr{B}^L_{F_w}), K) \subseteq \mathrm{WMin}(L(\mathscr{A}) + K, K).\end{aligned}$$

Thus the scheme from [48, Remark 5.5.5] remains valid when the image space of the considered vector problems is partially ordered by a nontrivial pointed closed convex cone $K \subseteq \mathbb{R}^k$.

6.3 Linear Vector Duality in Infinitely Dimensional Spaces

In this section we extend the investigations from the previous one to the infinitely dimensional case, as proposed in our paper [34]. As we shall see, some things can be easily generalized, while for others additional hypotheses are needed. Consider like in Chaps. 3–5 the Hausdorff locally convex vector spaces X, Z and V, with Z partially ordered by the convex cone $C \subseteq Z$ and V by the nontrivial pointed convex cone $K \subseteq V$. Let $L \in \mathscr{L}(X, V)$, $A \in \mathscr{L}(X, Z)$, $b \in Z$ and the convex cone $S \subseteq X$.

The primal *linear vector optimization problem* we consider now is

$$(PLI) \qquad \operatorname*{Min}_{x \in \mathscr{A}} Lx,$$

where

$$\mathscr{A} = \{x \in S : Ax - b \in C\}.$$

In case $X = \mathbb{R}^n$, $Z = \mathbb{R}^m$, $V = \mathbb{R}^k$, $S = \mathbb{R}^n_+$ and $C = \{0\}$, where the linear continuous mappings L and A can be identified with the matrices $L \in \mathbb{R}^{k \times n}$ and, respectively, $A \in \mathbb{R}^{m \times n}$, (*PLI*) becomes the classical linear vector optimization problem (*PLF*) investigated within Sect. 6.2.

Recall that an element $\bar{x} \in \mathscr{A}$ is said to be a *properly efficient solution in the sense of linear scalarization* to (*PLI*) if $L\bar{x} \in \mathrm{PMin}_{LS}(L(\mathscr{A}), K)$, i.e. there exists a $\lambda \in K^{*0}$ such that $\langle \lambda, L\bar{x} \rangle \leq \langle \lambda, Lx \rangle$ for all $x \in \mathscr{A}$, and the set of all the properly efficient solutions to (*PLI*) in the sense of linear scalarization is denoted by $\mathscr{P}\mathscr{E}_{LS}(PLI)$. An element $\bar{x} \in \mathscr{A}$ is said to be an *efficient solution* to (*PLI*) if $L\bar{x} \in \mathrm{Min}(L(\mathscr{A}), K)$, i.e. there exists no $x \in \mathscr{A}$ such that $Lx \leq_K L\bar{x}$, and the set of all the efficient solutions to (*PLI*) is denoted by $\mathscr{E}(PLI)$. A properly efficient solution $\bar{x}$ to (*PLI*) is also efficient to (*PLI*).

Remark 6.15. In general not all the efficient solutions to (*PLI*) are also properly efficient to it, as it is the case in the finitely dimensional framework of Sect. 6.2 (see Theorem 6.1). However, conditions sufficient to ensure the mentioned coincidence in the present setting can be derived from Theorems 3.4–3.5 or their consequences.

6.3.1 Vector Duals to the Linear Vector Optimization Problem

The vector dual problems assigned to (*PLF*) within Sect. 6.2.2 can be extended to the infinitely dimensional case, too, some straightforwardly. The *dual abstract optimization problem* to (*PLI*) is (cf. [48, 140])

$$(DLI^J) \qquad \operatorname*{Max}_{(\lambda,U)\in\mathscr{B}_I^J} h_I^J(\lambda,U),$$

where

$$\mathscr{B}_I^J = \Big\{(\lambda,U)\in K^{*0}\times\mathscr{L}(Z,V) : U^*\lambda\in C^*,\ (L-U\circ A)^*\lambda\in S^*\Big\}$$

and

$$h_I^J(\lambda,U) = Ub.$$

The vulnerability this vector dual problem presented in the framework of Sect. 6.2 in case $b = 0$ is inherited to the more general setting treated here, too.

The *vector Lagrange type dual* to (*PLI*) is (cf. [48, 140])

$$(DLI^L) \qquad \operatorname*{Max}_{(\lambda,z^*,v)\in\mathscr{B}_I^L} h_I^L(\lambda,z^*,v),$$

where

$$\mathscr{B}_I^L = \Big\{(\lambda,z^*,v)\in K^{*0}\times C^*\times V : \langle\lambda,v\rangle\le\langle z^*,b\rangle,\ L^*\lambda - A^*z^*\in S^*\Big\}$$

and

$$h_I^L(\lambda,z^*,v) = v.$$

The vector duals (DLF^I), (DLF^H), (DLF) and (DLF^D) were considered prior to [34] only in the finitely dimensional setting of Sect. 6.2, but they can be extended to the present framework, too, as follows. The generalization we propose for the vector dual inspired by Isermann's works is

$$(DLI^I) \qquad \operatorname*{Max}_{U\in\mathscr{B}_I^I} h_I^I(U),$$

where

$$\mathscr{B}_I^I = \Big\{U \in \mathscr{L}(Z,V) : \big((L - U \circ A)(S) + U(C)\big) \cap (-K) = \{0\}\Big\}$$

and

$$h_I^I(U) = Ub,$$

while the vector dual that extends (DLF^H) is

$$(DLI^H) \qquad \operatorname*{Max}_{U \in \mathscr{B}_I^H} h_I^H(U),$$

where

$$\mathscr{B}_I^H = \Big\{U \in \mathscr{L}(Z,V) : \big((L - U \circ A)(S) + U(C)\big) \cap (-K) = \{0\}\Big\}$$

and

$$h_I^H(U) = Ub + \text{PMin}_{LS}\big((L - U \circ A)(S) + U(C), K\big).$$

When $X = \mathbb{R}^n$, $S = \mathbb{R}^n_+$, $V = \mathbb{R}^k$, $Z = \mathbb{R}^m$ and $C = \{0\}$ these vector duals to (*PLI*) turn out to be exactly (DLF^I) and (DLF^H), respectively, taking also in consideration that (see Theorem 6.1) in that framework the properly efficient solutions of the vector minimization problem in the objective function of (DLI^H) coincide with the efficient ones of the same problem.

The generalizations to the present framework of (*DLF*) and (DLF^D) are

$$(DLI) \qquad \operatorname*{Max}_{(\lambda,U,v) \in \mathscr{B}_I} h_I(\lambda, U, v),$$

where

$$\mathscr{B}_I = \Big\{(\lambda, U, v) \in K^{*0} \times \mathscr{L}(Z,V) \times V : \langle \lambda, v\rangle = 0,\ U^*\lambda \in C^*,$$
$$(L - U \circ A)^*\lambda \in S^*\Big\}$$

and

$$h_I(\lambda, U, v) = Ub + v,$$

and, respectively,

$$(DLI^D) \qquad \operatorname*{Max}_{(\lambda,U,v) \in \mathscr{B}_I^D} h_I^D(\lambda, U, v),$$

where

$$\mathscr{B}_I^D = \Big\{(\lambda, U, v) \in K^{*0} \times \mathscr{L}(Z, V) \times V : \langle \lambda, v \rangle \leq 0,\ U^*\lambda \in C^*,$$
$$(L - U \circ A)^*\lambda \in S^*\Big\}$$

and

$$h_I^D(\lambda, U, v) = Ub + v.$$

Remark 6.16. If $(\lambda, U, v) \in \mathscr{B}_I$, one can easily note that $v \notin (K \cup (-K)) \setminus \{0\}$, while when $(\lambda, U, v) \in \mathscr{B}_I^D$ it follows that $v \notin K \setminus \{0\}$.

We begin with a result that generalizes in one direction Proposition 6.1, establishing thus a connection between the feasible elements of (DLI^J) and the ones of (DLI^I). A possible way to achieve also here an equivalence like in Proposition 6.1 would be by strongly separating the sets $(L - U \circ A)(S) + U(C)$ and $-K$. This can be done, under additional hypotheses, for instance by [140, Theorem 3.22], [117, Lemma 2.2] or [178, Theorem 11.4].

Proposition 6.7. *If $\lambda \in K^{*0}$ and $U \in \mathscr{L}(Z, V)$ fulfill $U^*\lambda \in C^*$ and $(L - U \circ A)^*\lambda \in S^*$, then $\big((L - U \circ A)(S) + U(C)\big) \cap (-K) = \{0\}$.*

Proof. Assume to the contrary that the conclusion is false. Then there exist $x \in S$ and $c \in C$ such that $0 \neq (L - U \circ A)x + Uc \in -K$. Consequently, $\langle \lambda, (L - U \circ A)x + Uc \rangle < 0$. But $\langle \lambda, (L - U \circ A)x + Uc \rangle = \langle (L - U \circ A)^*\lambda, x \rangle + \langle U^*\lambda, c \rangle$ and the hypotheses imply the nonnegativity of the both terms in the right-hand side of the last equality, so we reached the desired contradiction. □

Let us compare now the image sets of the vector duals assigned to (*PLI*) in this section. We begin with a consequence of Proposition 6.7.

Proposition 6.8. *One has $h_I^J(\mathscr{B}_I^J) \subseteq h_I^I(\mathscr{B}_I^I)$.*

Proof. Let $d \in h_I^J(\mathscr{B}_I^J)$. Thus, there exists $(\lambda, U) \in \mathscr{B}_I^J$ such that $d = Ub$. By Proposition 6.7 we obtain immediately that $U \in \mathscr{B}_I^I$. As $h_I^J(\lambda, U) = Ub = h_I^I(U)$, the conclusion follows. □

Proposition 6.9. *One has $h_I^I(\mathscr{B}_I^I) \subseteq h_I^H(\mathscr{B}_I^H)$.*

Proof. Let $d \in h^I(\mathscr{B}^I)$. Thus, there exists a $U \in \mathscr{B}_I^I$ such that $d = Ub$. But $\mathscr{B}_I^I$ and $\mathscr{B}_I^H$ coincide, thus $U \in \mathscr{B}_I^H$. Moreover, $(L - U \circ A)(0) + U(0) = 0$ and whenever $x \in S$ and $c \in C$ there holds $\langle \lambda, (L - U \circ A)x + Uc \rangle = \langle (L - U \circ A)^*\lambda, x \rangle + \langle U^*\lambda, c \rangle$ and this is nonnegative because $(\lambda, U) \in \mathscr{B}^J$. Consequently, $0 \in \text{PMin}_{LS}\big((L - U \circ A)(S) + U(C), K\big)$ and $d \in h_I^H(\mathscr{B}_I^H)$. □

Proposition 6.10. *One has $h_I^H(\mathscr{B}_I^H) \subseteq h_I(\mathscr{B}_I)$.*

Proof. Let $d \in h_I^H(\mathscr{B}_I^H)$. Thus, there exists a $U \in \mathscr{B}^H$ such that $d = Ub + v$, with $v \in \text{PMin}_{LS}\big((L - U \circ A)(S) + U(C), K\big)$. Then, there exist $\gamma \in K^{*0}$, $\overline{x} \in S$ and $\overline{c} \in C$ such that $v = (L - U \circ A)\overline{x} + U\overline{c}$ and

$$\langle \gamma, (L - U \circ A)\overline{x} + U\overline{c} \rangle \leq \langle \gamma, (L - U \circ A)x + Uc \rangle \ \forall x \in S \ \forall c \in C. \quad (6.3.14)$$

Taking in the right-hand side of (6.3.14) $c := \overline{c}$, it follows $\langle \gamma, (L - U \circ A)\overline{x} \rangle \leq \langle \gamma, (L - U \circ A)x \rangle$ for all $x \in S$. S being a cone, the existence of a point $\tilde{x} \in S$ for which $\langle \gamma, (L - U \circ A)\tilde{x} \rangle < 0$ would yield $\langle \gamma, (L - U \circ A)\overline{x} \rangle = -\infty$, that is impossible, so $\langle \gamma, (L - U \circ A)x \rangle \geq 0$ for all $x \in S$. Consequently, $(L - U \circ A)^*\gamma \in S^*$. As $0 \in S$, it follows also that $\langle \gamma, (L - U \circ A)\overline{x} \rangle \leq 0$, so $\langle \gamma, (L - U \circ A)\overline{x} \rangle = 0$.

Back to (6.3.14), taking now $x := \overline{x}$ one gets $\langle \gamma, U\overline{c} \rangle \leq \langle \gamma, Uc \rangle$ for all $c \in C$. Since C is a cone, too, the same argumentation as above leads to $U^*\gamma \in C^*$ and $\langle \gamma, U\overline{c} \rangle = 0$. Consequently, $\langle \gamma, (L - U \circ A)\overline{x} + U\overline{c} \rangle = \langle \gamma, v \rangle = 0$, so $(\gamma, U, v) \in \mathscr{B}_I$ and $h_I(\gamma, U, v) = d$. Therefore $d \in h_I(\mathscr{B}_I)$. □

Remark 6.17. By construction one has $h_I(\mathscr{B}_I) \subseteq h_I^D(\mathscr{B}_I^D)$.

Remark 6.18. Due to the fact that in the framework of Sect. 6.2 the vector duals we consider within this section become their counterparts considered there, the examples mentioned in Remarks 6.5, 6.6 and 6.8 can be invoked in order to show that the inclusions provided in Propositions 6.8–6.10 and Remark 6.17 do not turn in general into equalities.

Proposition 6.4 can be directly extended to the infinitely dimensional case, too.

Proposition 6.11. *One has* $h_I^D(\mathscr{B}_I^D) = h_I^L(\mathscr{B}_I^L)$.

Proof. "$\subseteq$" If $d \in h_I^D(\mathscr{B}_I^D)$, there exist $(\lambda, U, v) \in \mathscr{B}_I^D$ such that $d = h_I^D(\lambda, U, v) = Ub + v$. Taking $z^* := U^*\lambda$, one gets $\langle \lambda, d \rangle = \langle U^*\lambda, b \rangle + \langle \lambda, v \rangle \leq \langle z^*, b \rangle$, while $L^*\lambda - A^*z^* = (L - UA)^*\lambda \in S^*$. Consequently, $(\lambda, z^*, d) \in \mathscr{B}_I^L$ and, since $h_I^L(\lambda, z^*, d) = d$, it follows that $h_I^D(\mathscr{B}_I^D) \subseteq h_I^L(\mathscr{B}_I^L)$.

"$\supseteq$" If $d \in h_I^L(\mathscr{B}_I^L)$, there exist $(\lambda, z^*, d) \in \mathscr{B}_I^L$ such that $d = h_I^L(\lambda, z^*, d)$. As $\lambda \in K^{*0}$, there exists a $\tilde{\lambda} \in K$ such that $\langle \lambda, \tilde{\lambda} \rangle = 1$. Let $U \in \mathscr{L}(Z, V)$ be defined by $Uz := \langle z^*, z \rangle \tilde{\lambda}$ for $z \in Z$, and $v := d - Ub$. Then $\langle \lambda, v \rangle = \langle \lambda, d \rangle - \langle z^*, b \rangle \leq 0$ and $(L - UA)^*\lambda = L^*\lambda - A^*z^* \in S^*$. Consequently, $(\lambda, U, v) \in \mathscr{B}_I^D$ and, since $h_I^D(\lambda, U, v) = Ub + v = d$, it follows that $h_I^D(\mathscr{B}_I^D) \supseteq h_I^L(\mathscr{B}_I^L)$. □

Taking into consideration Propositions 6.8–6.10, Remark 6.17, Proposition 6.11 and Remark 6.18, one can conclude that the images of the feasible sets through their objective functions of the vector duals to (*PLI*) we dealt with respect the following inclusions chain

$$h_I^J(\mathscr{B}_I^J) \subseteq h_I^I(\mathscr{B}_I^I) \subsetneq h_I^H(\mathscr{B}_I^H) \subsetneq h_I(\mathscr{B}_I) \subsetneq h_I^D(\mathscr{B}_I^D) = h_I^L(\mathscr{B}_I^L), \quad (6.3.15)$$

extending thus (6.2.5) to infinitely dimensional spaces.

6.3.2 Duality Results for the Linear Vector Optimization Problem and Its Vector Duals

Let us prove now for the primal-dual pair of vector optimization problems (PLI) – (DLI) weak, strong and converse duality statements.

Theorem 6.16. *There exist no $x \in \mathscr{A}$ and $(\lambda, U, v) \in \mathscr{B}_I$ such that $Lx \leq_K Ub + v$.*

Proof. Assume the existence of $x \in \mathscr{A}$ and $(\lambda, U, v) \in \mathscr{B}_I$ such that $Lx \leq_K Ub+v$. Then $0 < \langle \lambda, Ub + v - Lx\rangle = \langle \lambda, Ub - Lx\rangle = \langle \lambda, Ub - U \circ Ax + U \circ Ax - Lx\rangle = \langle U^*\lambda, b - Ax\rangle - \langle (L - U \circ A)^*\lambda, x\rangle \leq 0$. As this cannot happen, the assumption we made is false. □

In order to prove strong and converse duality for (DLI) one needs additional hypotheses. The regularity conditions (RCV_i^G), $i \in \{1, 2, 3, 4\}$ become in this case

$$(RCV_1^I) \big| \; \exists x' \in S \text{ such that } Ax' - b \in \operatorname{int} C,$$

$$(RCV_2^I) \big| \; X \text{ and } Y \text{ are Fréchet spaces, } S \text{ and } C \text{ are closed and } b \in \operatorname{sqri}(A(S) - C),$$

$$(RCV_3^I) \big| \; \dim \operatorname{lin}(A(S) - C) < +\infty \text{ and } b \in \operatorname{ri}(A(S) - C),$$

and, respectively,

$$(RCV_4^I) \left| \begin{array}{l} S \text{ and } C \text{ are closed and for any } \lambda \in K^{*0} \\ \left\{ \bigcup\limits_{z^* \in C^*} (x^*, r) \in X^* \times \mathbb{R} : \langle z^*, b\rangle \leq r, \; x^* \in L^*\lambda - A^*z^* - S^* \right\} \\ \text{is closed in the topology } \omega(X^*, X) \times \mathscr{R}. \end{array} \right.$$

Remark 6.19. When X and Z are finitely dimensional, instead of the regularity condition (RCV_3^I) one can equivalently write $b \in A(\operatorname{ri} S) - \operatorname{ri} C$ and, moreover, in this condition one can replace the relative interiors of the cones which are actually orthants with the cones themselves.

Theorem 6.17. *If $\bar{x} \in \mathscr{PE}_{LS}(PLI)$ and one of the regularity conditions (RCV_i^I), $i \in \{1, 2, 3, 4\}$, is fulfilled, there exists $(\bar{\lambda}, \bar{U}, \bar{v}) \in \mathscr{E}(DLI)$ such that $L\bar{x} = \bar{U}b + \bar{v}$.*

Proof. Since $\bar{x}$ is properly efficient to (PLI), there exists a $\bar{\lambda} \in K^{*0}$ such that $\langle \bar{\lambda}, L\bar{x}\rangle \leq \langle \bar{\lambda}, Lx\rangle$ for all $x \in \mathscr{A}$. The fulfillment of any of the considered regularity conditions yields that for the scalar optimization problem

$$\inf_{x \in \mathscr{A}} \langle \bar{\lambda}, Lx\rangle$$

and its Lagrange dual

$$\sup_{z^* \in C^*} \inf_{x \in S} \left[\langle \bar{\lambda}, Lx\rangle + \langle z^*, b - Ax\rangle \right],$$

which can be equivalently written as

$$\sup_{\substack{z^* \in C^*, \\ L^*\bar{\lambda} - A^* z^* \in S^*}} \langle z^*, b \rangle,$$

there is strong duality, i.e. their optimal objective values coincide and the dual has an optimal solution, say $\bar{z}^* \in C^*$. Consequently, as $\bar{x}$ solves the primal problem, one gets $\langle \bar{\lambda}, L\bar{x} \rangle = \langle \bar{z}^*, b \rangle$, where $L^*\bar{\lambda} - A^*\bar{z}^* \in S^*$.

Because $\bar{\lambda} \in K^{*0}$, there exists a $\tilde{\lambda} \in K \setminus \{0\}$ such that $\langle \bar{\lambda}, \tilde{\lambda} \rangle = 1$. Let $\bar{U} \in \mathscr{L}(Z, V)$ be defined by $\bar{U}z := \langle \bar{z}^*, z \rangle \tilde{\lambda}$ for $z \in Z$, and $\bar{v} := L\bar{x} - \bar{U}b \in V$. Then $\langle \bar{\lambda}, \bar{v} \rangle = \langle \bar{\lambda}, L\bar{x} - \bar{U}b \rangle = \langle \bar{\lambda}, L\bar{x} \rangle - \langle \bar{z}^*, b \rangle = 0$, $\bar{U}^*\bar{\lambda} = \bar{z}^* \in C^*$ and $(L - \bar{U} \circ A)^*\bar{\lambda} = L^*\bar{\lambda} - A^*\bar{z}^* \in S^*$. Consequently, $(\bar{\lambda}, \bar{U}, \bar{v}) \in \mathscr{B}_I$ and $\bar{U}b + \bar{v} = \bar{U}b + L\bar{x} - \bar{U}b = L\bar{x}$. Assuming that $(\bar{\lambda}, \bar{U}, \bar{v})$ were not efficient to (*DLI*), i.e. the existence of another feasible solution $(\lambda, U, v) \in \mathscr{B}_I$ satisfying $\bar{U}b + \bar{v} \leq_K Ub + v$, it follows $L\bar{x} \leq_K Ub + v$, which contradicts Theorem 6.16. Consequently, $(\bar{\lambda}, \bar{U}, \bar{v}) \in \mathscr{E}(DLI)$ and $L\bar{x} = \bar{U}b + \bar{v}$. □

Like in the finitely dimensional case, a converse duality statement for (*DLI*) can be provided, too, but under additional hypotheses.

Theorem 6.18. *If $(\bar{\lambda}, \bar{U}, \bar{v}) \in \mathscr{E}(DLI)$, one of the regularity conditions (RCV_i^I), $i \in \{1, 2, 3, 4\}$, is fulfilled and $L(\mathscr{A}) + K$ is closed, there exists an $\bar{x} \in \mathscr{PE}_{LS}(PLI)$ such that $L\bar{x} = \bar{U}b + \bar{v}$.*

Proof. Let $\bar{d} := \bar{U}b + \bar{v}$ and suppose that $\bar{d} \notin L(\mathscr{A})$. Using Theorem 6.16 it follows easily that $\bar{d} \notin L(\mathscr{A}) + K$, too. Then Tuckey's separation theorem guarantees the existence of $\gamma \in V^* \setminus \{0\}$ and $a \in \mathbb{R}$ such that

$$\langle \gamma, \bar{d} \rangle < a < \langle \gamma, Lx + k \rangle \; \forall x \in \mathscr{A} \; \forall k \in K. \tag{6.3.16}$$

Assuming that $\gamma \notin K^*$ would yield the existence of a $k \in K$ for which $\langle \gamma, k \rangle < 0$. Taking into account that K is a cone, this implies a contradiction to (6.3.16), consequently $\gamma \in K^*$. Taking $k = 0$ in (6.3.16) it follows

$$\langle \gamma, \bar{d} \rangle < \langle \gamma, Lx \rangle \; \forall x \in \mathscr{A}. \tag{6.3.17}$$

On the other hand, one has $\langle \bar{\lambda}, \bar{d} \rangle = \langle \bar{\lambda}, \bar{U}b + \bar{v} \rangle = \langle \bar{U}^*\bar{\lambda}, b \rangle \leq \langle \bar{U}^*\bar{\lambda}, Ax \rangle$ for all $x \in \mathscr{A}$, so it holds

$$\langle \bar{\lambda}, Lx - \bar{d} \rangle \geq \langle (L - \bar{U} \circ A)^*\bar{\lambda}, x \rangle \geq 0 \; \forall x \in \mathscr{A}. \tag{6.3.18}$$

Now, taking $p := a - \langle \gamma, \bar{d} \rangle > 0$ it follows $\langle (r\bar{\lambda} + (1 - r)\gamma), \bar{d} \rangle = a - p + r(\langle \bar{\lambda}, \bar{d} \rangle - a + p)$ for all $r \in \mathbb{R}$. Note that there exists an $\bar{r} \in (0, 1)$ such that $\bar{r}(\langle \bar{\lambda}, \bar{d} \rangle - a + p) < p/2$ and $\bar{r}(\langle \bar{\lambda}, \bar{d} \rangle - a) > -p/2$, and let $\lambda := \bar{r}\bar{\lambda} + (1 - \bar{r})\gamma$. It is clear that $\lambda \in K^{*0}$. By (6.3.17) and (6.3.18) it follows $r\langle \bar{\lambda}, \bar{d} \rangle + (1 - r)a <$

$\langle r\bar{\lambda} + (1-r)\gamma, Lx\rangle$ for all $x \in \mathscr{A}$ and all $r \in (0,1)$, consequently

$$\langle \lambda, \bar{d}\rangle = \bar{r}\langle \bar{\lambda}, \bar{d}\rangle + (1-\bar{r})\langle \gamma, \bar{d}\rangle = \bar{r}\langle \bar{\lambda}, \bar{d}\rangle + (1-\bar{r})(a-p)$$
$$< \frac{p}{2} + \bar{r}(a-p) + (1-\bar{r})(a-p) = a - \frac{p}{2} < \langle \lambda, Lx\rangle \ \forall x \in \mathscr{A}.$$

Moreover, there exists a $\tilde{\lambda} \in K \setminus \{0\}$ such that $\langle \lambda, \tilde{\lambda}\rangle = 1$. Like in the proof of Theorem 6.17, the validity of any of the considered regularity conditions yields strong duality for the scalar optimization problem $\inf_{x\in\mathscr{A}}\langle \lambda, Lx\rangle$ and its Lagrange dual, i.e. there exists a $\bar{z}^* \in C^*$ with $L^*\lambda - A^*\bar{z}^* \in S^*$ for which $\inf_{x\in\mathscr{A}}\langle \lambda, Lx\rangle = \langle \bar{z}^*, b\rangle$.

Taking $U \in \mathscr{L}(Z,V)$ be defined by $Uz := \langle \bar{z}^*, z\rangle\tilde{\lambda}$, $z \in Z$. Then $U^*\lambda = \bar{z}^* \in C^*$ and $(L - U \circ A)^*\lambda \in S^*$. Consequently, the hyperplane $\mathscr{H} := \{Ub + v : v \in V, \langle \lambda, v\rangle = 0\}$, which is nothing but the set $\{w \in V : \langle \lambda, w\rangle = \langle \lambda, Ub\rangle\}$, is contained in $h_I(\mathscr{B}_I)$.

On the other hand, as $\langle \lambda, \bar{d}\rangle < \langle \bar{z}^*, b\rangle = \langle \lambda, Ub\rangle$, there exists a $\bar{k} \in K \setminus \{0\}$ such that $\langle \lambda, \bar{d} + \bar{k}\rangle = \langle \lambda, Ub\rangle$. Hence $\bar{d} + \bar{k} \in \mathscr{H} \subseteq h_I(\mathscr{B}_I)$. Noting that $\bar{d} \leq_K \bar{d} + \bar{k}$, we have just arrived to a contradiction to the maximality of $\bar{d}$ to the set $h_I(\mathscr{B}_I)$. Thus our initial supposition is false, consequently $\bar{d} \in L(\mathscr{A})$. Then there exists an $\bar{x} \in \mathscr{A}$ such that $L\bar{x} = \bar{d} = \bar{U}b + \bar{v}$. Using (6.3.18), it follows that $\bar{x} \in \mathscr{PE}_{LS}(PLI)$. □

Regarding necessary and optimality conditions for the primal-dual pair of vector optimization problems $(PLI) - (DLI)$ we make the following observation.

Remark 6.20. If $\bar{x} \in \mathscr{A}$ and $(\bar{\lambda}, \bar{U}, \bar{v}) \in \mathscr{B}$ fulfill $L\bar{x} = \bar{U}b + \bar{v}$, then the *complementarity conditions* $\langle (L - \bar{U} \circ A)^*\bar{\lambda}, \bar{x}\rangle = 0$ and $\langle \bar{U}^*\bar{\lambda}, A\bar{x} - b\rangle = 0$ are fulfilled.

Remark 6.21. In the framework of Sect. 6.2 the set $L(\mathscr{A}) + K$ is closed. Thus, a natural question is when is the closedness of this set guaranteed in more general settings. Let us investigate what happens when $X = \mathbb{R}^n$, $Z = \mathbb{R}^m$ and $V = \mathbb{R}^k$ and we have also the convex cones $S \subseteq \mathbb{R}^n$, $C \subseteq \mathbb{R}^m$ and the nontrivial pointed convex cone $K \subseteq \mathbb{R}^k$. Then $L \in \mathbb{R}^{k\times n}$, $A \in \mathbb{R}^{m\times n}$ and $b \in \mathbb{R}^m$. If S and C are closed and $A^{-1}(b + \operatorname{ri} C) \neq \emptyset$, then via [178, Theorem 6.7] it follows that $\mathscr{A}$ is closed. Moreover, in this case [178, Corollary 8.3.3] yields $0^+\mathscr{A} = 0^+(A^{-1}(b + C)) \cap 0^+(S) = \big(0^+(A^{-1}(C)) + 0^+(A^{-1}(b))\big) \cap S$, where the last equality follows via [178, Corollary 9.1.2]. Employing [178, Corollary 8.3.4] one gets $0^+(A^{-1}(C)) = A^{-1}(0^+C)$, consequently $0^+\mathscr{A} = (A^{-1}(C) + A^{-1}(0)) \cap S = A^{-1}(C) \cap S$. If $\ker L \cap A^{-1}(C) \cap S = \{0\}$, then via [178, Theorem 9.1] it follows that $L(\mathscr{A})$ is closed and $L(0^+\mathscr{A}) = 0^+L(\mathscr{A})$, consequently $0^+L(\mathscr{A}) = L(A^{-1}(C) \cap S)$. Note that $0 \in L(A^{-1}(C) \cap S)$. For $(\lambda, U, v) \in \mathscr{B}_I$, one gets via Proposition 6.7 that $(L - UA)(S) \cap (-K) = \{0\}$, which yields $(L - UA)(A^{-1}(C) \cap S) \cap (-K) = \{0\}$. Assuming moreover that K is closed (then $K^{*0} = \operatorname{int} K^*$) and $L(A^{-1}(C) \cap S) \subseteq (L - UA)(A^{-1}(C) \cap S)$, it follows that $0^+L(\mathscr{A}) \cap (-K) = \{0\}$ and, finally, [178, Corollary 9.1.2] yields that $L(\mathscr{A}) + K$ is closed. One can note that several additional hypotheses, that are automatically fulfilled or may be skipped in the framework of Sect. 6.2, were necessary in order to guarantee the desired outcome even if we worked in finitely dimensional spaces. Other sufficient conditions that ensure the closedness of $L(\mathscr{A}) + K$ can be found for instance in [221, Theorem 1.1.8] or [220, Corollary 3.12].

Combining Proposition 6.10 and Theorem 6.16, one can easily provide the weak duality statement for (*PLI*) and (DLI^H), too.

Theorem 6.19. *There exist no* $x \in \mathscr{A}$, $U \in \mathscr{B}_I^H$ *and* $v \in \mathrm{PMin}_{LS}\big((L - U \circ A)(S) + U(C), K\big)$ *such that* $Lx \leqq_K Ub + v$.

Strong duality for (DLI^H) can be proven under the same hypotheses as for (*DLI*), too.

Theorem 6.20. *If* $\bar{x} \in \mathscr{P}\mathscr{E}_{LS}(PLI)$ *and one of the regularity conditions* (RCV_i^I), $i \in \{1, 2, 3, 4\}$, *is fulfilled, there exists a* $\bar{U} \in \mathscr{E}(DLI^H)$ *such that* $L\bar{x} = \bar{U}b + \bar{v}$, *where* $\bar{v} \in \mathrm{PMin}_{LS}\big((L - \bar{U} \circ A)(S) + \bar{U}(C), K\big)$.

Proof. Like in the proof of Theorem 6.17, the proper efficiency of $\bar{x}$ to (*PLI*) delivers a $\bar{\lambda} \in K^{*0}$ and the fulfillment of any of the considered regularity conditions a $\bar{z}^* \in C^*$ such that $\langle \bar{\lambda}, L\bar{x} \rangle = \langle \bar{z}^*, b \rangle$ and $L^*\bar{\lambda} - A^*\bar{z}^* \in S^*$. As $\bar{\lambda} \in K^{*0}$, there exists a $\tilde{\lambda} \in K \setminus \{0\}$ such that $\langle \bar{\lambda}, \tilde{\lambda} \rangle = 1$. Taking $\bar{U} \in \mathscr{L}(Z, V)$ be defined by $\bar{U}z := \langle \bar{z}^*, z \rangle \tilde{\lambda}$, $z \in Z$, Proposition 6.7 yields then $\bar{U} \in \mathscr{B}_I^H$.

Taking $\bar{v} := L\bar{x} - \bar{U}b$, one gets $\bar{v} = (L - \bar{U} \circ A)\bar{x} + \bar{U}(A\bar{x} - b) \in (L - \bar{U} \circ A)(S) + \bar{U}(C)$. One has $\langle \bar{\lambda}, \bar{v} \rangle = \langle \bar{\lambda}, L\bar{x} - \bar{U}b \rangle = \langle \bar{z}^*, b \rangle - \langle \bar{\lambda}, \langle \bar{z}^*, b \rangle k \rangle = 0$ and $\langle \bar{\lambda}, (L - \bar{U} \circ A)x + \bar{U}c \rangle \geq 0$ for all $x \in S$ and $c \in C$. Consequently, $\bar{v} \in \mathrm{PMin}_{LS}\big((L - \bar{U} \circ A)(S) + \bar{U}(C), K\big)$.

Assuming that $\bar{U}$ were not efficient to (DLI^H), i.e. the existence of another feasible solution $U \in \mathscr{B}^H$ satisfying $\bar{U}b + \bar{v} \leq_K Ub + v$ for a $v \in \mathrm{PMin}_{LS}\big((L - U \circ A)(S) + U(C), K\big)$, it follows $L\bar{x} \leq_K Ub + v$, which contradicts Theorem 6.19. Consequently, $\bar{U} \in \mathscr{E}(DLI^H)$ and $L\bar{x} = \bar{U}b + \bar{v}$. □

Remark 6.22. It remains an open question whether a converse duality theorem for (DLI^H) is valid under hypotheses similar to the ones of Theorem 6.18. Moreover, one can provide duality statements for (DLI^I), too, but we skip them here since even in the finitely dimensional case of Sect. 6.2 such statements are valid for this vector dual only for $b \neq 0$, when they coincide with the ones for the corresponding Lagrange type vector dual problem.

Finally, let us compare the sets of maximal elements of the image sets of the vector duals we assigned to (*PLI*).

Theorem 6.21. *It holds*

$$\mathrm{Max}(h_I^J(\mathscr{B}_I^J), K) \subseteq \mathrm{Max}(h_I(\mathscr{B}_I), K) = \mathrm{Max}(h_I^L(\mathscr{B}_I^L), K)$$

and the inclusion becomes equality when $b \neq 0$.

Proof. Assume the existence of a $d \in \text{Max}(h_I(\mathscr{B}_I), K) \setminus \text{Max}(h_I^L(\mathscr{B}_I^L), K)$. Then there exist a $\bar{d} \in h_I^L(\mathscr{B}_I^L)$, such that $d \leq_K \bar{d}$, and $(\lambda, z^*, \bar{d}) \in \mathscr{B}_I^L$ such that $\bar{d} = h_I^L(\lambda, z^*, \bar{d})$ and $\langle \lambda, \bar{d} \rangle = \langle z^*, b \rangle$. There exists also a $\tilde{\lambda} \in K \setminus \{0\}$ such that $\langle \lambda, \tilde{\lambda} \rangle = 1$. Let $U \in \mathscr{L}(Z, V)$ be defined by $Uz := \langle z^*, z \rangle \tilde{\lambda}$, for $z \in Z$. Then $U^*\lambda = z^* \in C^*$. Moreover, $(L - U \circ A)^*\lambda = L^*\lambda - A^*z^* \in S^*$. Taking $v := \bar{d} - Ub$, one gets $\langle \lambda, v \rangle = 0$. Consequently, $(\lambda, U, v) \in \mathscr{B}_I$ and thus $\bar{d} \in h_I(\mathscr{B}_I)$. But since $d \in \text{Max}(h_I(\mathscr{B}_I), K)$ and $d \leq_K \bar{d}$ a contradiction is attained, therefore $\text{Max}(h_I(\mathscr{B}_I), K) \subseteq \text{Max}(h_I^L(\mathscr{B}_I^L), K)$.

Take now $d \in \text{Max}(h_I^L(\mathscr{B}_I^L), K)$. Then there exists $(\lambda, z^*, d) \in \mathscr{B}_I^L$ such that $\langle \lambda, d \rangle \leq \langle z^*, b \rangle$. From the maximality of d in $h_I^L(\mathscr{B}_I^L)$ it follows that one actually has $\langle \lambda, d \rangle = \langle z^*, b \rangle$. Defining U and v like above, one can directly verify that $(\lambda, U, v) \in \mathscr{B}_I$ and $d \in h_I(\mathscr{B}_I)$. By (6.3.15) it follows that $d \in \text{Max}(h_I(\mathscr{B}_I), K)$, hence $\text{Max}(h_I^L(\mathscr{B}_I^L), K) \subseteq \text{Max}(h_I(\mathscr{B}_I), K)$, too.

Therefore $\text{Max}(h_I(\mathscr{B}_I), K) = \text{Max}(h_I^L(\mathscr{B}_I^L), K)$ and the rest follows from [48, Theorem 4.5.2]. □

Remark 6.23. From Theorem 6.20 one can conclude that when one of the regularity conditions (RCV_i^I), $i \in \{1, 2, 3, 4\}$, is fulfilled one has $\text{PMin}_{LS}(L(\mathscr{A}), K) \subseteq \text{Max}(h_I^H(\mathscr{B}_I^H), K)$. It remains an open challenge to find out under which hypotheses does this inclusion turn into an equality and also to see that in general $\text{Max}(h_I^H(\mathscr{B}_I^H), K)$ actually coincides with the maximal sets of the image sets considered within Theorem 6.21. In the next statement we show that in the framework considered in Remark 6.21 under an additional hypothesis $\text{Max}(h_I^H(\mathscr{B}_I^H), K)$ is larger than its counterparts, generalizing thus [48, Proposition 5.5.7].

Theorem 6.22. *Let $X = \mathbb{R}^n$, $Z = \mathbb{R}^m$ and $V = \mathbb{R}^k$. Take $S \subseteq \mathbb{R}^n$ and $C \subseteq \mathbb{R}^m$ to be convex cones and the nontrivial pointed closed convex cone $K \subseteq \mathbb{R}^k$, as well as $L \in \mathbb{R}^{k \times n}$, $A \in \mathbb{R}^{m \times n}$ and $b \in \mathbb{R}^m$. If $(L - UA)(S) + U(C)$ is closed, it holds*

$$\text{Max}(h_I(\mathscr{B}_I), K) \subseteq \text{Max}(h_I^H(\mathscr{B}_I^H), K).$$

Proof. Assume the existence of a $d \in \text{Max}(h_I(\mathscr{B}_I), K) \setminus h_I^H(\mathscr{B}_I^H)$. Then there exists an element $(\lambda, U, v) \in \mathscr{B}_I$ such that $d = Ub + v$ and $U \in \mathscr{B}_I^H$, as well as $v \notin (L - UA)(S) + U(C)$.

Assuming that there exist $x \in S$ and $c \in C$ such that $(L - UA)x + Uc \leq_K v$ yields $0 \leq \lambda^\top((L - UA)x + Uc) < \lambda^\top v = 0$, that is a contradiction. Consequently, $((L - UA)(S) + U(C) - v) \cap (-K) = \emptyset$. Then Proposition 6.5 yields the existence of a $\bar{\lambda} \in \text{int}\, K^*$ such that $\bar{\lambda}^\top v < 0$, $(L - UA)^\top \bar{\lambda} \in S^*$ and $U^\top \bar{\lambda} \in C^*$.

Then $\lambda + \bar{\lambda} \in \text{int}\, K^*$ and $\langle \lambda + \bar{\lambda}, v \rangle < 0$, hence there exists a $\tilde{v} \in V$ such that $v \leq_K \tilde{v}$ and $\langle \lambda + \bar{\lambda}, \tilde{v} \rangle = 0$. Moreover, $(L - UA)^\top(\lambda + \bar{\lambda}) \in S^*$ and $U^\top(\lambda + \bar{\lambda}) \in C^*$, therefore $(\lambda + \bar{\lambda}, U, \tilde{v}) \in \mathscr{B}_I$. Then $d = Ub + v \leq_K Ub + \tilde{v}$, that contradicts the maximality of d in $h_I(\mathscr{B}_I)$. Hence $d \in h_I^H(\mathscr{B}_I^H)$ and there exist $\bar{x} \in S$ and $\bar{c} \in C$ such that $v = (L - UA)\bar{x} + U\bar{c} \in \text{PMin}_{LS}\big((L - U \circ A)(S) + U(C), K\big)$. If $Ub + v \notin \text{Max}(h_I^H(\mathscr{B}_I^H), K)$ and there exists an element $w \in h_I^H(\mathscr{B}_I^H)$ such that

$Ub + v \leqq_K w$, Proposition 6.10 yields $w \in h_I(\mathcal{B}_I)$ and the maximality of d in $h_I(\mathcal{B}_I)$ is contradicted again. □

Remark 6.24. Working in the framework of Theorem 6.22, one can work analogously to Remark 6.21 to identify hypotheses that guarantee the closedness of $(L - UA)(S) + U(C)$. Note also that in the setting of Sect. 6.2 the set $(L - UA)(S) + U(C)$ is closed.

Remark 6.25. From Theorems 6.17, 6.18, 6.20 and 6.21 one can conclude that when one of the regularity conditions (RCV_i^I), $i \in \{1, 2, 3, 4\}$, is fulfilled and $L(\mathcal{A}) + K$ is closed the following inclusion scheme holds

$$\begin{aligned}\operatorname{Max}(h_I^J(\mathcal{B}_I^J), K) &\subseteq \operatorname{PMin}_{LS}(L(\mathcal{A}), K) = \operatorname{Max}(h_I(\mathcal{B}_I), K)\\ &= \operatorname{Max}(h_I^L(\mathcal{B}_I^L), K) \subseteq \operatorname{Max}(h_I^H(\mathcal{B}_I^H), K),\end{aligned} \tag{6.3.19}$$

and the first inclusion becomes an equality when $b \neq 0$. Because of (6.3.15) we believe that the second inclusion in (6.3.19) is fulfilled as an equality under the hypotheses mentioned above, too, but a proof of this fact is still unknown.

6.3.3 Wolfe and Mond-Weir Type Linear Vector Duality

Other vector dual problems that can be attached to the primal linear vector optimization problem (*PLI*) can be obtained by particularizing the Wolfe and Mond-Weir type vector duals assigned to a general constrained vector optimization problem in Sects. 5.3.2 and 5.4.2. Due to the continuity of the involved functions, the mentioned vector duals to (*PLI*) of Lagrange type and Fenchel-Lagrange type coincide (see for instance Remark 5.30 for more on this), thus we shall consider here only the first ones.

In order to formulate them, we have to see what becomes in this framework the constraint involving a subdifferential that appears in all of them. For $u \in S$, $\lambda \in K^{*0}$ and $z^* \in C^*$, one has

$$\begin{aligned}\partial\big((\lambda L) + (z^*(b - A\cdot)) + \delta_S\big)(u) = \big\{&x^* \in X^* : x^* \in L^*\lambda - A^*z^* - S^*,\\ &\langle L^*\lambda - A^*z^* - x^*, u\rangle = 0\big\},\end{aligned}$$

therefore $0 \in \partial\big((\lambda L) + z^*(b - A\cdot) + \delta_S\big)(u)$ if and only if $L^*\lambda - A^*z^* \in S^*$ and $\langle L^*\lambda - A^*z^* - x^*, u\rangle = 0$.

Given these considerations, the vector dual to (*PLI*) that is a special case of (DVC_W^L) turns out to be

$$(DLI_W) \qquad \operatorname*{Max}_{(\lambda, z^*, u, r) \in \mathcal{B}_{IW}} h_{IW}(\lambda, z^*, u, r),$$

where

$$\mathscr{B}_{I_W} = \Big\{(\lambda, z^*, v, r) \in K^{*0} \times C^* \times S \times (K \setminus \{0\}) : L^*\lambda - A^*z^* \in S^*,$$
$$\langle L^*\lambda - A^*z^*, u\rangle = 0\Big\}$$

and

$$h_{I_W}(\lambda, z^*, u, r) = Lu + \frac{\langle z^*, b - Au\rangle}{\langle \lambda, r\rangle} r.$$

Analogously, the vector dual to (*PLI*) that is a special case of (DVC_M^L) turns out to be

$$(DLI_M) \qquad \underset{(\lambda, z^*, u) \in \mathscr{B}_{I_M}}{\text{Max}} h_{I_M}(\lambda, z^*, u),$$

where

$$\mathscr{B}_{I_M} = \Big\{(\lambda, z^*, u) \in K^{*0} \times C^* \times \mathscr{A} : L^*\lambda - A^*z^* \in S^*, \ \langle L^*\lambda - A^*z^*, u\rangle = 0\Big\}$$

and

$$h_{I_M}(\lambda, z^*, u) = Lu,$$

while the one arising from (DVC_{MW}^L) is

$$(DLI_{MW}) \qquad \underset{(\lambda, z^*, u) \in \mathscr{B}_{I_{MW}}}{\text{Max}} h_{I_{MW}}(\lambda, z^*, u),$$

where

$$\mathscr{B}_{I_{MW}} = \Big\{(\lambda, z^*, u) \in K^{*0} \times C^* \times S : L^*\lambda - A^*z^* \in S^*, \langle L^*\lambda - A^*z^*, u\rangle = 0,$$
$$\langle z^*, (b - Au)\rangle \geq 0\Big\}$$

and

$$h_{I_{MW}}(\lambda, z^*, u) = Lu.$$

On the other hand, the vector dual to (*PLI*) that is a special case of (DVC_L^W) turns out to be

$$(DLI^W) \qquad \underset{(\lambda, z^*, u, v) \in \mathscr{B}_{I^W}}{\text{Max}} h_{I^W}(\lambda, z^*, u, v),$$

where

$$\mathscr{B}_{IW} = \Big\{(\lambda, z^*, u, v) \in K^{*0} \times C^* \times S \times V : L^*\lambda - A^*z^* \in S^*,$$
$$\langle L^*\lambda - A^*z^*, u\rangle = 0,\ \langle \lambda, v\rangle \leq \langle z^*, b\rangle\Big\}$$

and

$$h_{IW}(\lambda, z^*, u, v) = v.$$

Analogously, the vector dual to (*PLI*) that is obtained as a special case of (DVC_L^M) is

$$(DLI^M) \qquad \operatorname*{Max}_{(\lambda,z^*,u,v)\in\mathscr{B}_{IM}} h_{IM}(\lambda, z^*, u, v),$$

where

$$\mathscr{B}_{IM} = \Big\{(\lambda, z^*, u, v) \in K^{*0} \times C^* \times \mathscr{A} \times V : L^*\lambda - A^*z^* \in S^*,$$
$$\langle L^*\lambda - A^*z^*, u\rangle = 0,\ \langle \lambda, v - Lu\rangle \leq 0\Big\}$$

and

$$h_{IM}(\lambda, z^*, u) = v,$$

while the one arising from (DVC_L^{MW}) is

$$(DLI^{MW}) \qquad \operatorname*{Max}_{(\lambda,z^*,u,v)\in\mathscr{B}_{IMW}} h_{IMW}(\lambda, z^*, u, v),$$

where

$$\mathscr{B}_{IMW} = \Big\{(\lambda, z^*, u, v) \in K^{*0} \times C^* \times S \times V : L^*\lambda - A^*z^* \in S^*,$$
$$\langle L^*\lambda - A^*z^*, u\rangle = 0,\ \langle z^*, b - Au\rangle \geq 0, \langle \lambda, v - Lu\rangle \leq 0\Big\}$$

and

$$h_{IMW}(\lambda, z^*, u, v) = v.$$

In order to provide vector dual problems to (*PLI*) that share the same image sets as the ones given above and rely on the formulation of the classical contributions to this matter, one can consider, like in the proof of Theorems 6.17 or 6.20, for $(\lambda, z^*, u) \in K^{*0} \times C^* \times S$, the linear continuous maping $U \in \mathscr{L}(Z, V)$, defined by $Uz := \langle z^*, z\rangle\tilde{\lambda}$, $z \in Z$, where $\tilde{\lambda} \in K \setminus \{0\}$ fulfilling $\langle \lambda, \tilde{\lambda}\rangle = 1$ exists because $\lambda \in K^{*0}$. Then $z^* = U^*\lambda$ and $L^*\lambda - A^*z^*$ becomes $(L - U \circ A)^*\lambda$, and the vector duals provided above can be correspondingly modified, as follows.

The vector duals to (PLI) that are special cases of (DVG_W) turn out to become

$$(DLI_W) \qquad \operatorname*{Max}_{(\lambda,U,u,r)\in\mathscr{B}_{I_W}} h_{I_W}(\lambda, U, u, r),$$

where

$$\mathscr{B}_{I_W} = \Big\{(\lambda, U, v, r) \in K^{*0} \times \mathscr{L}(Z, V) \times S \times (K \setminus \{0\}) : U^*\lambda \in C^*,$$
$$(L - U \circ A)^*\lambda \in S^*, \ \langle (L - U \circ A)^*\lambda, u\rangle = 0\Big\}$$

and

$$h_{I_W}(\lambda, U, u, r) = Lu + \frac{\langle U^*\lambda, b - Au\rangle}{\langle \lambda, r\rangle} r,$$

$$(DLI_M) \qquad \operatorname*{Max}_{(\lambda,U,u)\in\mathscr{B}_{I_M}} h_{I_M}(\lambda, U, u),$$

where

$$\mathscr{B}_{I_M} = \Big\{(\lambda, U, u) \in K^{*0} \times \mathscr{L}(Z, V) \times \mathscr{A} : U^*\lambda \in C^*,$$
$$(L - U \circ A)^*\lambda \in S^*, \ \langle (L - U \circ A)^*\lambda, u\rangle = 0\Big\}$$

and

$$h_{I_M}(\lambda, U, u) = Lu,$$

and, respectively

$$(DLI_{MW}) \qquad \operatorname*{Max}_{(\lambda,U,u)\in\mathscr{B}_{I_{MW}}} h_{I_{MW}}(\lambda, U, u),$$

where

$$\mathscr{B}_{I_{MW}} = \Big\{(\lambda, U, u) \in K^{*0} \times \mathscr{L}(Z, V) \times S : U^*\lambda \in C^*, \ (L - U \circ A)^*\lambda \in S^*,$$
$$\langle (L - U \circ A)^*\lambda, u\rangle = 0, \ \langle U^*\lambda, (b - Au)\rangle \geq 0\Big\}$$

and

$$h_{I_{MW}}(\lambda, U, u) = Lu.$$

On the other hand, the vector duals to (*PLI*) that are special cases of (DVG^W) turn into

$$(DLI^W) \qquad \underset{(\lambda,U,u,v)\in\mathscr{B}_{IW}}{\text{Max}} h_{IW}(\lambda,U,u,v),$$

where

$$\mathscr{B}_{IW} = \Big\{(\lambda,U,u,v) \in K^{*0} \times \mathscr{L}(Z,V) \times S \times V : U^*\lambda \in C^*,$$
$$(L - U \circ A)^*\lambda \in S^*,\ \langle (L - U \circ A)^*\lambda, u\rangle = 0,\ \langle \lambda, v\rangle \leq \langle U^*\lambda, b\rangle\Big\}$$

and

$$h_{IW}(\lambda,U,u,v) = v,$$

$$(DLI^M) \qquad \underset{(\lambda,U,u,v)\in\mathscr{B}_{IM}}{\text{Max}} h_{IM}(\lambda,U,u,v),$$

where

$$\mathscr{B}_{IM} = \Big\{(\lambda,U,u,v) \in K^{*0} \times \mathscr{L}(Z,V) \times \mathscr{A} \times V : U^*\lambda \in C^*,$$
$$(L - U \circ A)^*\lambda \in S^*,\ \langle (L - U \circ A)^*\lambda, u\rangle = 0,\ \langle \lambda, v - Lu\rangle \leq 0\Big\}$$

and

$$h_{IM}(\lambda,U,u) = v,$$

and, respectively,

$$(DLI^{MW}) \qquad \underset{(\lambda,U,u,v)\in\mathscr{B}_{IMW}}{\text{Max}} h_{IMW}(\lambda,U,u,v),$$

where

$$\mathscr{B}_{IMW} = \Big\{(\lambda,U,u,v) \in K^{*0} \times \mathscr{L}(Z,V) \times S \times V : U^*\lambda \in C^*,\ (L - U \circ A)^*\lambda \in S^*,$$
$$\langle (L - U \circ A)^*\lambda, u\rangle = 0,\ \langle U^*\lambda, b - Au\rangle \geq 0, \langle \lambda, v - Lu\rangle \leq 0\Big\}$$

and

$$h_{IMW}(\lambda,U,u,v) = v.$$

The weak and strong duality assertions for the primal linear vector optimization problem (*PLI*) and these vector duals to it follow from the general case.

Furthermore, the image sets of these vector duals and their maximal sets can be compared with the others we assigned to (*PLI*) within this section, taking also into consideration the general inclusions proven in Sect. 5.5.

In the finitely dimensional framework considered in Sect. 6.2, the vector duals considered in this subsection should be correspondingly modified and the duality statements hold, like for the other vector duals to (*PLF*) investigated there, without assuming the fulfillment of any regularity condition. Moreover, note that the case $b = 0$ produces no trouble to the vector dual problems considered within this subsection, too.

6.3.4 Duality with Respect to Weakly Efficient Solutions

In this subsection we deliver vector duality statements for the classical linear vector optimization problem in infinitely dimensional spaces and its vector dual problems with respect to weakly efficient solutions. To the framework considered in the rest of this section we add the hypotheses that $\text{qi}\, K \neq \emptyset$ and K is closed. The primal linear vector optimization problem is in this case

$$(PLI_w) \qquad \underset{x \in \mathscr{A}}{\text{WMin}}\, Lx,$$

where

$$\mathscr{A} = \{x \in S : Ax - b \in C\}.$$

Recall that an element $\bar{x} \in \mathscr{A}$ is said to be a *weakly efficient solution* to (PLI_w) if $L\bar{x} \in \text{WMin}(L(\mathscr{A}), K)$, i.e. there exists a $\lambda \in K^* \setminus \{0\}$ such that $\langle \lambda, L\bar{x} \rangle \leq \langle \lambda, Lx \rangle$ for all $x \in \mathscr{A}$, and the set of all the weakly efficient solutions to (PLI_w) is denoted by $\mathscr{W}\mathscr{E}(PLI_w)$.

We begin with a result that extends Proposition 6.7, establishing as we shall see a connection between the feasible sets of elements of the vector dual problems with respect to weakly efficient solutions we assign to (PLI_w). Note that in this case we have actually an equivalence like in Proposition 6.6.

Proposition 6.12. *If $U \in \mathscr{L}(Z, V)$ and the pair $(K, (L - U \circ A)(S) + U(C))$ has the property (QC), there exists a $\lambda \in K^* \setminus \{0\}$ fulfilling $U^*\lambda \in C^*$ and $(L - U \circ A)^*\lambda \in S^*$ if and only if $((L - U \circ A)(S) + U(C)) \cap (-\text{qi}\, K) = \emptyset$.*

Proof. "$\Rightarrow$" Assume to the contrary that the conclusion is false. Then there exist $x \in S$ and $c \in C$ such that $(L - U \circ A)x + Uc \in -\text{qi}\, K$. Consequently, $\langle \lambda, (L - U \circ A)x + Uc \rangle < 0$. But $\langle \lambda, (L - U \circ A)x + Uc \rangle = \langle (L - U \circ A)^*\lambda, x \rangle + \langle U^*\lambda, c \rangle$ and the hypotheses imply the nonnegativity of the both terms in the right-hand side of the last equality, so we reached the desired contradiction.

"$\Leftarrow$" The hypothesis yields $0 \notin (L - U \circ A)(S) + U(C)) + \text{qi}\, K = \text{qi}(K + (L - U \circ A)(S) + U(C))$, due to (QC). But the set $K + (L - U \circ A)(S) + U(C)$ is convex and it contains 0. One can apply then Lemma 1.2, which guarantees the existence of a $\lambda \in V^* \setminus \{0\}$ satisfying $\langle \lambda, 0 \rangle \leq \langle \lambda, v+k \rangle$ for all $v \in K + (L - U \circ A)(S) + U(C)$ and all $k \in K$. As $0 \in (L - U \circ A)(S) + U(C)$ and K is a cone, it follows that $\lambda \in K^* \setminus \{0\}$. Analogously, as $0 \in (L - U \circ A)(S) \cap U(C) \cap K$, one gets $\langle (L - U \circ A)^* \lambda, x \rangle \geq 0$ for all $x \in S$, and $\langle U^* \lambda, z \rangle \geq 0$ for all $z \in C$. Consequently, $U^* \lambda \in C^*$ and $(L - U \circ A)^* \lambda \in S^*$. □

Remark 6.26. In case K has a nonempty interior it needs not be closed for the investigations performed within this subsection and the property (QC) is automatically fulfilled for the pair $(K, (L - U \circ A)(S) + U(C))$ whenever $U \in \mathscr{L}(Z, V)$. Proposition 6.12 remains valid in that case, too, the only important modification in its proof being the usage of Eidelheit's separation statement for separating the sets $(L - U \circ A)(S) + U(C))$ and K.

The vector dual problems assigned to (PLF_w) within Sect. 6.2.4 can be extended to the infinitely dimensional case, too. The *dual abstract optimization problem* to (PLI_w) is (cf. [48, 140])

$$(DLI_w^J) \qquad \underset{(\lambda, U) \in \mathscr{B}_{I_w}^J}{\text{WMax}} \ h_{I_w}^J(\lambda, U),$$

where

$$\mathscr{B}_{I_w}^J = \left\{ (\lambda, U) \in (K^* \setminus \{0\}) \times \mathscr{L}(Z, V) : U^* \lambda \in C^*, \ (L - U \circ A)^* \lambda \in S^* \right\}$$

and

$$h_{I_w}^J(\lambda, U) = Ub,$$

while the generalization we propose for the vector dual inspired by Isermann's works is

$$(DLI_w^I) \qquad \underset{U \in \mathscr{B}_{I_w}^I}{\text{WMax}} \ h_{I_w}^I(U),$$

where

$$\mathscr{B}_{I_w}^I = \left\{ U \in \mathscr{L}(Z, V) : \big((L - U \circ A)(S) + U(C)\big) \cap (-\operatorname{qi} K) = \emptyset \right\}$$

and

$$h_{I_w}^I(U) = Ub.$$

The vulnerability these vector dual problems presented in the framework of Sect. 6.2.4 in case $b = 0$ is inherited to the more general setting treated here, too.

The *vector Lagrange type dual* to (PLI_w) is (cf. [48, 140])

$$(DLI_w^L) \qquad \underset{(\lambda,z^*,v)\in\mathscr{B}_{I_w}^L}{\mathrm{WMax}} \; h_{I_w}^L(\lambda,z^*,v),$$

where

$$\mathscr{B}_{I_w}^L = \Big\{(\lambda,z^*,v)\in(K^*\setminus\{0\})\times C^*\times V : \langle\lambda,v\rangle\le\langle z^*,b\rangle,\ L^*\lambda - A^*z^*\in S^*\Big\}$$

and

$$h_{I_w}^L(\lambda,z^*,v) = v,$$

while the vector dual with respect to weakly efficient solutions modeled after (DLI^H) is

$$(DLI_w^H) \qquad \underset{U\in\mathscr{B}_{I_w}^H}{\mathrm{WMax}} \; h_{I_w}^H(U),$$

where

$$\mathscr{B}_{I_w}^H = \Big\{U\in\mathscr{L}(Z,V) : \big((L-U\circ A)(S)+U(C)\big)\cap(-\operatorname{qi} K)=\emptyset\Big\}$$

and

$$h_{I_w}^H(U) = Ub + \mathrm{WMin}\big((L-U\circ A)(S)+U(C),K\big).$$

The generalizations to the present framework of (DLF_w) and (DLF_w^D) are

$$(DLI_w) \qquad \underset{(\lambda,U,v)\in\mathscr{B}_{I_w}}{\mathrm{WMax}} \; h_{I_w}(\lambda,U,v),$$

where

$$\mathscr{B}_{I_w} = \Big\{(\lambda,U,v)\in(K^*\setminus\{0\})\times\mathscr{L}(Z,V)\times V : \langle\lambda,v\rangle=0,$$
$$U^*\lambda\in C^*,\ (L-U\circ A)^*\lambda\in S^*\Big\}$$

and

$$h_{I_w}(\lambda,U,v) = Ub + v,$$

and, respectively,

$$(DLI_w^D) \qquad \underset{(\lambda,U,v)\in\mathscr{B}_{I_w}^D}{\mathrm{WMax}} \; h_{I_w}^D(\lambda,U,v),$$

where

$$\mathscr{B}^D_{I_w} = \Big\{(\lambda, U, v) \in (K^* \setminus \{0\}) \times \mathscr{L}(Z, V) \times V : \langle \lambda, v \rangle \leq 0,$$
$$U^*\lambda \in C^*,\ (L - U \circ A)^*\lambda \in S^*\Big\}$$

and

$$h^D_{I_w}(\lambda, U, v) = Ub + v.$$

Remark 6.27. If $(\lambda, U, v) \in \mathscr{B}_{I_w}$, one can easily note that $v \notin \operatorname{qi} K \cup (-\operatorname{qi} K)$, while when $(\lambda, U, v) \in \mathscr{B}^D_{I_w}$ it follows that $v \notin \operatorname{qi} K$.

An inclusion chain similar to the ones given for their counterparts with respect to efficient solutions in (6.3.15) holds for these vector duals to (PLI_w), too, extending thus the one given in [48, Section 5.5] for only some of them in case $K = \mathbb{R}^k_+$. Therefore, assuming that for all $U \in \mathscr{B}^I_{I_w}$ the pair $(K, (L - U \circ A)(S) + U(C))$ has the property (QC), one obtains

$$h^J_{I_w}(\mathscr{B}^J_{I_w}) = h^I_{I_w}(\mathscr{B}^I_{I_w}) \subsetneq h^H_{I_w}(\mathscr{B}^H_{I_w}) \subsetneq h_{I_w}(\mathscr{B}_{I_w}) \subsetneq h^D_{I_w}(\mathscr{B}^D_{I_w}) = h^L_{I_w}(\mathscr{B}^L_{I_w}). \tag{6.3.20}$$

For the primal-dual pair of vector optimization problems $(PLI_w) - (DLI_w)$ weak, strong and converse duality statements can be proven similarly to their counterparts from Sect. 6.3.2, employing where necessary the modifications performed in the finitely dimensional case for the same purpose.

Theorem 6.23. *There exist no $x \in \mathscr{A}$ and $(\lambda, U, v) \in \mathscr{B}_{I_w}$ such that $Lx <_K Ub + v$.*

Theorem 6.24. *If $\bar{x} \in \mathscr{W}\mathscr{E}(PLI_w)$ and one of the regularity conditions (RCV^I_i), $i \in \{1, 2, 3, 4\}$, is fulfilled, there exists $(\bar{\lambda}, \bar{U}, \bar{v}) \in \mathscr{W}\mathscr{E}(DLI_w)$ such that $L\bar{x} = \bar{U}b + \bar{v}$.*

Theorem 6.25. *When $(\bar{\lambda}, \bar{U}, \bar{v}) \in \mathscr{W}\mathscr{E}(DLI_w)$, $L(\mathscr{A}) + K$ is closed and one of the regularity conditions (RCV^I_i), $i \in \{1, 2, 3, 4\}$, is fulfilled, then $\bar{U}b + \bar{v} \in$* WMin$(L(\mathscr{A}) + K, K)$.

Remark 6.28. If $\bar{x} \in \mathscr{A}$ and $(\bar{\lambda}, \bar{U}, \bar{v}) \in \mathscr{B}_w$ fulfill $L\bar{x} = \bar{U}b + \bar{v}$, then the *complementarity conditions* $\langle (L - \bar{U} \circ A)^*\bar{\lambda}, \bar{x}\rangle = 0$ and $\langle \bar{U}^*\bar{\lambda}, A\bar{x} - b\rangle = 0$ are fulfilled.

Note that Remarks 6.19, 6.21 and 6.22 remain valid in the framework of this subsection, too.

For the primal-dual pair of vector optimization problems $(PLI_w) - (DLI^H_w)$, we deliver weak and strong duality statements, too.

Theorem 6.26. *There exist no* $x \in \mathscr{A}$, $U \in \mathscr{B}_{I_w}^H$ *and* $v \in \mathrm{WMin}\big((L - U \circ A)(S) + U(C), K\big)$ *such that* $Lx <_K Ub + v$.

Theorem 6.27. *If* $\bar{x} \in \mathscr{W}\mathscr{E}(PLI_w)$ *and one of the regularity conditions* (RCV_i^I), $i \in \{1, 2, 3, 4\}$, *is fulfilled, there exist* $\bar{U} \in \mathscr{W}\mathscr{E}(DLI_w^H)$ *and* $\bar{v} \in \mathrm{WMin}\big((L - \bar{U} \circ A)(S) + \bar{U}(C), K\big)$ *such that* $L\bar{x} = \bar{U}b + \bar{v}$.

Regarding the sets of weakly maximal elements of the vector duals we considered within this subsection, one can show the following statement by following the proof of Theorem 6.21.

Theorem 6.28. *It holds*

$$\mathrm{WMax}(h_{I_w}^J(\mathscr{B}_{I_w}^J), K) \subseteq \mathrm{WMax}(h_{I_w}(\mathscr{B}_{I_w}), K) = \mathrm{WMax}(h_{I_w}^L(\mathscr{B}_{I_w}^L), K)$$

and the inclusion becomes equality when $b \neq 0$.

Remark 6.29. From Theorem 6.27 one can conclude that when one of the regularity conditions (RCV_i^I), $i \in \{1, 2, 3, 4\}$, is fulfilled one has $\mathrm{WMin}(L(\mathscr{A}), K) \subseteq \mathrm{WMax}(h_{I_w}^H(\mathscr{B}_{I_w}^H), K)$. It remains an open challenge to find out under which hypotheses does this inclusion turn into an equality and also to compare in general $\mathrm{WMax}(h_{I_w}^H(\mathscr{B}_{I_w}^H), K)$ with the maximal sets of the image sets considered within Theorem 6.28.

Remark 6.30. From Theorems 6.24, 6.25, 6.27 and 6.28 one can conclude that when one of the regularity conditions (RCV_i^I), $i \in \{1, 2, 3, 4\}$, is fulfilled, the pair $(K, (L - U \circ A)(S) + U(C))$ has the property (QC) for all $U \in \mathscr{B}_{I_w}^I$ and $L(\mathscr{A}) + K$ is closed the following inclusion scheme holds in case $b \neq 0$

$$\mathrm{WMin}(L(\mathscr{A}), K) \subseteq \mathrm{WMax}(h_{I_w}^J(\mathscr{B}_{I_w}^J), K) = \mathrm{WMax}(h_{I_w}^I(\mathscr{B}_{I_w}^I), K)$$
$$= \mathrm{WMax}(h_{I_w}(\mathscr{B}_{I_w}), K) = \mathrm{WMax}(h_{I_w}^L(\mathscr{B}_{I_w}^L), K) \subseteq \mathrm{WMin}(L(\mathscr{A}) + K, K),$$

while if $b = 0$ one has

$$\mathrm{WMax}(h_{I_w}^J(\mathscr{B}_{I_w}^J), K) = \mathrm{WMax}(h_{I_w}^I(\mathscr{B}_{I_w}^I), K) \subseteq \mathrm{WMin}(L(\mathscr{A}), K)$$
$$\subseteq \mathrm{WMax}(h_{I_w}(\mathscr{B}_{I_w}), K) = \mathrm{WMax}(h_{I_w}^L(\mathscr{B}_{I_w}^L), K) \subseteq \mathrm{WMin}(L(\mathscr{A}) + K, K).$$

6.4 Vector Duality for Vector Semidefinite Optimization Problems

In this section we deal by means of vector duality with a vector optimization problem consisting in the vector minimization of a matrix function with respect to the cone of the symmetric positive semidefinite matrices subject to both geometric and semidefinite inequality constraints. Let the nonempty set $S \subseteq \mathbb{R}^n$ and the matrix functions $F : \mathbb{R}^n \to \mathcal{S}^k$ and $H : \mathbb{R}^n \to \mathcal{S}^m$. For $i, j \in \{1, \dots, k\}$, denote by $f_{ij} : \mathbb{R}^n \to \mathbb{R}$ the function defined as $f_{ij}(x) = (F(x))_{ij}$. Recall that the scalar product of two matrices $A, B \in \mathcal{S}^k$ is defined as $\langle A, B \rangle = \mathrm{Tr}(A^\top B)$. The primal *semidefinite vector optimization problem* we consider now is

$$(PVS) \qquad \operatorname*{Min}_{x \in \mathcal{A}} F(x),$$

where

$$\mathcal{A} = \{x \in S : H(x) \in -\mathcal{S}^m_+\},$$

where the vector minimization is done with respect to the cone $\mathcal{S}^k_+$.

Recall that an element $\bar{x} \in \mathcal{A}$ is said to be a *properly efficient solution in the sense of linear scalarization* to (PVS) if $F(\bar{x}) \in \mathrm{PMin}_{LS}(F(\mathcal{A}), \mathcal{S}^k_+)$, i.e. there exists a $\Lambda \in \hat{\mathcal{S}}^k_+$ such that $\mathrm{Tr}\left(\Lambda^\top F(\bar{x})\right) \leq \mathrm{Tr}\left(\Lambda^\top F(x)\right)$ for all $x \in \mathcal{A}$, and the set of all the properly efficient solutions to (PVS) in the sense of linear scalarization is denoted by $\mathcal{PE}_{LS}(PVS)$. An element $\bar{x} \in \mathcal{A}$ is said to be an *efficient solution* to (PVS) if $F(\bar{x}) \in \mathrm{Min}(F(\mathcal{A}), \mathcal{S}^k_+)$, i.e. there exists no $x \in \mathcal{A}$ such that $F(x) \preceq_k F(\bar{x})$, and the set of all the efficient solutions to (PVS) is denoted by $\mathcal{E}(PVS)$. A properly efficient solution $\bar{x}$ to (PVS) is also efficient to (PVS).

Remark 6.31. Similar vector optimization problems were considered, for instance, in [111, 112], with all the involved functions taken cone-convex and differentiable, without the geometric constraint $x \in S$ and by considering finitely many similar semidefinite inequality constraints. Besides delivering optimality conditions regarding the ideal efficient points to considered vector optimization problems, some investigations via duality were performed for them, too, Lagrange and Wolfe dual problems being assigned to the attached scalarized problems. Moreover, a Lagrange type vector dual was proposed in [111], but with a different construction than the ones considered within this work.

The vector dual problem we assign to (PVS) is inspired by the ones proposed in Sects. 6.2 and 6.3 for primal linear optimization problems and by the ones considered in [200, 201] for vector optimization problems whose image spaces were partially ordered by the corresponding nonnegative orthants, being

$$(DVS) \qquad \operatorname*{Max}_{(\Lambda, Q, P, V) \in \mathcal{B}_S} H_S(\Lambda, Q, P, V),$$

where

$$\mathcal{B}_S = \Bigg\{(\Lambda, Q, P, V) \in \hat{\mathcal{S}}^k_+ \times \mathcal{S}^m_+ \times (\mathbb{R}^n)^{k \times k} \times \mathbb{R}^{k \times k} : P = (p_{ij})_{i,j=1,\dots,k},$$
$$p_{ij} \in \mathrm{dom}\, f_{ij}^* \ \forall i, j \in \{1, \dots, k\} \text{ s.t. } \Lambda_{ij} \neq 0,$$
$$-\sum_{i,j=1}^{k} \Lambda_{ij} p_{ij} \in \mathrm{dom}(QH)^*_S, \mathrm{Tr}(\Lambda^\top V) = 0\Bigg\}$$

and, for $i, j = 1, \ldots, k$,

$$(H_S(\Lambda, Q, P, V))_{ij}$$

$$= V_{ij} - \begin{cases} f_{ij}^*(p_{ij}) + \frac{1}{z(\Lambda)\Lambda_{ij}}(QH)_S^*\left(-\sum_{i,j=1}^{k} \Lambda_{ij} p_{ij}\right), & \text{if } \Lambda_{ij} \neq 0, \\ 0, & \text{otherwise,} \end{cases}$$

where $z(\Lambda)$ denotes the number of nonzero entries of the matrix Λ.

Remark 6.32. Like for the other vector dual problems we considered within this work, one can replace in $\mathscr{B}_S$ the constraint equality $\text{Tr}(\Lambda^\top V) = 0$ by $\text{Tr}(\Lambda^\top V) \leq 0$, obtaining thus a vector dual problem to (*PVS*) with a larger feasible set and, consequently, image set.

Remark 6.33. If $(\Lambda, Q, P, V) \in \mathscr{B}_S$, one can easily note that $V \notin (\mathscr{S}_+^k \cup (-\mathscr{S}_+^k)) \setminus \{0\}$.

Now let us formulate the weak duality statement for (*PVS*) and (*DVS*).

Theorem 6.29. *There exist no $x \in \mathscr{A}$ and $(\Lambda, Q, P, V) \in \mathscr{B}_S$ such that $F(x) \preceq_k H_S(\Lambda, Q, P, V)$.*

Proof. Assume the existence of $x \in \mathscr{A}$ and $(\Lambda, Q, P, V) \in \mathscr{B}_S$ such that $F(x) \preceq_k H_S(\Lambda, Q, P, V)$. Then $0 > \text{Tr}\left(\Lambda^\top(F(x) - H_S(\Lambda, Q, P, V))\right) = \sum_{i,j=1}^k \Lambda_{ij}(f_{ij}(x) + f_{ij}^*(p_{ij})) + (QH)_S^*\left(-\sum_{i,j=1}^k \Lambda_{ij} p_{ij}\right) \geq \left(\sum_{i,j=1}^k \Lambda_{ij} p_{ij}^\top x - \text{Tr}(Q^\top H(x)) - \delta_S(x) - \left(\sum_{i,j=1}^k \Lambda_{ij} p_{ij}\right)^\top x \geq 0\right.$ because $x \in \mathscr{A}$. As this cannot happen, the assumption we made is false. □

In order to prove strong duality for (*DVS*) one needs additional hypotheses. The regularity conditions (RCV_i^G), $i \in \{1, 2, 3, 4\}$ become in this case

$$(RCV_1^S) \big| \; \exists x' \in S \text{ such that } H(x') \in -\hat{\mathscr{S}}_+^m,$$

$$(RCV_2^S) \big| \; 0 \in \text{ri}(H(S) - C),$$

which is obtained as a special case of both (RCV_2^G) and (RCV_3^G) due to the fact that we work here in finitely dimensional spaces, and, respectively,

$$(RCV_4^S) \Big| \; S \text{ is closed and for any } \Lambda \in \hat{\mathscr{S}}_+^k \text{ epi}(\Lambda F)^* + \bigcup_{Q \in \mathscr{S}_+^m} \text{epi}(QH)_S^* \text{ is closed.}$$

Theorem 6.30. *If S is a convex set, f_{ij}, $i, j = 1, \ldots, k$, are convex functions, H is $\mathscr{S}_+^m$-convex, $\bar{x} \in \mathscr{PE}_{LS}(PVS)$ and one of the regularity conditions (RCV_i^S), $i \in \{1, 2, 4\}$, is fulfilled, there exists $(\bar{\Lambda}, \bar{Q}, \bar{P}, \bar{V}) \in \mathscr{E}(DVS)$ such that $F(\bar{x}) = H_S(\bar{\Lambda}, \bar{Q}, \bar{P}, \bar{V})$.*

Proof. Since $\bar{x}$ is properly efficient to *(PVS)*, there exists a $\bar{\Lambda} \in \hat{\mathscr{S}}^k_+$ such that $\mathrm{Tr}\left(\Lambda^\top F(\bar{x})\right) \leq \mathrm{Tr}\left(\Lambda^\top F(x)\right)$ for all $x \in \mathscr{A}$. The fulfillment of any of the considered regularity conditions yields strong duality for the scalarized optimization problem attached to *(PVS)*

$$\inf_{x \in \mathscr{A}} \mathrm{Tr}\left(\Lambda^\top F(x)\right)$$

and its Fenchel-Lagrange dual

$$\sup_{\substack{Q \in \mathscr{S}^m_+, \\ T \in \mathbb{R}^n}} \Big\{ -(\Lambda F)^*(T) - (QH)^*_S(-T) \Big\},$$

thus the latter has the optimal solutions $\bar{Q}$ and $\bar{T}$ and

$$\mathrm{Tr}\left(\Lambda^\top F(\bar{x})\right) = -(\bar{\Lambda} F)^*(\bar{T}) - (\bar{Q} H)^*_S(-\bar{T}).$$

Because f_{ij}, $i, j = 1, \ldots, k$, are convex functions defined on $\mathbb{R}^n$ with full domain they are continuous, too, consequently there exist $\tilde{p}_{ij} \in \mathbb{R}^n$, $i, j = 1, \ldots, k$, such that $\tilde{p}_{ij} = 0$ if $\bar{\Lambda}_{ij} = 0$,

$$(\bar{\Lambda} F)^*(\bar{T}) = \sum_{\substack{i,j=1, \\ \bar{\Lambda}_{ij} \neq 0}}^k (\bar{\Lambda}_{ij} f_{ij})^*(\tilde{p}_{ij}) = \sum_{\substack{i,j=1, \\ \bar{\Lambda}_{ij} \neq 0}}^k \bar{\Lambda}_{ij} f^*_{ij}\left(\frac{\tilde{p}_{ij}}{\bar{\Lambda}_{ij}}\right)$$

and $\sum_{i,j=1}^k \tilde{p}_{ij} = \bar{T}$. For $i, j \in \{1, \ldots, k\}$ take $\bar{p}_{ij} = \tilde{p}_{ij} / \bar{\Lambda}_{ij}$ and

$$\bar{V}_{ij} = f_{ij}(\bar{x}) + f^*_{ij}(\bar{p}_{ij}) + \frac{1}{z(\bar{\Lambda}) \bar{\Lambda}_{ij}} (\bar{Q} H)^*_S \left(-\sum_{i,j=1}^k \tilde{p}_{ij} \right)$$

if $\bar{\Lambda}_{ij} \neq 0$ and $\bar{p}_{ij} = \tilde{p}_{ij}$ and $\bar{V}_{ij} = f_{ij}(\bar{x})$ otherwise. Then

$$\begin{aligned} &\mathrm{Tr}(\bar{\Lambda}^\top \bar{V}) \\ &= \sum_{\substack{i,j=1, \\ \bar{\Lambda}_{ij} \neq 0}}^k \bar{\Lambda}_{ij} \left(f_{ij}(\bar{x}) + f^*_{ij}(\bar{p}_{ij}) + \left(\frac{1}{z(\bar{\Lambda}) \bar{\Lambda}_{ij}} \right) (\bar{Q} H)^*_S \left(-\sum_{i,j=1}^k \bar{\Lambda}_{ij} \bar{p}_{ij} \right) \right) = 0. \end{aligned}$$

Consequently, after denoting $\bar{P} = (\bar{p}_{ij})_{i,j=1,\ldots,k}$, one notices that $(\bar{\Lambda}, \bar{Q}, \bar{P}, \bar{V}) \in \mathscr{B}_S$. Assuming that $(\bar{\Lambda}, \bar{Q}, \bar{P}, \bar{V}) \notin \mathscr{E}(DVS)$, employing Theorem 6.29 yields a contradiction, therefore $(\bar{\Lambda}, \bar{Q}, \bar{P}, \bar{V}) \in \mathscr{E}(DVS)$. □

The corresponding statement giving necessary and sufficient optimality conditions for the primal-dual pair of problems $(PVS) - (DVS)$ follows.

Theorem 6.31. *(a) If S is a convex set, f_{ij}, $i, j = 1, \dots, k$, are convex functions, H is $\mathscr{S}_+^m$-convex, $\bar{x} \in \mathscr{PE}_{LS}(PVS)$ and one of the regularity conditions (RCV_i^S), $i \in \{1,2,4\}$, is fulfilled, there exists $(\bar{\Lambda}, \bar{Q}, \bar{P}, \bar{V}) \in \mathscr{E}(DVS)$ such that*

(i) $F(\bar{x}) = H_S(\bar{\Lambda}, \bar{Q}, \bar{P}, \bar{V})$;
(ii) $f_{ij}(\bar{x}) + f_{ij}^*(\bar{p}_{ij}) = \bar{p}_{ij}^\top \bar{x}$ *when* $\bar{\Lambda}_{ij} \neq 0$;
(iii) $(\bar{Q}H)_S^*\left(-\sum\limits_{i,j=1}^{k} \bar{\Lambda}_{ij}\bar{p}_{ij}\right) = -\left(\sum\limits_{i,j=1}^{k} \bar{\Lambda}_{ij}\bar{p}_{ij}\right)^\top \bar{x}$;
(iv) $\mathrm{Tr}(\bar{Q}^\top H(\bar{x})) = 0$;
(v) $\mathrm{Tr}(\bar{\Lambda}^\top \bar{V}) = 0$;

(b) Assume that $\bar{x} \in \mathscr{A}$ and $(\bar{\Lambda}, \bar{Q}, \bar{P}, \bar{V}) \in \hat{\mathscr{S}}_+^k \times \mathscr{S}_+^m \times (\mathbb{R}^n)^{k\times k} \times \mathbb{R}^{k\times k}$ fulfill the relations $(i) - (v)$, where $\bar{P} = (\bar{p}_{ij})_{i,j=1,\dots,k}$. Then $\bar{x} \in \mathscr{PE}_{LS}(PVS)$ and $(\bar{\Lambda}, \bar{Q}, \bar{P}, \bar{V}) \in \mathscr{E}(DVS)$.

Proof. (a) The existence of a $(\bar{\Lambda}, \bar{Q}, \bar{P}, \bar{V}) \in \mathscr{E}(DVS)$, where $\bar{P} = (\bar{p}_{ij})_{i,j=1,\dots,k}$, such that $F(\bar{x}) = H_S(\bar{\Lambda}, \bar{Q}, \bar{P}, \bar{V})$ is guaranteed by Theorem 6.30. The relations (i) and (v) are thus satisfied. Moreover,

$$\begin{aligned}&\mathrm{Tr}(\bar{\Lambda}^\top \bar{V})\\ &= \sum_{\substack{i,j=1,\\ \bar{\Lambda}_{ij}\neq 0}}^{k} \bar{\Lambda}_{ij}\left(f_{ij}(\bar{x}) + f_{ij}^*(\bar{p}_{ij}) + \left(\frac{1}{z(\bar{\Lambda})\bar{\Lambda}_{ij}}\right)(\bar{Q}H)_S^*\left(-\sum_{i,j=1}^{k}\bar{\Lambda}_{ij}\bar{p}_{ij}\right)\right),\end{aligned}$$

and this is actually equal to 0. On the other hand, the Young-Fenchel inequality yields $f_{ij}(\bar{x}) + f_{ij}^*(\bar{p}_{ij}) \geq \bar{p}_{ij}^\top \bar{x}$ and

$$(\bar{Q}H)_S^*\left(-\sum_{i,j=1}^{k}\bar{\Lambda}_{ij}\bar{p}_{ij}\right) + \mathrm{Tr}\left(\bar{Q}^\top H(\bar{x})\right) \geq \left(-\sum_{i,j=1}^{k}\bar{\Lambda}_{ij}\bar{p}_{ij}\right)^\top \bar{x},$$

which, taking into consideration the equality from above, imply $\mathrm{Tr}(\bar{Q}^\top H(\bar{x})) \geq 0$. But $\mathrm{Tr}(\bar{Q}^\top H(\bar{x})) \leq 0$ because $Q \in \mathscr{S}_+^m$ and $H(\bar{x}) \in -\mathscr{S}_+^m$, consequently both Young-Fenchel inequalities are fulfilled as equalities and $\mathrm{Tr}(\bar{Q}^\top H(\bar{x})) = 0$, hence relations $(ii) - (iv)$ are fulfilled, too.

(b) From (ii) it follows that $\bar{p}_{ij} \in \mathrm{dom}\, f_{ij}^*$ for all $i, j \in \{1, \dots, k\}$ such that $\bar{\Lambda}_{ij} \neq 0$, while (iii) yields $-\sum_{i,j=1}^{k}\bar{\Lambda}_{ij}\bar{p}_{ij} \in \mathrm{dom}\,\mathrm{dom}(QH)_S^*$. Because of (v), it follows that $(\bar{\Lambda}, \bar{Q}, \bar{P}, \bar{V}) \in \mathscr{B}_S$.

Multiplying (ii) for f_{ij} with $\bar{\Lambda}_{ij}$ and summing up these relations and also $(iii) - (iv)$ one obtains

$$\sum_{\substack{i,j=1,\\ \bar{\Lambda}_{ij}\neq 0}}^{k} \bar{\Lambda}_{ij}\left(f_{ij}(\bar{x}) + f_{ij}^{*}(\bar{p}_{ij}) + \left(\frac{1}{z(\bar{\Lambda})\bar{\Lambda}_{ij}}\right)(\bar{Q}H)_{S}^{*}\left(-\sum_{i,j=1}^{k} \bar{\Lambda}_{ij}\bar{p}_{ij}\right)\right) = 0,$$

that yields because of the strong duality for the scalarized optimization problem attached to (PVS)

$$\inf_{x\in\mathcal{A}} \operatorname{Tr}\left(\Lambda^{\top} F(x)\right)$$

and its Fenchel-Lagrange dual that $\bar{x} \in \mathcal{PE}_{LS}(PVS)$. The efficiency of $(\bar{\Lambda}, \bar{Q}, \bar{P}, \bar{V})$ to (DVS) follows immediately by (i) and Theorem 6.29. □

Remark 6.34. A possible way to generalize the investigations made within this subsection may be by means of the *K-semidefinite cone* introduced in [85], that has as special case when $K = \mathbb{R}^n$ the corresponding cone of positive semidefinite matrices.

Chapter 7
Monotone Operators Approached via Convex Analysis

7.1 Historical Overview and Motivation

The monotone operators started being intensively investigated during the 1960's by authors like Browder, Brézis or Minty, and it did not take much time until their connections with convex analysis were noticed by Rockafellar, Gossez and others. The fact that the (convex) subdifferential of a proper, convex and lower semicontinuous function is a maximally monotone operator was one of the reasons for connecting these at a first sight maybe unrelated research fields. One of the most important challenges of the next decades was to identify a function that could be associated to a monotone operator in order to help investigating it by means of convex analysis, in addition to the previously used methods belonging to fixed point theory and equilibrium problems. Such functions were proposed by Coodey, Simons or Krauss, but the real breakthrough was brought by Fitzpatrick's function, introduced in [86], neglected for more than a decade and independently rediscovered in the early 2000's by Martínez-Legaz and Théra, and Burachik and Svaiter, respectively. Shortly afterwards, the Fitzpatrick family of representative functions was introduced, offering new tools for approaching the monotone operators via convex analysis. Since then, the number of papers where different aspects of monotone operators were investigated, especially by means of convex analysis, has increased in a spectacular manner, due to authors like Bauschke, Borwein, Boţ, Marques Alves, Martínez-Legaz, Penot, Simons, Svaiter, Voisei, Yao, Zălinescu and some of the already mentioned ones, besides the new results many older statements being rediscovered or improved in this way.

Perhaps the most famous problem regarding monotone operators concerns the maximality of the sum of two maximally monotone operators. Different hypotheses that guarantee the mentioned outcome were successfully proposed for the case the space on which the mentioned monotone operators are defined on is reflexive, but it is still unknown whether they work or not if the space is a general Banach one. Other interesting problems involving monotone operators regard their surjectivity

S.-M. Grad, *Vector Optimization and Monotone Operators via Convex Duality*, Vector Optimization, DOI 10.1007/978-3-319-08900-3_7

properties, the properties of their domains and ranges, the relations between different classes of them, their extensions etc. The investigations on monotone operators have led to advances back in convex analysis, too, let us mention here only the notions of Fenchel totally unstable functions (cf. [21, 190]) or sets that are closed regarding others (cf. [42, 45]). Moreover, the algorithms for finding zeros of (combinations of) monotone operators were successfully employed for solving convex optimization problems, too.

The sum of the ranges of two monotone operators defined on Banach spaces is usually larger than the range of their sum. Under some additional conditions these sets are almost equal, i.e. their interiors and closures coincide. Brézis and Haraux brought the first contributions in this directions in [60] and since then determining when the sum of the ranges of two monotone operators is almost equal in the sense mentioned above to the range of their sum is known as the *Brézis-Haraux approximation* problem, being treated in works like [9, 70, 72, 73, 171, 176, 190]. There is a rich literature on the applications of the Brézis-Haraux approximation, let us mention here only the ones for variational inequality problems, Hammerstein equations and Neumann problem (cf. [60]), complementarity problems (cf. [70]), generalized equations of maximally monotone type (cf. [171]) and Bregman and projection algorithms. Our contributions to this topic, summarized in Sect. 7.3 and originally published in [35, 40, 42, 44], concern Brézis-Haraux type approximation statements for the sum of a monotone operator with the composition with a linear mapping of another one, where the involved spaces are general Banach ones. When particularizing the involved operators to subdifferentials of proper, convex and lower semicontinuous functions, some statements from [70, 176] are corrected and extended, respectively.

Problems arising from fields like inverse problems, Fenchel-Rockafellar and Singer-Toland duality schemes, Clarke-Ekeland least action principle (cf. [5]), variational inequalities (cf. [19]), Schrödinger equations and others (cf. [4]) can be modelled to lead to the surjectivity or the identification of zeros of a combination of monotone operators. These, together with the known surjectivity properties of a monotone operator, let us mention just the classical ones due to Minty and Rockafellar (see, for instance, [190]), respectively, motivated the investigations regarding the ranges of combinations of monotone operators whose outcomes were published in recent works such as [162, 163, 177, 190, 222]. In Sect. 7.4 we present, following our paper [30], weak closedness type conditions involving representative functions that equivalently characterize or guarantee the surjectivity of a sum of a maximally monotone operator with a translation of another one. Particularizing then these results for the zeros of the mentioned sum and for the case when the involved monotone operators are subdifferentials, we improved several recent statements from the literature.

Similarities and connections between monotone operators and bifunctions were noticed in the seminal paper [12], followed by works like [116, 135, 160], where the latter were investigated mostly by means of equilibrium problems and different maximality or boundedness results for them were provided. On the other hand, we proposed in [33] a way to deal with the maximal monotonicity of the bifunctions by means of representative functions and this path was followed in very recent papers like [2, 136]. In Sect. 7.5 we attach to a bifunction two functions which are then used for approaching the maximal monotonicity of the bifunction by means of convex analysis. We succeeded to extend in this way to general Banach spaces some results known in the literature only for reflexive ones. Moreover, we provided positive answers to some recently posed conjectures from [135, 136].

7.2 Preliminaries on Monotone Operators

Before proceeding with our investigations on monotone operators, we present some notions and preliminary results used later in the exposition, following [19,21,65,86, 104, 161, 172, 173, 190, 221] and some of the references therein.

7.2.1 Monotone Operators

Within this chapter, unless otherwise mentioned, the involved spaces will be considered to be Banach spaces, equipped with norms usually denoted by $\|\cdot\|$, while the norm on its dual space is denoted by $\|\cdot\|_*$. Let X and Y be nontrivial real Banach spaces. We present first the definition of a monotone operator, followed by ones of different properties the latter can have.

Definition 7.1. A multifunction $T : X \rightrightarrows X^*$ is called a *monotone operator* provided that for any $x, y \in X$ one has $\langle y^* - x^*, y - x\rangle \geq 0$ whenever $x^* \in T(x)$ and $y^* \in T(y)$.

Having a monotone operator $T : X \rightrightarrows X^*$, its *domain* is the set $D(T) = \{x \in X : T(x) \neq \emptyset\}$, its *range* is $R(T) = \cup\{T(x) : x \in X\}$, while its *graph* is $G(T) = \{(x, x^*) : x \in X, x^* \in T(x)\}$. One can also consider the monotone operator $-T : X \rightrightarrows X^*$ whose graph is $G(-T) = \{(x, x^*) \in X \times X^* : (x, -x^*) \in G(T)\}$.

Definition 7.2. The monotone operator $T : X \rightrightarrows X^*$ is called *maximal* when its graph is not properly included in the graph of any other monotone operator $T' : X \rightrightarrows X^*$.

The next class of monotone operators was introduced in [104] and afterwards it was shown that it coincides in the maximality case with some other ones considered in various circumstances in the literature.

Definition 7.3. A monotone operator $T : X \rightrightarrows X^*$ is called of *type* (D) provided that each element of its *monotone closure* operator $\overline{T} : X^{**} \rightrightarrows X^*$,

$$G(\overline{T}) = \big\{(x^{**}, x^*) \in X^{**} \times X^* : \langle x^{**} - y, x^* - y^*\rangle \geq 0\ \forall (y, y^*) \in G(T)\big\}$$

is the limit in the weak*× strong topology of $X^{**} \times X^*$ of a bounded net $\{(x_i, x_i^*)_i\} \subseteq G(T)$.

Remark 7.1. The monotone closure is not the only closure of a monotone operator considered in the literature. Another one can be found, for instance, in [62].

Remark 7.2. In reflexive Banach spaces every maximally monotone operator is of type (D) and coincides with its closure operator. On the other hand, not every monotone operator of type (D) is maximal, as the example presented in [104, Remarques 2, p.376] shows. Note also that according to [173], $\operatorname{cl} R(T) = \operatorname{cl} R(\overline{T})$ for any monotone operator $T : X \rightrightarrows X^*$.

Another class of monotone operators we consider within this work is the following one, originally introduced in [60], but mentioned in the literature under different names like *star-monotone operators* (see [171]), 3^**-monotone operators* (cf. [70, 176, 217]) and (BH)*-operators* (in [72, 73]).

Definition 7.4. A monotone operator $T : X \rightrightarrows X^*$ is said to be *rectangular* if for all $x^* \in R(T)$ and $x \in D(T)$ there is some $\beta(x^*, x) \in \mathbb{R}$ such that $\inf_{(y,y^*)\in G(T)} \langle x^* - y^*, x - y\rangle \geq \beta(x^*, x)$.

Example 7.1. The subdifferential of a proper, convex and lower semicontinuous function defined on X is a classical example for all these classes of monotone operators. In [104, Théoréme 3.1] it was proven that it is a monotone operator of type (D), according to [217] (see also [190]) it is rectangular, while its maximal monotonicity was proven for the first time in [179]. However, one can find in the literature (see, for instance, [9, 10, 104, 190]) also examples of monotone operators belonging to the mentioned classes that are not subdifferentials. Moreover, in [10, Example 5.4] one can find a maximally monotone operator that is not rectangular, while in [10, Example 3.3] a rectangular monotone operator that is not maximal is mentioned.

Remark 7.3. One of the most important maximally monotone operators is the *duality map*

$$\mathcal{J} : X \rightrightarrows X^*,$$

$$\mathcal{J}(x) = \partial\Big(\frac{1}{2}\|\cdot\|^2\Big)(x) = \Big\{x^* \in X^* : \|x\|^2 = \|x^*\|_*^2 = \langle x^*, x\rangle\Big\}, x \in X,$$

that can be used, for instance, as noted below, for formulating a maximality criterium for a monotone operator.

The following statements from [19] and [176], respectively, will be used later in our investigations.

Lemma 7.1. *When X is a reflexive Banach space, a monotone operator $T : X \rightrightarrows X^*$ is maximal if and only if the mapping $T(x + \cdot) + \mathcal{J}(\cdot)$ is surjective for all $x \in X$.*

Lemma 7.2. *Given the monotone operator of type (D) $T : X \rightrightarrows X^*$ and the nonempty subset $E \subseteq X^*$ such that for any $x^* \in E$ there is some $x \in X$ fulfilling* $\inf_{(y,y^*)\in G(T)}\langle x^* - y^*, x - y\rangle > -\infty$, *one has* $E \subseteq \text{cl}(R(T))$ *and* $\text{int}(E) \subseteq R(\overline{T})$.

7.2.2 Representative Functions

In order to deal with monotone operators by means of convex analysis, different functions were attached to them in the literature. The one that has facilitated the most important progresses in this direction is the one introduced by Fitzpatrick in [86].

Definition 7.5. The *Fitzpatrick function* attached to the monotone operator $T : X \rightrightarrows X^*$ is

$$\varphi_T : X \times X^* \to \overline{\mathbb{R}},\ \varphi_T(x, x^*) = \sup\big\{\langle y^*, x\rangle + \langle x^*, y\rangle - \langle y^*, y\rangle : y^* \in Ty\big\}.$$

The Fitzpatrick function attached to any monotone operator is convex and weak-weak* lower semicontinuous. Moreover, using it one can show that a monotone operator $T : X \rightrightarrows X^*$ is rectangular if and only if $D(T) \times R(T) \subseteq \text{dom}\,\varphi_T$. Note also that in [9] one can find interesting connections between rectangular monotone operators and almost convex sets (that are called there nearly convex). The function $\psi_T := \overline{\text{co}}(c + \delta_{G(T)})$, where the closure is considered in the strong topology, is very well connected to the Fitzpatrick function. On $X \times X^*$ we have $\psi_T^{*\top} = \varphi_T$ and, when X is a reflexive Banach space, one also has $\varphi_T^{*\top} = \psi_T$. If $T : X \rightrightarrows X^*$ is maximally monotone, then $\varphi_T \geq c$ and $G(T) = \{(x, x^*) \in X \times X^* : \varphi_T(x, x^*) = \langle x^*, x\rangle\}$. These properties of the Fitzpatrick function motivate attaching to monotone operators other functions, as follows.

Definition 7.6. Given the monotone operator $T : X \rightrightarrows X^*$, a convex and strong-strong lower semicontinuous function $h_T : X \times X^* \to \overline{\mathbb{R}}$ fulfilling $h_T \geq c$ and $G(T) \subseteq \{(x, x^*) \in X \times X^* : h_T(x, x^*) = c(x, x^*)\}$ is said to be a *representative function* of T. The set $\mathscr{F}_T$ of all the representative functions of the monotone operator T is said to be the *Fitzpatrick family* of T.

Note that if $G(T) \neq \emptyset$ (in particular if T is maximally monotone), then every representative function of T is proper. It follows immediately that $\varphi_T, \psi_T \in \mathscr{F}_T$. If $f : X \to \overline{\mathbb{R}}$ is a proper, convex and lower semicontinuous function, then the function $(x, x^*) \mapsto f(x) + f^*(x^*)$ is a representative function of the maximally monotone operator $\partial f : X \rightrightarrows X^*$ and we call it the *Fenchel representative function* (cf. [30]). If f is moreover sublinear, the only representative function associated to ∂f is the Fenchel one, which coincides in this case with the Fitzpatrick function of ∂f. Some properties of maximally monotone operators and representative functions attached to them that we need further follow (cf. [65]).

Lemma 7.3. *Let $T : X \rightrightarrows X^*$ be a maximally monotone operator and $h_T \in \mathscr{F}_T$. Then*

(i) $\varphi_T(x,x^*) \leq h_T(x,x^*) \leq \psi_T(x,x^*)$ *for all* $(x,x^*) \in X \times X^*$*;*
(ii) The restriction of $h_T^{*\top}$ *to* $X \times X^*$ *is also a representative function of* T*;*
(iii) $\{(x,x^*) \in X \times X^* : h_T(x,x^*) = c(x,x^*)\} = \{(x,x^*) \in X \times X^* : h_T^{*\top}(x,x^*) = c(x,x^*)\} = G(T)$.

Given the monotone operator $T : X \rightrightarrows X^*$ with $G(T) \neq \emptyset$ and $h_T \in \mathscr{F}_T$, denote by $\hat{h}_T : X \times X^* \to \overline{\mathbb{R}}$ the function defined as $\hat{h}_T(x,x^*) = h_T(x,-x^*)$, $x \in X$, $x^* \in X^*$. Note that $\hat{f}_T$ is proper, convex and strong-strong lower semicontinuous, too and $\hat{h}_T(x,x^*) \geq -\langle x^*,x\rangle$ and $\hat{h}_T^*(x^*,x) = h_T^*(x^*,-x)$ for all $x \in X$ and all $x^* \in X^*$.

Let us now give two maximality criteria for monotone operators involving convex functions, the first one, following [65, Theorem 3.1] and [172, Proposition 2.1], in reflexive spaces, the other one originally given in [161, Theorem 3.1] with the hypothesis $0 \in \operatorname{sqri}\big(\operatorname{Pr}_X(\operatorname{dom} h)\big)$ and generalized by translation arguments as given below in [33].

Lemma 7.4. *Let X be reflexive. If $h : X \times X^* \to \overline{\mathbb{R}}$ is a proper, convex and lower semicontinuous function with $h \geq c$, then the monotone operator $\{(x,x^*) \in X \times X^* : h(x,x^*) = c(x,x^*)\}$ is maximal if and only if $h^{*\top} \geq c$.*

Lemma 7.5. *Let $h : X \times X^* \to \overline{\mathbb{R}}$ be a proper and convex function with $h \geq c$ and $h^{*\top} \geq c$ on $X \times X^*$. If* $\operatorname{sqri}\operatorname{Pr}_X(\operatorname{dom} h) \neq \emptyset$*, then the operator $\{(x,x^*) \in X \times X^* : h^*(x^*,x) = c(x,x^*)\}$ is maximally monotone.*

7.3 Brézis-Haraux Type Approximations

We give in this section some results concerning the so-called *Brézis-Haraux type approximation* of the range of the sum of a monotone operator with a monotone operator composed with a linear continuous mapping, following our papers [35, 40, 42, 44]. These results are then particularized by taking for the monotone operators the subdifferentials of some proper, convex and lower semicontinuous functions.

7.3.1 Brézis-Haraux Type Approximations for Sums of Rectangular Monotone Operators

Consider two monotone operators $S : X \rightrightarrows X^*$ and $T : Y \rightrightarrows Y^*$ and a linear continuous mapping $A : X \to Y$. It is known that $S + A^* \circ T \circ A$ is a monotone operator and under certain conditions it is maximally monotone (see [42, 44, 171, 172], for instance). The construction $S + A^* \circ T \circ A$ encompasses at least two important special cases. Taking S to be the *zero operator* defined as $S(x) = \{0\}$ for all $x \in X$, the results we give provide their counterparts for the composition of a monotone operator with a linear continuous mapping, while when $X = Y$ and A is the identity mapping of X one obtains corresponding results regarding the sum of two monotone operators. We show first that $S + A^* \circ T \circ A$ is rectangular when S and T are rectangular monotone operators.

Theorem 7.1. *If the monotone operators S and T are rectangular, then $S + A^* \circ T \circ A$ is rectangular, too.*

Proof. If $D(S + A^* \circ T \circ A) = \emptyset$, the conclusion arises trivially. Otherwise take $w^* \in R(S + A^* \circ T \circ A)$, i.e. there are some $w \in X$ and $x^*, z^* \in X^*$ such that $x^* \in S(w)$, $z^* \in A^* \circ T \circ A(w)$ and $w^* = x^* + z^*$. Let $x \in D(S + A^* \circ T \circ A)$. We have

$$\inf_{(y,y^*)\in G(S+A^*\circ T\circ A)} \langle w^* - y^*, x - y\rangle = \inf_{\substack{(y,u^*)\in G(S),\\ (y,v^*)\in G(A^*\circ T\circ A),\\ u^*+v^*=y^*}} \langle x^* + z^* - (u^* + v^*), x - y\rangle$$

$$\geq \inf_{(y,u^*)\in G(S)} \langle x^* - u^*, x - y\rangle + \inf_{(y,v^*)\in G(A^*\circ T\circ A)} \langle z^* - v^*, x - y\rangle. \tag{7.3.1}$$

As $z^* \in A^* \circ T \circ A(w)$, there is some $r^* \in T \circ A(w)$ such that $z^* = A^* r^*$. Clearly, $r^* \in R(T)$. Denote $u = Ax \in D(T)$. When $v^* \in A^* \circ T \circ A(y)$ there is some $s^* \in T \circ A(y)$ such that $v^* = A^* s^*$. We have

$$\inf_{(y,v^*)\in G(A^*\circ T\circ A)} \langle z^* - v^*, x - y\rangle = \inf_{(y,s^*)\in G(T\circ A)} \langle A^* r^* - A^* s^*, x - y\rangle$$

$$= \inf_{(y,s^*)\in G(T\circ A)} \langle r^* - s^*, A(x - y)\rangle \geq \inf_{(v,s^*)\in G(T)} \langle r^* - s^*, u - v\rangle \geq \beta(r^*, u) \in \mathbb{R},$$

since T is rectangular. As S is also rectangular, (7.3.1) yields that $S + A^* \circ T \circ A$ is rectangular, too. □

Remark 7.4. Taking $X = Y$ and A to be the identity mapping of X, one rediscovers as a special case of Theorem 7.1 the result given in [9, Lemma 11], i.e. that the sum of two rectangular monotone operators is rectangular, too.

The next statement provides a Brézis-Haraux type approximation of the range of $S + A^* \circ T \circ A$ through the ranges of the monotone operators S and T.

Theorem 7.2. *If the monotone operators S and T are rectangular and $S + A^* \circ T \circ A$ is of type (D), one has*

(i) $\operatorname{cl} R(S + A^* \circ T \circ A) = \operatorname{cl}(R(S) + A^*(R(T))) = \operatorname{cl} R(\overline{S + A^* \circ T \circ A})$;
(ii) $\operatorname{int} R(S + A^* \circ T \circ A) \subseteq \operatorname{int}(R(S) + A^*(R(T))) \subseteq \operatorname{int} R(\overline{S + A^* \circ T \circ A})$.

Proof. As the monotone operator $S + A^* \circ T \circ A$ is of type (D) its domain is nonempty, thus $D(S) \cap D(A^* \circ T \circ A) \neq \emptyset$. By Theorem 7.1 we obtain that it is rectangular, too.

Take $x^* \in R(S + A^* \circ T \circ A)$. Then there exist $x \in D(S + A^* \circ T \circ A)$ and $y^*, z^* \in X^*$ such that $x^* = y^* + z^*$, $y^* \in S(x)$ and $z^* \in A^* \circ T \circ A(x)$. Obviously $z^* \in A^*(R(T))$, thus $x^* = y^* + z^* \in R(S) + A^*(R(T))$. Consequently $R(S + A^* \circ T \circ A) \subseteq R(S) + A^*(R(T))$ and the same inclusion exists also between the closures, respectively the interiors, of these sets.

Let now $x^* \in R(S) + A^*(R(T))$, thus there are some $x_1^* \in R(S)$, $x_2^* \in R(A^* \circ T \circ A)$ and $z^* \in R(T)$ such that $x^* = x_1^* + x_2^*$ and $x_2^* = A^* z^*$. Taking an $x \in D(S + A^* \circ T \circ A)$ there holds

$$\inf_{(y,y^*) \in G(S+A^* \circ T \circ A)} \langle x^* - y^*, x - y \rangle = \inf_{\substack{(y,u^*) \in G(S), \\ (y,v^*) \in G(A^* \circ T \circ A), \\ u^* + v^* = y^*}} \langle x_1^* + x_2^* - (u^* + v^*), x - y \rangle$$

$$\geq \inf_{(y,u^*) \in G(S)} \langle x_1^* - u^*, x - y \rangle + \inf_{(y,v^*) \in G(A^* \circ T \circ A)} \langle x_2^* - v^*, x - y \rangle > -\infty,$$

as both S and $A^* \circ T \circ A$ are rectangular. Applying Lemma 7.2 for $E = R(S) + A^*(R(T))$ and $S + A^* \circ T \circ A$, we obtain that $R(S) + A^*(R(T)) \subseteq \operatorname{cl} R(\overline{S + A^* \circ T \circ A})$ and $\operatorname{int}(R(S) + A^*(R(T))) \subseteq R(\overline{S + A^* \circ T \circ A})$. Taking into consideration what we have already proven above, (i) and (ii) follow. □

Remark 7.5. Taking $X = Y$ and A to be the identity mapping of X, one rediscovers as a special case of Theorem 7.2 the result given in [70, Theorem 3.1] and [176, Theorem 1].

When X is moreover reflexive the inequalities in Theorem 7.2(ii) turn into equalities and we get a more accurate Brézis-Haraux approximation of the range of $S + A^* \circ T \circ A$.

Theorem 7.3. *If the Banach space X is moreover reflexive, the monotone operators S and T are rectangular and $S + A^* \circ T \circ A$ is maximally monotone, one has*

(i) $\operatorname{cl}(R(S) + A^*(R(T))) = \operatorname{cl} R(S + A^* \circ T \circ A)$;
(ii) $\operatorname{int} R(S + A^* \circ T \circ A) = \operatorname{int}(R(S) + A^*(R(T)))$.

Proof. As X is reflexive, the maximally monotone operator $S + A^* \circ T \circ A$ is of type (D), too, and $S + A^* \circ T \circ A = \overline{S + A^* \circ T \circ A}$. The conclusion follows via Theorem 7.2. □

Remark 7.6. Taking $X = Y$ and A to be the identity mapping of X, one rediscovers as a special case of Theorem 7.3 the result given in [70, Corollary 3.1] and [176, Corollary 1].

7.3.2 *Brézis-Haraux Type Approximations for Sums of Subdifferentials*

Now we turn our attention to the most famous example for many classes of monotone operators, namely the subdifferential of a proper, convex and lower semicontinuous function. Let the proper, convex and lower semicontinuous functions $f : X \to \overline{\mathbb{R}}$ and $g : Y \to \overline{\mathbb{R}}$, and the linear continuous mapping $A : X \to Y$ fulfilling the feasibility condition $A(\operatorname{dom} f) \cap \operatorname{dom} g \neq \emptyset$. Like in Sects. 2.2.3, 2.3.3, and the other places where we dealt with unconstrained optimization problems, let us note that valuable special cases of the results presented in the following can be obtained by taking $X = Y$ and A to be the identity mapping of X and, respectively, when f is the zero function. Before giving a Brézis-Haraux type statement involving ranges of subdifferentials, we introduce the following regularity condition inspired from (RC_4^U)

$$(RCM^{BH}) \mid \operatorname{epi} f^* + (A^* \times \operatorname{id}_{\mathbb{R}})(\operatorname{epi} g^*) \text{ is closed in the topology } \omega(X^*, X) \times \mathscr{R}.$$

Theorem 7.4. *If (RCM^{BH}) is valid, then one has*

(i) $\operatorname{cl}(R(\partial f) + A^*(R(\partial g))) = \operatorname{cl} R(\partial f + A^* \circ \partial g \circ A) = \operatorname{cl} R(\partial(f + g \circ A))$;
(ii) $\operatorname{int} R(\partial(f + g \circ A)) = \operatorname{int} R(\partial f + A^* \circ \partial g \circ A) \subseteq \operatorname{int}(R(\partial f) + A^*(R(\partial g))) \subseteq \operatorname{int} D(\partial(f^* \Box A^* g^*)) = \operatorname{int} D(\partial(f + g \circ A)^*)$.

Proof. As f, g and $f + g \circ A$ are proper, convex and lower semicontinuous, by Example 7.1 we know that $\partial(f + g \circ A)$ is a monotone operator of type (D), while ∂f and ∂g are rectangular.

By Corollary 2.14 we know that (RCM^{BH}) implies $\partial f + A^* \circ \partial g \circ A = \partial(f + g \circ A)$, therefore $\partial f + A^* \circ \partial g \circ A$ is maximally monotone operator of type (D), too.

Applying Theorem 7.2 for $S = \partial f$ and $T = \partial g$ we get

$$\operatorname{cl}(R(\partial f) + A^*(R(\partial g))) = \operatorname{cl} R(\partial f + A^* \circ \partial g \circ A) = \operatorname{cl} R(\partial(f + g \circ A)),$$

i.e. (i), and

$$\operatorname{int} R(\partial f + A^* \circ \partial g \circ A) \subseteq \operatorname{int}(R(\partial f) + A^*(R(\partial f))) \subseteq \operatorname{int} R(\overline{\partial f + A^* \circ \partial g \circ A}),$$

which becomes

$$\operatorname{int} R(\partial(f + g \circ A)) \subseteq \operatorname{int}(R(\partial f) + A^*(R(\partial g))) \subseteq \operatorname{int} R(\overline{\partial(f + g \circ A)}). \tag{7.3.2}$$

From Sect. 2.2.3 one can deduce that under (RCM^{BH}) it holds $(f + g \circ A)^* = f^* \Box A^* g^*$, by [104, Théoréme 3.1] we get $R(\overline{\partial(f + g \circ A)}) = D(\partial(f + g \circ A)^*) = D(\partial(f^* \Box A^* g^*))$. Combining this with (7.3.2) one gets (ii). □

Remark 7.7. Similar results to the ones in Theorem 7.4 have been obtained for the case when $X = Y$ and A is the identity mapping of X in [176, Corollary 2] and [70, Corollary 3.2], under the hypothesis that $\bigcup_{t>0} t(\text{dom} f - \text{dom} g)$ is a closed linear subspace of X. However, some of the results obtained there are not true in general Banach spaces. In [176] it is claimed that the mentioned hypotheses yield $\text{int}(R(\partial f) + R(\partial g)) = \text{int} D(\partial(f^* \Box g^*))$, while according to [70] they imply that $\text{int}(R(\partial f) + R(\partial g)) = \text{int} D(\partial(f + g)^*)$. However, as the situation depicted in Example 7.2, which is due to Fitzpatrick and was brought into our attention by [173, Example 2.21], shows, these conclusions can be false when working in nonreflexive Banach spaces.

Example 7.2. Take $X = c_0$, the space of the real sequences converging to 0, which is a nonreflexive Banach space with the usual norm $\|x\| = \sup_{n\geq 1} |x_n|$ for $x = (x_n)_{n\geq 1} \in c_0$, and let $f, g : c_0 \to \mathbb{R}$, with f taking everywhere the value 0 and $g(x) = \|x\| + \|x - e_1\|$, for all $x \in c_0$, where $e_1 = (1, 0, 0, \ldots) \in c_0$. Both functions f and g are proper, convex and continuous and the regularity condition required in [70, 176] is fulfilled. Moreover for any $x \in c_0$ one has $\partial g(x) = \partial\| \cdot \|(x) + \partial\| \cdot -e_1\|(x)$. The dual space of c_0 is ℓ^1, which consists of all the sequences $y = (y_n)_{n\geq 1}$ such that $\|y\|_* = \sum_{n=1}^{+\infty} |y_n| < +\infty$. Denote by F the set of sequences in ℓ^1 having finitely many nonzero entries and by B^* the closed unit ball in ℓ^1.

It is known that $\| \cdot \|^*(y) = 0$ if $\|y\|_* \leq 1$ and $\| \cdot \|^*(y) = +\infty$ otherwise, which leads to $\partial\| \cdot \|(x) = B^*$ if $x = 0$, $\partial\| \cdot \|(e_1) = \{e_1\}$, $\partial\| \cdot \|(-e_1) = \{-e_1\}$ and $\partial\| \cdot \|(x) = \{y \in \ell^1 : \|y\|_* \leq 1, \langle y, x\rangle = \|x\|\} \subseteq F$, otherwise, where we note that $e_1 \in \ell^1$, too. Moreover, we have $\partial\| \cdot -e_1\|(x) = \partial\| \cdot \|(x - e_1)$ for any $x \in c_0$. Further one gets $\partial g(0) = -e_1 + B^*$ and $\partial g(e_1) = e_1 + B^*$. Otherwise, i.e. if $x \in c_0 \backslash \{0, e_1\}$, $\partial g(x) \subseteq F$. Therefore

$$R(\partial g) \subseteq (-e_1 + B^*) \cup (e_1 + B^*) \cup F. \tag{7.3.3}$$

Since $\text{int} R(\partial g)$ includes $\text{int} B^* \pm e_1$, assuming it convex yields $0 = 1/2(e_1 - e_1) \in \text{int} R(\partial g)$. Hence there exists a neighborhood of 0, say U, completely included in $R(\partial g)$. Take some $\lambda > 0$ sufficiently small such that

$$\nu(\lambda) = \left(0, \frac{\lambda}{2^2}, \frac{\lambda}{2^3}, \frac{\lambda}{2^4}, \ldots\right) \in U.$$

Thus $\nu(\lambda) \in R(\partial g)$. One can check that $\|\nu(\lambda) \pm e_1\|_* = 1 + \frac{\lambda}{2} > 1$, so, taking into consideration (7.3.3), $\nu(\lambda)$ must be in F. It is clear that this does not happen, thus we reached a contradiction. Therefore $\text{int} R(\partial g)$ is not convex, unlike $\text{int} \overline{R(\partial g)}$, whose convexity follows via [189, Theorem 20].

On the other hand, the relations claimed in [70, 176] to be valid and mentioned in Remark 7.7 become both now $\text{int} R(\partial g) = \text{int} D(\partial g^*)$, which is equivalent, via [104, Théoréme 3.1], to $\text{int} R(\partial g) = \text{int} \overline{R(\partial g)}$. But, as we have seen above, this does not happen for f and g as selected above, thus the allegations concerning the interior of the sum of the ranges of two subdifferentials in [70, 176] are false.

In the light of Remark 7.7 and Example 7.2, let us give below the consequence of Theorem 7.4 for the case $X = Y$ and A is the identity mapping of X which corrects and generalizes, by asking the fulfillment of a weaker regularity condition, [176, Corollary 2] and [70, Corollary 3.2].

Corollary 7.1. *Let f and g be two proper, convex and lower semicontinuous functions on the Banach space X with extended real values such that* $\operatorname{dom} f \cap \operatorname{dom} g \neq \emptyset$. *Assuming that*

$$\operatorname{epi} f^* + \operatorname{epi} g^* \textit{is closed in the product topology } \omega(X^*, X) \times \mathscr{R},$$

one has

(i) $\operatorname{cl}(R(\partial f) + R(\partial g)) = \operatorname{cl} R(\partial f + \partial g) = \operatorname{cl} R(\partial(f + g))$;
(ii) $\operatorname{int} R(\partial f + \partial g) = \operatorname{int} R(\partial(f + g)) \subseteq \operatorname{int}(R(\partial f) + R(\partial g)) \subseteq \operatorname{int} D(\partial(f^* \Box g^*)) = \operatorname{int} D(\partial((f + g)^*))$.

Remark 7.8. Considering moreover that the Banach space X is reflexive, Theorem 7.3 yields that the inclusions in Corollary 7.1*(ii)* turn into equalities.

7.3.3 Applications of the Brézis-Haraux Type Approximations

Besides the fields of applications of the Brézis-Haraux type approximations mentioned before (see, for instance, [60, 171]), we present below two concrete ways to apply the results we provided within this section.

7.3.3.1 Existence of a Solution to an Optimization Problem

Let the proper, convex and lower semicontinuous functions $f : X \to \overline{\mathbb{R}}$ and $g : Y \to \overline{\mathbb{R}}$ and the linear continuous mapping $A : X \to Y$ such that $A(\operatorname{dom} f) \cap \operatorname{dom} g \neq \emptyset$.

Theorem 7.5. *Assume that* (RCM^{BH}) *is satisfied and moreover that* $0 \in \operatorname{int}(R(\partial f) + A^*(R(\partial g)))$. *Then there exists a neighborhood V of 0 in X^* such that for all $x^* \in V$ there exists an $\bar{x} \in \operatorname{dom} f \cap A^{-1}(\operatorname{dom} g)$ for which*

$$f(\bar{x}) + g(A\bar{x}) - \langle x^*, \bar{x}\rangle = \min_{x \in X} \left[f(x) + g(Ax) - \langle x^*, x\rangle \right].$$

Proof. By Theorem 7.4 we have $\operatorname{int}(R(\partial f) + A^*(R(\partial g))) \subseteq \operatorname{int} D(\partial(f^* \Box A^* g^*))$, thus $0 \in \operatorname{int} D(\partial(f^* \Box A^* g^*))$, i.e. there is a neighborhood V of 0 in X^* such that $V \subseteq D(\partial(f^* \Box A^* g^*)) = D(\partial((f + g \circ A)^*))$.

Let $x^* \in V$. The properties of the subdifferential yield that there is an $\bar{x} \in \operatorname{dom} f \cap A^{-1}(\operatorname{dom} g)$ such that $(f + g \circ A)^*(x^*) + (f + g \circ A)^{**}(\bar{x}) = \langle x^*, \bar{x}\rangle$.

As $f + g \circ A$ is a proper, convex and lower semicontinuous function we have $(f + g \circ A)^{**} = f + g \circ A$, hence the equality stated above becomes

$$f(\bar{x})+g(A\bar{x})-\langle x^*,\bar{x}\rangle = -(f+g\circ A)^*(x^*) = -\max_{x\in X}\{\langle x^*,x\rangle - f(x)-g(Ax)\},$$

yielding thus the conclusion. □

Remark 7.9. Under the hypotheses of Theorem 7.5, (RCM^{BH}) is equivalent to

$$\inf_{x\in X}\left[f(x)+g(Ax)-\langle x^*,x\rangle\right] = \max_{y^*\in Y^*}\left\{-f^*(x^*-A^*y^*)-g^*(y^*)\right\}\ \forall x^*\in X^*.$$

Thus one may notice that the conclusion of the mentioned statement can be refined in the sense that the outcome is something that may be called *locally stable total Fenchel duality*, i.e. the situation where both the primal and the dual problem have optimal solutions and their values coincide for small enough linear perturbations of the objective function of the primal problem. Let us notice moreover that as $0 \in V$, for $x^* = 0$ we obtain also the Fenchel total duality statement, too.

7.3.3.2 Existence of a Solution to a Complementarity Problem

Consider now X to be a reflexive Banach space, let $C \subseteq X$ be a closed convex cone and $S : X \rightrightarrows X^*$ a maximally monotone operator. In the following we will show that Theorem 7.3 can guarantee under certain hypotheses the existence of a solution to the complementarity problem (cf. [70])

$$(CP) \qquad \begin{cases} x\in C,\ x^*\in C^*,\\ \langle x^*,x\rangle = 0,\\ x^*\in S(x).\end{cases}$$

But before we can prove the mentioned statement we have to mention a recent result of ours, originally given in [42, 44]. Recall that the sum of two maximally monotone operators is always a monotone operator that in general fails to be maximal and the problem of finding hypotheses that guarantee its maximality has been firstly solved in [180].

Lemma 7.6. *Given two maximally monotone operators $S, T : X \rightrightarrows X^*$, if the condition*

$$(RCM^M)\left|\begin{array}{l}\{(x^*+y^*,x,y,r) : \varphi_S^*(x^*,x)+\varphi_T^*(y^*,y)\le r\}\ \textit{is closed}\\ \textit{regarding the subspace}\ X^*\times\Delta_X\times\mathbb{R},\end{array}\right.$$

is fulfilled then $S + T$ is a maximally monotone operator, too.

Proof. Fix first some $z \in X$ and $z^* \in X^*$. We prove that there is always an $\bar{x} \in X$ such that $z^* \in (S+T)(\bar{x}+z) + \mathscr{J}(\bar{x})$. Consider the functions $f, g : X \times X^* \to \overline{\mathbb{R}}$, defined by

$$f(x,x^*) = \inf_{y^* \in X^*} \left[\varphi_S(x+z, x^*+z^*-y^*) + \varphi_T(x+z, y^*)\right] - \langle x^*+z^*, z\rangle$$

and

$$g(x,x^*) = \frac{1}{2}\|x\|^2 + \frac{1}{2}\|x^*\|_*^2 - \langle z^*, x\rangle,\ (x,x^*) \in X \times X^*.$$

Let us calculate the conjugates of f and g. For any $(w^*, w) \in X^* \times X$ we have

$$f^*(w^*,w) = \sup_{\substack{x \in X,\\ x^* \in X^*}} \Big\{\langle w^*, x\rangle + \langle x^*, w\rangle - \inf_{y^* \in X^*} \left[\varphi_S(x+z, x^*+z^*-y^*)\right.$$

$$\left.+\varphi_T(x+z, y^*)\right] + \langle x^*+z^*, z\rangle\Big\} = \sup_{\substack{x \in X,\\ x^*, y^* \in X^*}} \{\langle w^*, x\rangle + \langle x^*, w\rangle + \langle x^*+z^*, z\rangle$$

$$-\varphi_S(x+z, x^*+z^*-y^*) - \varphi_T(x+z, y^*)\} = \sup_{\substack{u \in X,\\ u^*, y^* \in X^*}} \{\langle w^*, u-z\rangle + \langle u^*+y^*$$

$$-z^*, w\rangle + \langle u^*+y^*, z\rangle - \varphi_S(u,u^*) - \varphi_T(u,y^*)\} = \sup_{\substack{u \in X,\\ u^*, y^* \in X^*}} \{\langle w^*, u\rangle + \langle u^*+y^*,$$

$$w+z\rangle - \varphi_S(u,u^*) - \varphi_T(u,y^*)\} - \langle w^*, z\rangle - \langle z^*, w\rangle.$$

Considering the function $F : X \times X \times X^* \times X^* \to \overline{\mathbb{R}}$, $F(a,b,a^*,b^*) = \varphi_S(a,a^*) + \varphi_T(b,b^*)$ and the linear mappings $A : X \times X^* \times X^* \to X \times X \times X^* \times X^*$, $A(a,a^*,b^*) = (a,a,a^*,b^*)$ and $M : X^* \times X \to X^* \times X \times X, M(a^*,a) = (a^*,a,a)$, we have that

$$f^*(w^*,w) = (F \circ A)^*(M(w^*, w+z)) - \langle w^*, z\rangle - \langle z^*, w\rangle\ \forall (w^*,w) \in X^* \times X.$$

Because $F^* : X^* \times X^* \times X \times X \to \overline{\mathbb{R}}$, $F^*(a^*,b^*,a,b) = \varphi_S^*(a^*,a) + \varphi_T^*(b^*,b)$ and $A^* : X^* \times X^* \times X \times X \to X^* \times X \times X$, $A^*(a^*,b^*,a,b) = (a^*+b^*,a,b)$, one has

$$A^* \times \mathrm{id}_{\mathbb{R}}(\mathrm{epi}(F^*)) = \{(a^*+b^*,a,b,r) : \varphi_S^*(a^*,a) + \varphi_T^*(b^*,b) \le r\}.$$

Knowing that $\operatorname{Im} M \times \mathbb{R} = X^* \times \Delta_X \times \mathbb{R}$, the regularity condition (RCM^M) is equivalent to saying that $A^* \times \mathrm{id}_{\mathbb{R}}(\mathrm{epi}(F^*))$ is closed regarding the subspace $\operatorname{Im} M \times \mathbb{R}$. So, by Theorem 2.10, we have that for any $(w^*,w) \in X^* \times X$ it holds

$$\begin{aligned}&(F \circ A)^*(M(w^*, w+z))\\ &\quad = \min\big\{F^*(a^*,b^*,a,b) : (a^*+b^*,a,b) = (w^*, w+z, w+z)\big\}.\end{aligned}$$

Back to f^*, one gets immediately that for any $(w^*, w) \in X^* \times X$

$$f^*(w^*, w) = \min_{a^*+b^*=w^*} \left[\varphi_S^*(a^*, w+z) + \varphi_T^*(b^*, w+z)\right] - \langle w^*, z\rangle - \langle z^*, w\rangle.$$

Regarding g^*, the conjugate of g, for any $(w^*, w) \in X^* \times X$ one has

$$\begin{aligned} g^*(w^*, w) &= \sup_{\substack{x\in X,\\ x^*\in X^*}} \left\{\langle w^*, x\rangle + \langle x^*, w\rangle - \frac{1}{2}\|x\|^2 - \frac{1}{2}\|x^*\|_*^2 + \langle z^*, x\rangle\right\} \\ &= \sup_{x\in X} \left\{\langle w^* + z^*, x\rangle - \frac{1}{2}\|x\|^2\right\} + \sup_{x^*\in X^*} \left\{\langle x^*, w\rangle - \frac{1}{2}\|x^*\|_*^2\right\} \\ &= \frac{1}{2}\|w^* + z^*\|_*^2 + \frac{1}{2}\|w\|^2. \end{aligned}$$

For any $(x, x^*) \in X \times X^*$ and $y^* \in X^*$, by Lemma 7.3 one gets

$$\begin{aligned} &\varphi_S(x+z, x^*+z^*-y^*) + \varphi_T(x+z, y^*) - \langle x^*+z^*, z\rangle + g(x, x^*) \geq \\ &\langle x^*+z^*-y^*, x+z\rangle + \langle y^*, x+z\rangle - \langle x^*+z^*, z\rangle + \\ &\frac{1}{2}\|x\|^2 + \frac{1}{2}\|x^*\|_*^2 - \langle z^*, x\rangle = \frac{1}{2}\|x\|^2 + \frac{1}{2}\|x^*\|_*^2 + \langle x^*, x\rangle \geq 0. \end{aligned}$$

Taking in the left-hand side the infimum subject to all $y^* \in X^*$, we get $f(x, x^*)+g(x, x^*) \geq 0$. Thus $\inf_{(x,x^*)\in X\times X^*}[f(x, x^*)+g(x, x^*)] \geq 0$. Because of the convexity of f and g and since the latter is continuous Fenchel's duality theorem (cf. [48, Theorem 3.3.7]) guarantees the existence of a pair $(\bar{x}^*, \bar{x}) \in X^* \times X$ such that

$$\begin{aligned} \inf_{(x,x^*)\in X\times X^*} [f(x, x^*) + g(x, x^*)] &= \max_{(x^*,x)\in X^*\times X} \{-f^*(x^*, x) - g^*(-x^*, -x)\} \\ &= -f^*(\bar{x}^*, \bar{x}) - g^*(-\bar{x}^*, -\bar{x}). \end{aligned}$$

Using the result from above, one gets $f^*(\bar{x}^*, \bar{x}) + g^*(-\bar{x}^*, -\bar{x}) \leq 0$. So there are some $\bar{a}^*$ and $\bar{b}^*$ in X^* such that $\bar{a}^* + \bar{b}^* = \bar{x}^*$ and

$$\varphi_S^*(\bar{a}^*, \bar{x}+z) + \varphi_T^*(\bar{b}^*, \bar{x}+z) - \langle \bar{x}^*, z\rangle - \langle z^*, \bar{x}\rangle + \frac{1}{2}\|-\bar{x}^* + z^*\|_*^2 + \frac{1}{2}\|-\bar{x}\|^2 \leq 0.$$

Taking into account that $\bar{a}^* + \bar{b}^* = \bar{x}^*$, we get

$$\begin{aligned} 0 \geq &\left(\varphi_S^*(\bar{a}^*, \bar{x}+z) - \langle \bar{a}^*, \bar{x}+z\rangle\right) + \left(\varphi_T^*(\bar{b}^*, \bar{x}+z) - \langle \bar{b}^*, \bar{x}+z\rangle\right) \\ &+ \left(\langle \bar{x}^* - z^*, \bar{x}\rangle + \frac{1}{2}\|\bar{x}^* - z^*\|_*^2 + \frac{1}{2}\|\bar{x}\|^2\right) \geq 0, \end{aligned}$$

where the last inequality comes from Lemma 7.3. Thus the inequalities above must hold as equalities, hence

$$\varphi_S^*(\bar{a}^*, \bar{x} + z) = \langle \bar{a}^*, \bar{x} + z \rangle, \ \ \varphi_T^*(\bar{b}^*, \bar{x} + z) = \langle \bar{b}^*, \bar{x} + z \rangle,$$

and

$$\langle \bar{a}^* + \bar{b}^* - z^*, \bar{x} \rangle + \frac{1}{2}\|\bar{a}^* + \bar{b}^* - z^*\|_*^2 + \frac{1}{2}\|\bar{x}\|^2 = 0.$$

These three equalities are equivalent, due to Lemma 7.3, to $\bar{a}^* \in S(\bar{x} + z)$, $\bar{b}^* \in T(\bar{x} + z)$ and, respectively,

$$z^* - \bar{a}^* - \bar{b}^* \in \partial \frac{1}{2}\|\cdot\|^2(\bar{x}) = \mathcal{J}(\bar{x}).$$

Summing these three relations up, one gets

$$z^* - \bar{a}^* - \bar{b}^* + \bar{a}^* + \bar{b}^* \in (S + T)(\bar{x} + z) + \mathcal{J}(\bar{x}).$$

As z and z^* have been arbitrarily chosen, the conclusion follows via Lemma 7.1. □

Remark 7.10. The regularity condition (RCM^M) we gave in Lemma 7.6 is the weakest in the literature that guarantees the maximal monotonicity of the sum of two maximally monotone operators. For a review on more restrictive regularity conditions that deliver the same outcome the reader is referred to [44]. Note moreover that in [21, Theorem 25.4] one can find another weak regularity condition for this, that is formulated via arbitrary representative functions attached to the involved maximally monotone operators, while in [42, Theorem 1] and [21, Theorem 25.1] (see also [38]) weak hypotheses that guarantee the maximal monotonicity of the sum of a maximally monotone operator with another one that is composed with a linear continuous mapping are provided.

Now we are ready to formulate the announced assertion regarding the existence of a solution to (CP).

Theorem 7.6. *Suppose that the monotone operator S is maximal and rectangular, the regularity condition*

$$(RCM^C) \left| \begin{array}{l} \{(x^* + y^*, x, y, r) : (x^*, x, r) \in \operatorname{epi}(\varphi_S^*), y \in C, y^* \in -C^*\} \text{ is closed} \\ \text{regarding the subspace } X^* \times \Delta_X \times \mathbb{R}, \end{array} \right.$$

is satisfied and $0 \in \operatorname{int}(R(S) - C^*)$. *Then the complementarity problem* (CP) *admits a solution.*

Proof. Recall first that $\delta_C^* = \delta_{-C^*}$ and $N_C(x) = \{y^* \in -C^* : \langle y^*, x\rangle = 0\}$ for all $x \in C$. Moreover, $R(N_C) = -C^*$ since $R(N_C) \subseteq -C^* = N_C(0)$.

The Fitzpatrick function attached to N_C is, when $(x, x^*) \in X \times X^*$,

$$\varphi_{N_C}(x, x^*) = \sup_{(y,y^*)\in G(N_C)} \{\langle y^*, x\rangle + \langle x^*, y\rangle - \langle y^*, y\rangle\}$$

$$= \sup_{\substack{y\in C, y^*\in -C^*,\\ \langle y^*, y\rangle = 0}} \{\langle y^*, x\rangle + \langle x^*, y\rangle\} = \begin{cases} 0, & \text{if } x \in C, x^* \in -C^*,\\ +\infty, & \text{otherwise,}\end{cases}$$

while its conjugate at $(z^*, z) \in X^* \times X$ is

$$\varphi_{N_C}^*(z^*, z) = \sup_{\substack{x\in C,\\ x^*\in -C^*}} \{\langle z^*, x\rangle + \langle x^*, z\rangle\} = \begin{cases} 0, & \text{if } z \in C,\ z^* \in -C^*,\\ +\infty, & \text{otherwise.}\end{cases}$$

As (RCM^C) is actually (RCM^M) for S and N_C, the maximality of the monotone operator $S + N_C$ is secured via Lemma 7.6, so by Theorem 7.3 one gets

$$\operatorname{int}(R(S) - C^*) = \operatorname{int}(R(S) + R(N_C)) = \operatorname{int} R(S + N_C).$$

Then we get $0 \in \operatorname{int} R(S + N_C)$, thus $0 \in R(S + N_C)$, i.e. there exists an $x \in C$ such that $0 \in (S + N_C)(x)$. Thus we found an $x^* \in S(x)$ such that $-x^* \in -N_C(x)$, which, since $N_C(x) \subseteq C^*$, yields that (x, x^*) is a solution to (CP). □

7.4 Surjectivity Results Involving the Sum of Two Maximally Monotone Operators

In this section we approach by means of convex analysis different surjectivity problems involving maximally monotone operators defined on a reflexive Banach space, following our paper [30]. First we deliver characterizations via closedness type regularity conditions involving representative functions of the surjectivity of the sum of a maximally monotone operator with a translation of another one. Besides particularizing them for some valuable special cases, we derive from these equivalences regularity conditions for guaranteeing the surjectivity of the sum of two maximally monotone operators and different particular instances of it that are weaker than their previous counterparts from the literature.

7.4.1 Surjectivity Results for the Sum of Two Maximally Monotone Operators

Let X be a reflexive Banach space and S and T be two maximally monotone operators defined on X. Before giving the first main statement of this subsection, the following observation is necessary.

Remark 7.11. Let $p \in X$ and $p^* \in X^*$. Then $p^* \in R(S(p+\cdot)+T(\cdot))$ if and only if $(p, p^*) \in G(S) - G(-T)$.

Theorem 7.7. *Let $p \in X$ and $p^* \in X^*$. The following statements are equivalent*

(i) $p^* \in R(S(p+\cdot)+T(\cdot))$*;*

(ii) *for all $h_S \in \mathscr{F}_S$ and all $h_T \in \mathscr{F}_T$ one has* $\operatorname{dom} h_S \cap (\operatorname{dom} \hat{h}_T + (p, p^*)) \neq \emptyset$ *and the function $h_S^* \Box \big(\hat{h}_T^* + \langle (p^*, p), (\cdot,\cdot)\rangle\big)$ is lower semicontinuous at (p^*, p) and exact at (p^*, p);*

(iii) *there exist $h_S \in \mathscr{F}_S$ and $h_T \in \mathscr{F}_T$ fulfilling* $\operatorname{dom} h_S \cap (\operatorname{dom} \hat{h}_T + (p, p^*)) \neq \emptyset$ *such that the function $h_S^* \Box \big(\hat{h}_T^* + \langle (p^*, p), (\cdot,\cdot)\rangle\big)$ is lower semicontinuous at (p^*, p) and exact at (p^*, p).*

Proof. Note first that the assertion "*(ii)* $\Rightarrow$ *(iii)*" is immediate and one also has

$$\big(\hat{h}_T(\cdot - p, \cdot - p^*)\big)^* = \hat{h}_T^* + \langle p^*, \cdot\rangle + \langle \cdot, p\rangle. \tag{7.4.4}$$

"*(iii)* $\Rightarrow$ *(i)*" Proposition 2.1 yields the equivalence of *(iii)* to

$$\big(h_S + \hat{h}_T(\cdot - p, \cdot - p^*)\big)^*(p^*, p) = \min_{u^* \in X^*, u \in X} \big[h_S^*(p^* - u^*, p - u) + \hat{h}_T^*(u^*, u) + \langle p^*, u\rangle + \langle u^*, p\rangle\big] \tag{7.4.5}$$

Denoting by $(\bar{u}^*, \bar{u}) \in X^* \times X$ the point where this minimum is attained, we obtain, via Lemma 7.3,

$$\begin{aligned}\big(h_S + \hat{h}_T(\cdot - p, \cdot - p^*)\big)^*(p^*, p) &= h_S^*(p^* - \bar{u}^*, p - \bar{u}) + \hat{h}_T^*(\bar{u}^*, \bar{u}) + \langle p^*, \bar{u}\rangle \\ &+ \langle \bar{u}^*, p\rangle \geq \langle p^* - \bar{u}^*, p - \bar{u}\rangle - \langle \bar{u}^*, \bar{u}\rangle + \langle p^*, \bar{u}\rangle + \langle \bar{u}^*, p\rangle = \langle p^*, p\rangle.\end{aligned} \tag{7.4.6}$$

But Lemma 7.3 also yields for every $x \in X$ and $x^* \in X^*$

$$\begin{aligned}(h_S + \hat{h}_T(\cdot - p, \cdot - p^*))(x, x^*) &\geq \langle x^*, x\rangle + \langle -(x^* - p^*), x - p\rangle \\ &= \langle x^*, p\rangle + \langle p^*, x\rangle - \langle p^*, p\rangle,\end{aligned}$$

thus $\langle p^*, p\rangle \geq \langle x^*, p\rangle + \langle p^*, x\rangle - (h_S + \hat{h}_T(\cdot - p, \cdot - p^*))(x, x^*)$. Consequently,

$$\big(h_S + \hat{h}_T(\cdot - p, \cdot - p^*)\big)^*(p^*, p) \leq \langle p^*, p\rangle. \tag{7.4.7}$$

Together with (7.4.6) this yields

$$\big(h_S + \hat{h}_T(\cdot - p, \cdot - p^*)\big)^*(p^*, p) = \langle p^*, p\rangle,$$

and consequently the inequalities invoked to obtain (7.4.6) must be fulfilled as equalities. Therefore

$$h_S^*(p^* - \bar{u}^*, p - \bar{u}) = \langle p^* - \bar{u}^*, p - \bar{u}\rangle \text{ and } \hat{h}_T^*(\bar{u}^*, \bar{u}) = \langle -\bar{u}^*, \bar{u}\rangle. \tag{7.4.8}$$

Having these, Lemma 7.3 yields then $p^* - \bar{u}^* \in S(p - \bar{u})$ and $\bar{u}^* \in T(-\bar{u})$, followed by $p^* \in S(p - \bar{u}) + T(-\bar{u})$, i.e. $p^* \in R(S(p + \cdot) + T(\cdot))$.

"$(i) \Rightarrow (ii)$" Whenever $h_S \in \mathscr{F}_S$, $h_T \in \mathscr{F}_T$, (i) yields, via Remark 7.11, $(p, p^*) \in \operatorname{dom} h_S - \operatorname{dom} \hat{h}_T$, i.e. $\operatorname{dom} h_S \cap (\operatorname{dom} \hat{h}_T + (p^*, p)) \neq \emptyset$.

For every $h_S \in \mathscr{F}_S$, $h_T \in \mathscr{F}_T$, $u \in X$ and $u^* \in X^*$ we have $h_S^*(p^* - u^*, p - u) + \hat{h}_T^*(u^*, u) + \langle (p^*, p), (u, u^*)\rangle \geq \langle p^* - u^*, p - u\rangle - \langle u^*, u\rangle + \langle p^*, u\rangle + \langle u^*, p\rangle = \langle p^*, p\rangle$, consequently, $h_S^* \square \big(\hat{h}_T^* + \langle (p^*, p), (\cdot, \cdot)\rangle\big)\,(p^*, p) \geq \langle p^*, p\rangle$ and, since the function in the right-hand side is strong-strong continuous its value at (p^*, p) must be also smaller than $\overline{h_S^* \square \big(\hat{h}_T^* + \langle (p^*, p), (\cdot, \cdot)\rangle\big)}(p^*, p)$. But from [21, Theorem 7.6] we know, via (7.4.4), that one has $\overline{h_S^* \square \big(\hat{h}_T^* + \langle (p^*, p), (\cdot, \cdot)\rangle\big)} = (h_S + \hat{h}_T(-(p^*, p) + (\cdot, \cdot)))^*$ and since (7.4.7) always holds, it follows that $\overline{h_S^* \square \big(\hat{h}_T^* + \langle (p^*, p), (\cdot, \cdot)\rangle\big)}\,(p^*, p) \leq \langle p^*, p\rangle$. Consequently,

$$h_S^* \square \big(\hat{h}_T^* + \langle (p^*, p), (\cdot, \cdot)\rangle\big)(p^*, p) \geq \overline{h_S^* \square \big(\hat{h}_T^* + \langle (p^*, p), (\cdot, \cdot)\rangle\big)}(p^*, p) = \langle p^*, p\rangle. \tag{7.4.9}$$

Since $p^* \in R(S(p + \cdot) + T(\cdot))$, there exist $(\bar{u}^*, \bar{u}) \in X^* \times X$ fulfilling (7.4.8). Then $h_S^*(p^* - \bar{u}^*, p - \bar{u}) + \hat{h}_T^*(\bar{u}^*, \bar{u}) + \langle (p^*, p), (\bar{u}, \bar{u}^*)\rangle = \langle p^*, p\rangle$, i.e. $h_S^* \square \big(\hat{h}_T^* + \langle (p^*, p), (\cdot, \cdot)\rangle\big)(p^*, p) = h_S^*(p^* - \bar{u}^*, p - \bar{u}) + \hat{h}_T^*(\bar{u}^*, \bar{u}) + \langle (p^*, p), (\bar{u}, \bar{u}^*)\rangle = \langle p^*, p\rangle$, therefore the exactness of the infimal convolution in (ii) is proven, while its lower semicontinuity follows via (7.4.9). □

From Theorem 7.7 we obtain immediately the following surjectivity result.

Corollary 7.2. *For $p \in X$, one has $R(S(p + \cdot) + T(\cdot)) = X^*$ if and only if*

$$\left|\begin{array}{l} \forall p^* \in X^* \forall h_S \in \mathscr{F}_S \forall h_T \in \mathscr{F}_T \text{ one has } \operatorname{dom} h_S \cap (\operatorname{dom} \hat{h}_T + (p, p^*)) \neq \emptyset \text{ and} \\ h_S^* \square \big(\hat{h}_T^* + \langle (p^*, p), (\cdot, \cdot)\rangle\big) \text{ is lower semicontinuous at } (p^*, p) \text{ and exact at } (p^*, p), \end{array}\right.$$

and this is further equivalent to

$$\left|\begin{array}{l}\forall p^* \in X^* \exists h_S \in \mathscr{F}_S \exists h_T \in \mathscr{F}_T \text{ with } \operatorname{dom} h_S \cap (\operatorname{dom} \hat{h}_T + (p, p^*)) \neq \emptyset \text{ such that} \\ h_S^* \square (\hat{h}_T^* + \langle\langle (p^*, p), (\cdot, \cdot)\rangle\rangle) \text{ is lower semicontinuous at } (p^*, p) \text{ and exact at } (p^*, p).\end{array}\right.$$

Inspired by Corollary 7.2 we are able to introduce a sufficient condition that guarantees the surjectivity of $S(p + \cdot) + T(\cdot)$ for a given $p \in X$.

Theorem 7.8. *Let $p \in X$. Then $R(S(p + \cdot) + T(\cdot)) = X^*$ if*

$$(RCM^S) \left|\begin{array}{l}\forall p^* \in X^* \exists h_S \in \mathscr{F}_S \exists h_T \in \mathscr{F}_T \text{ with } \operatorname{dom} h_S \cap (\operatorname{dom} \hat{h}_T + (p, p^*)) \neq \emptyset \\ \text{such that } h_S^* \square (\hat{h}_T^* + \langle\langle (p^*, p), (\cdot, \cdot)\rangle\rangle) \text{ is lower semicontinuous} \\ \text{on } X^* \times \{p\} \text{ and exact at } (p^*, p).\end{array}\right.$$

Next we characterize the surjectivity of the monotone operator $S + T$ via a condition involving representative functions. The first statement follows directly from Theorem 7.7, while the second one is a direct consequence.

Theorem 7.9. *Let $p^* \in X^*$. The following statements are equivalent*

(i) $p^* \in R(S + T)$;

(ii) *for all $h_S \in \mathscr{F}_S$ and $h_T \in \mathscr{F}_T$ one has $\operatorname{dom} h_S \cap (\operatorname{dom} \hat{h}_T + (0, p^*)) \neq \emptyset$ and the function $h_S^* \square (\hat{h}_T^* + \langle p^*, \cdot\rangle)$ is lower semicontinuous at $(p^*, 0)$ and exact at $(p^*, 0)$;*

(iii) *there exist $h_S \in \mathscr{F}_S$ and $h_T \in \mathscr{F}_T$ with $\operatorname{dom} h_S \cap (\operatorname{dom} \hat{h}_T + (0, p^*)) \neq \emptyset$ such that the function $h_S^* \square (\hat{h}_T^* + \langle p^*, \cdot\rangle)$ is lower semicontinuous at $(p^*, 0)$ and exact at $(p^*, 0)$.*

Corollary 7.3. *One has $R(S + T) = X^*$ if and only if*

$$\left|\begin{array}{l}\forall p^* \in X^* \forall h_S \in \mathscr{F}_S \forall h_T \in \mathscr{F}_T \text{ one has } \operatorname{dom} h_S \cap (\operatorname{dom} \hat{h}_T + (0, p^*)) \neq \emptyset \text{ and} \\ h_S^* \square (\hat{h}_T^* + \langle p^*, \cdot\rangle) \text{ is lower semicontinuous at } (p^*, 0) \text{ and exact at } (p^*, 0),\end{array}\right.$$

and this is further equivalent to

$$\left|\begin{array}{l}\forall p^* \in X^* \exists h_S \in \mathscr{F}_S \exists h_T \in \mathscr{F}_T \text{ with } \operatorname{dom} h_S \cap (\operatorname{dom} \hat{h}_T + (0, p^*)) \neq \emptyset \text{ such that} \\ h_S^* \square (\hat{h}_T^* + \langle p^*, \cdot\rangle) \text{ is lower semicontinuous at } (p^*, 0) \text{ and exact at } (p^*, 0).\end{array}\right.$$

Inspired by Corollary 7.3 we are able to introduce a sufficient condition that guarantees the surjectivity of $S + T$.

Theorem 7.10. *One has $R(S + T) = X^*$ if*

$$(RCM^J) \left|\begin{array}{l}\forall p^* \in X^* \exists h_S \in \mathscr{F}_S \exists h_T \in \mathscr{F}_T \text{ with } \operatorname{dom} h_S \cap (\operatorname{dom} \hat{h}_T + (0, p^*)) \neq \emptyset \\ \text{such that } h_S^* \square (\hat{h}_T^* + \langle p^*, \cdot\rangle) \text{ is lower semicontinuous on } X^* \times \{0\} \\ \text{and exact at } (p^*, 0).\end{array}\right.$$

Remark 7.12. In the literature there were given other regularity conditions guaranteeing the surjectivity of $S + T$, namely, for fixed $h_S \in \mathscr{F}_S$ and $h_T \in \mathscr{F}_T$,

- (cf. [163, Corollary 2.7]) $\operatorname{dom} h_T = X \times X^*$;
- (cf. [190, Theorem 30.2]) $\operatorname{dom} h_S - \operatorname{dom} \hat{h}_T = X \times X^*$;
- (cf. [222, Corollary 4]) $\{0\} \times X^* \subseteq \operatorname{sqri}(\operatorname{dom} h_S - \operatorname{dom} \hat{h}_T)$.

It is obvious that the first one implies the second, whose fulfillment yields the validity of the third condition. This one yields that for any $x^*, p^* \in X^*$ one has

$$\big(h_S + \hat{h}_T(\cdot, \cdot - p^*)\big)^*(x^*, 0) = \min_{u^* \in X^*, u \in X} \big[h_S^*(x^* - u^*, -u) + \hat{h}_T^*(u^*, u) + \langle p^*, u\rangle\big],$$

which is equivalent, when $\operatorname{dom} h_S \cap (\operatorname{dom} \hat{h}_T + (0, p^*)) \neq \emptyset$ (condition automatically fulfilled when any of the three regularity conditions given above is satisfied), to the fact that whenever $p^* \in X^*$ the function $h_S^* \square \big(\hat{h}_T^* + \langle p^*, \cdot\rangle\big)$ is lower semicontinuous at $(x^*, 0)$ and exact at $(x^*, 0)$ for all $x^* \in X^*$. It is obvious that this implies (RCM^J) and below we present a situation where (RCM^J) holds, unlike the conditions cited from the literature for the surjectivity of $S + T$.

Example 7.3. Let $X = \mathbb{R}$ and consider the maximally monotone operators $S, T : \mathbb{R} \rightrightarrows \mathbb{R}$ defined by

$$S(x) = \begin{cases} \{0\}, & \text{if } x > 0, \\ (-\infty, 0], & \text{if } x = 0, \\ \emptyset, & \text{otherwise}, \end{cases} \quad \text{and } T(x) = \begin{cases} \mathbb{R}, & \text{if } x = 0, \\ \emptyset, & \text{otherwise}, \end{cases} \quad x \in \mathbb{R}.$$

They are actually subdifferentials of proper, convex and lower-semicontinuous functions, which are also sublinear, namely $S = N_{[0,+\infty)}$ and $T = N_{\{0\}}$. Obviously, $R(S + T) = \mathbb{R}$ and the Fitzpatrick families of both S and T contain only the corresponding Fitzpatrick function, i.e. $\varphi_S = \delta_{[0,+\infty)\times(-\infty,0]} = \varphi_S^{*\top}$ and $\varphi_T = \delta_{\{0\}\times\mathbb{R}} = \varphi_T^{*\top}$.

Then $\operatorname{dom} \varphi_S - \operatorname{dom} \hat{\varphi}_T = \mathbb{R}_+ \times \mathbb{R}$, where $\mathbb{R}_+ = [0, +\infty)$, and it is obvious that $\{0\} \times \mathbb{R}$ is not included in $\operatorname{sqri}(\operatorname{dom} \varphi_S - \operatorname{dom} \hat{\varphi}_T) = (0, +\infty) \times \mathbb{R}$. Consequently, the three conditions mentioned in Remark 7.12 fail in this situation. On the other hand, for $p^*, x, x^* \in \mathbb{R}$ one has

$$\varphi_S^* \square \big(\hat{\varphi}_T^* + \langle p^*, \cdot\rangle\big)(x^*, x) = \begin{cases} 0, & \text{if } x \geq 0, \\ +\infty, & \text{if } x < 0, \end{cases}$$

and this function is lower semicontinuous on $\mathbb{R} \times \mathbb{R}_+$ and exact at all $(x^*, x) \in \mathbb{R} \times \mathbb{R}_+$. Consequently, (RCM^J) is valid in this case.

Remark 7.13. When one of h_S and h_T is continuous, the condition (RCM^J) is automatically fulfilled. It is known (see for instance [190]) that the domain of the Fitzpatrick function attached to the duality map $\mathscr{J}$, which is a maximally monotone operator, is the whole product space $X \times X^*$. By [221, Theorem 2.2.20] it follows that $\varphi_{\mathscr{J}}$ is continuous, thus by Corollary 7.2 we obtain that $S(p + \cdot) + \mathscr{J}(\cdot)$ is surjective, whenever $p \in X$. In this way we rediscover a known property of the maximally monotone operators, already mentioned in Lemma 7.1, used for instance for verifying the maximal monotonicity of the sum of two monotone operators under certain hypotheses, as done for instance in [42, 44]. Moreover, via Corollary 7.3 one gets that $S + \mathscr{J}$ is surjective, rediscovering Rockafellar's classical surjectivity theorem for maximally monotone operators (see for instance [190, Theorem 29.5]).

Remark 7.14. One can notice via (7.4.4) that (7.4.5) can be rewritten when $p^* = 0$ and $p = 0$ as

$$\inf_{x \in X, x^* \in X^*} \left[h_S(x, x^*) + \hat{h}_T(x, x^*)\right] = \max_{u^* \in X^*, u \in X} \left\{ - h_S^*(-u^*, -u) - \hat{h}_T^*(u^*, u)\right\}, \tag{7.4.10}$$

i.e. there is strong duality for the convex optimization problem formulated above in the left-hand side of (7.4.10) and its Fenchel dual problem. When $(\bar{u}, \bar{u}^*) \in X \times X^*$ is an optimal solution to the dual problem, i.e. the point where the maximum in the right-hand side of (7.4.10) is attained, one obtains $\bar{u}^* \in S(\bar{u})$ and $-\bar{u}^* \in T(\bar{u})$. Employing now Lemma 7.3, we obtain $h_S(\bar{u}, \bar{u}^*) = h_S^*(-\bar{u}^*, -\bar{u}) = \langle \bar{u}^*, \bar{u} \rangle$ and $\hat{h}_T(\bar{u}, \bar{u}^*) = \hat{h}_T^*(\bar{u}^*, \bar{u}) = -\langle \bar{u}^*, \bar{u} \rangle$, therefore

$$h_S(\bar{u}, \bar{u}^*) + \hat{h}_T(\bar{u}, \bar{u}^*) = h_S^*(-\bar{u}^*, -\bar{u}) + \hat{h}_T^*(\bar{u}^*, \bar{u}) = 0.$$

Thus, the infimum in the left-hand side of (7.4.10) is attained, i.e. the primal optimization problem given there has an optimal solution, too, so total duality holds for the primal-dual pair of optimization problems in discussion. Therefore we can note for this special kind of optimization problems the coincidence of the strong and total Fenchel duality.

Remark 7.15. Given $p \in X$ and $p^* \in X^*$, the function $h_S^* \Box \big(\hat{h}_T^* + \langle (p^*, p), (\cdot, \cdot) \rangle\big)$ can be replaced in Theorem 7.7(*ii*)–(*iii*) with $\big(h_S^* - \langle (p^*, p), (\cdot, \cdot) \rangle\big) \Box \hat{h}_T^*$ without altering the statement. The other conditions considered afterwards within this section can be correspondingly rewritten, too.

Remark 7.16. The results given within this subsection can be extended for the sum of a maximally monotone operator with another one composed with a linear mapping, as considered in Sect. 7.3. However, because even in the case treated here the results are quite complicated we chose to work in the present framework. Another possible direction of generalization of the results provided in this subsection is for the situation when the involved Banach spaces are not necessarily reflexive, possibly by exploiting ideas and techniques from [161, 162]. Last but not least, it should be possible to obtain Lemma 7.6 as a consequence of Theorem 7.10 and taking into consideration Remark 7.13.

7.4.2 Special Cases

7.4.2.1 Zeros of Sums of Monotone Operators

An important consequence of Theorem 7.9 is the following statement, where we provide equivalent characterizations by means of representative functions of the situation when 0 lies in the range of $S + T$.

Corollary 7.4. *One has* $0 \in R(S+T)$ *if and only if*

$$\left| \begin{array}{l} \forall h_S \in \mathscr{F}_S \forall h_T \in \mathscr{F}_T \text{ one has } \operatorname{dom} h_S \cap \operatorname{dom} \hat{h}_T \neq \emptyset \text{ and the function} \\ h_S^* \Box \hat{h}_T^* \text{ is lower semicontinuous at } (0,0) \text{ and exact at } (0,0), \end{array} \right.$$

and this is further equivalent to

$$\left| \begin{array}{l} \exists h_S \in \mathscr{F}_S \exists h_T \in \mathscr{F}_T \text{ with } \operatorname{dom} h_S \cap \operatorname{dom} \hat{h}_T \neq \emptyset \text{ such that the function} \\ h_S^* \Box \hat{h}_T^* \text{ is lower semicontinuous at } (0,0) \text{ and exact at } (0,0). \end{array} \right.$$

From Corollary 7.4 one can deduce a sufficient condition which ensures that $0 \in R(S+T)$.

Corollary 7.5. *One has* $0 \in R(S+T)$ *if*

$$(RCM^Z) \left| \begin{array}{l} \exists h_S \in \mathscr{F}_S \exists h_T \in \mathscr{F}_T \text{ with } \operatorname{dom} h_S \cap \operatorname{dom} \hat{h}_T \neq \emptyset \text{ such that} \\ h_S^* \Box \hat{h}_T^* \text{ is lower semicontinuous on } X^* \times \{0\} \text{ and exact at } (0,0). \end{array} \right.$$

Remark 7.17. The problem of guaranteeing that $0 \in R(S+T)$ and furthermore of finding a solution of this equation has received a large interest in the literature because of both theoretical and practical reasons. In [19, Theorem 4.5] the condition $(0,0) \in \operatorname{core}(\operatorname{co} G(S) - \operatorname{co} G(-T))$ is shown to imply $0 \in R(S+T)$, while in [222, Lemma 1] the same result is achieved under the assumption $(0,0) \in \operatorname{sqri}(\operatorname{dom} h_S - \operatorname{dom} \hat{h}_T)$. Following similar arguments to the ones in Remark 7.12 one can show that both these conditions yield the validity of (RCM^Z). Checking the situation from Example 7.3, we see that the second condition fails, while (RCM^Z) is fulfilled. As $\operatorname{core}(\operatorname{co} G(S) - \operatorname{co} G(-T)) = \operatorname{int}(\mathbb{R}_+ \times (-\mathbb{R}_+) - \{0\} \times \mathbb{R}) = (0,+\infty) \times \mathbb{R}$ does not contain $(0,0)$, it is straightforward that (RCM^Z) is indeed weaker than both conditions mentioned above.

7.4.2.2 Surjectivity Results Involving Normal Cones

Let $U \subseteq X$ be a nonempty closed convex set. Its normal cone N_U is a maximally monotone operator whose only representative function (cf. [8, Corollary 5.9]) is the Fenchel one, namely $h_{N_U}(x,x^*) = \delta_U(x) + \sigma_U(x^*)$, $(x,x^*) \in X \times X^*$.

From Theorem 7.7 and its consequences we obtain by taking $T = N_U$ the following results.

Corollary 7.6. *Let $p \in X$. Then $R(S(p+\cdot) + N_U(\cdot)) = X^*$ if and only if*

$$\left| \begin{array}{l} \forall p^* \in X^* \forall h_S \in \mathscr{F}_S \text{ one has } \operatorname{dom} h_S \cap (U \times \operatorname{dom} \sigma_{-U} + (p, p^*)) \neq \emptyset \text{ and the} \\ \text{function } (y^*, y) \mapsto \inf\limits_{x \in -U, x^* \in X^*} \left[(h_S^* - \langle (p^*, p), (\cdot, \cdot) \rangle)(y^* - x^*, y - x) + \sigma_U(x^*)\right] \\ \text{is lower semicontinuous at } (p^*, p) \text{ and the infimum within} \\ \text{is attained when } (y^*, y) = (p^*, p), \end{array} \right.$$

and this is further equivalent to

$$\left| \begin{array}{l} \forall p^* \in X^* \exists h_S \in \mathscr{F}_S \text{ with } \operatorname{dom} h_S \cap (U \times \operatorname{dom} \sigma_{-U} + (p, p^*)) \neq \emptyset \text{ the} \\ \text{function } (y^*, y) \mapsto \inf\limits_{x \in -U, x^* \in X^*} \left[(h_S^* - \langle (p^*, p), (\cdot, \cdot) \rangle)(y^* - x^*, y - x) + \sigma_U(x^*)\right] \\ \text{is lower semicontinuous at } (p^*, p) \text{ and the infimum within is attained} \\ \text{when } (y^*, y) = (p^*, p). \end{array} \right.$$

Corollary 7.7. *Let $p \in X$. Then $R(S(p+\cdot) + N_U(\cdot)) = X^*$ if*

$$\left| \begin{array}{l} \forall p^* \in X^* \exists h_S \in \mathscr{F}_S \text{ with } \operatorname{dom} h_S \cap (U \times \operatorname{dom} \sigma_{-U} + (p, p^*)) \neq \emptyset \text{ the} \\ \text{function } (y^*, y) \mapsto \inf\limits_{x \in -U, x^* \in X^*} \left[(h_S^* - \langle (p^*, p), (\cdot, \cdot) \rangle)(y^* - x^*, y - x) + \sigma_U(x^*)\right] \\ \text{is lower semicontinuous on } X^* \times \{p\} \text{ and the infimum within} \\ \text{is attained when } (y^*, y) = (p^*, p). \end{array} \right.$$

Corollary 7.8. *One has $0 \in R(S + N_U)$ if*

$$(RCM^N) \left| \begin{array}{l} \exists h_S \in \mathscr{F}_S \text{ with } \operatorname{dom} h_S \cap (U \times \operatorname{dom} \sigma_{-U}) \neq \emptyset \text{ such that the function} \\ (y^*, y) \mapsto \inf\limits_{x \in U} [(h_S^*(\cdot, y + x) \Box \sigma_U)(y^*)] \text{ is lower semicontinuous} \\ \text{on } X^* \times \{0\} \text{ and the infimum within is attained when } (y^*, y) = (0, 0). \end{array} \right.$$

Remark 7.18. In [19, Corollary 5.7] it is stated that the regularity condition $0 \in \operatorname{core} \operatorname{co}(D(S) - U)$ yields $0 \in R(S + N_U)$. Similarly to the considerations from Remarks 7.12 and 7.17 one can notice that this condition is indeed stronger than (RCM^N).

Not without importance is the question how can one equivalently characterize the surjectivity of a maximally monotone operator via its representative functions. To proceed to answering it, take $U = X$. Then $T = N_X$, i.e. $T(x) = \{0\}$ for all $x \in X$, and the Fenchel representative function of N_X is $(x, x^*) \mapsto \delta_X(x) + \sigma_X(x^*) = \delta_{\{0\}}(x^*)$. Then $S + T = S$ and the surjectivity of S can be characterized, via Corollary 7.6, as follows.

Corollary 7.9. *One has $R(S) = X^*$ if and only if*

$$\left| \begin{array}{l} \forall p^* \in X^* \forall h_S \in \mathscr{F}_S \text{ the function } y^* \mapsto -\left(h_S^*(y^*, \cdot)\right)^*(p^*) \text{ is lower} \\ \text{semicontinuous at } p^* \text{ and } \exists x \in X \text{ such that } p^* \in (\partial h_S^*(p^*, \cdot))(x), \end{array} \right.$$

and this is further equivalent to

$$\left| \begin{array}{l} \forall p^* \in X^* \exists h_S \in \mathscr{F}_S \text{ the function } y^* \mapsto -\big(h_S^*(y^*,\cdot)\big)^*(p^*) \text{ is lower} \\ \text{semicontinuous at } p^* \text{ and } \exists x \in X \text{ such that } p^* \in (\partial h_S^*(p^*,\cdot))(x). \end{array} \right.$$

Proof. Corollary 7.6 asserts the equivalence of the surjectivity of the maximally monotone operator S to the lower semicontinuity at $(p^*, 0)$ of the function

$$(y^*, y) \mapsto \inf_{x \in X, x^* \in X^*} \big[(h_S^* - \langle p^*, \cdot \rangle)(y^* - x^*, y + x) + \sigma_X(x^*)\big]$$

concurring with the attainment of the infimum within when $(y^*, y) = (p^*, 0)$, for every $p^* \in X^*$. Taking a closer look at this function, we note that it can be simplified to $(y^*, y) \mapsto \inf_{x \in X} \big[h_S^*(y^*, y + x) - \langle p^*, y + x \rangle\big]$, which can be further reduced to $y^* \mapsto -\big(h_S^*(y^*, \cdot)\big)^*(p^*)$.

For $p^* \in X^*$, the attainment of the infimum from above when $(y^*, y) = (p^*, 0)$ means actually the existence of an $x \in X$ such that $h_S^*(p^*, x) - \langle p^*, x \rangle = -\big(h_S^*(p^*, \cdot)\big)^* (p^*)$, which is nothing but $p^* \in (\partial h_S^*(p^*, \cdot))(x)$. □

Remark 7.19. In [163, Corollary 2.2] it is shown that S is surjective if $\operatorname{dom}(\varphi_S) = X \times X^*$. This result can be obtained as a consequence of Corollary 7.9 knowing that the characterizations provided there for $R(S) = X^*$ are fulfilled when φ_S is continuous.

Remark 7.20. Since determining when $0 \in R(S)$ is important even beyond optimization, using Corollary 7.9 one can provide the following regularity condition for guaranteeing this

$$\left| \begin{array}{l} \exists h_S \in \mathscr{F}_S \text{ the function } y^* \mapsto -\big(h_S^*(y^*,\cdot)\big)^*(0) \text{ is lower} \\ \text{semicontinuous and } \exists x \in X \text{ such that } p^* \in (\partial h_S^*(0,\cdot))(x). \end{array} \right.$$

7.4.2.3 Surjectivity Results Involving Subdifferentials

Let now the proper, convex and lower semicontinuous functions $f, g : X \to \overline{\mathbb{R}}$. Take first $T = \partial g$ and consider for it the Fenchel representative function. Then Corollary 7.2 yields the following statement.

Corollary 7.10. *Let $p \in X$. Then $R(S(p + \cdot) + \partial g(\cdot)) = X^*$ if and only if*

$$\left| \begin{array}{l} \forall p^* \in X^* \forall h_S \in \mathscr{F}_S \text{ one has } \operatorname{dom} h_S \cap (\operatorname{dom} g \times (-\operatorname{dom} g^*) + (p, p^*)) \neq \emptyset \\ \text{and the function } h_S^* \square \big(g(-\cdot) + g^*(\cdot) + \langle (p^*, p), (\cdot, \cdot) \rangle\big) \text{ is} \\ \text{lower semicontinuous at } (p^*, p) \text{ and exact at } (p^*, p), \end{array} \right.$$

and this is further equivalent to

$$\left| \begin{array}{l} \forall p^* \in X^* \exists h_S \in \mathscr{F}_S \text{ with } \operatorname{dom} h_S \cap (\operatorname{dom} g \times (-\operatorname{dom} g^*) + (p, p^*)) \neq \emptyset \\ \text{such that the function } h_S^* \square \big(g(-\cdot) + g^*(\cdot) + \langle (p^*, p), (\cdot, \cdot) \rangle \big) \text{ is} \\ \text{lower semicontinuous at } (p^*, p) \text{ and exact at } (p^*, p). \end{array} \right.$$

Remark 7.21. In [163, Proposition 2.9] it was proven that when g and g^* are real valued the monotone operator $S(p + \cdot) + \partial g(\cdot)$ is surjective whenever $p \in X$. This statement can be rediscovered as a consequence of Corollary 7.10, too. Using [221, Proposition 2.1.6] one obtains that g and g^* are continuous under the mentioned hypotheses. Then the Fenchel representative function of ∂g is continuous and this yields the fulfillment of the regularity condition from Corollary 7.10. Consequently, $S(p + \cdot) + \partial g(\cdot)$ is surjective whenever $p \in X$.

The other statements involving two maximally monotone operators given above can be particularized for this special case, too. However, we give here only a consequence of Corollary 7.5.

Corollary 7.11. *One has* $0 \in R(S + \partial g)$ *if*

$$\left| \begin{array}{l} \exists h_S \in \mathscr{F}_S \text{ with } \operatorname{dom} h_S \cap (\operatorname{dom} g \times (-\operatorname{dom} g^*)) \neq \emptyset \text{ such that the function} \\ h_S^* \square \big(g(-\cdot) + g^*(\cdot)\big) \text{ is lower semicontinuous on } X^* \times \{0\} \text{ and exact at } (0,0). \end{array} \right.$$

Take now also $S = \partial f$, to which we associate the Fenchel representative function, too. Let the function $\hat{g} : X \to \overline{\mathbb{R}}$, $\hat{g}(x) = g(-x)$. Corollary 7.2 yields the following result.

Corollary 7.12. *Let* $p \in X$. *If* $\operatorname{dom} f \cap (p + \operatorname{dom} g) \neq \emptyset$, *then* $R(\partial f(p + \cdot) + \partial g(\cdot)) = X^*$ *if and only if*

$$\left| \begin{array}{l} \forall p^* \in X^* \text{ one has } \operatorname{dom} f^* \cap (p^* - \operatorname{dom} g^*) \neq \emptyset, \text{ the function } f \square (\hat{g} + p^*) \\ \text{is lower semicontinuous at } p \text{ and exact at } p \text{ and the function} \\ f^* \square (g^* + p) \text{ is lower semicontinuous at } p^* \text{ and exact at } p^*. \end{array} \right.$$

Moreover, from Corollary 7.11 one can deduce the following statement.

Corollary 7.13. *One has* $0 \in R(\partial f + \partial g)$ *if* $\operatorname{dom} f \cap \operatorname{dom} g \neq \emptyset$, $\operatorname{dom} f^* \cap (-\operatorname{dom} g^*) \neq \emptyset$ *and*

$$\left| \begin{array}{l} f \square \hat{g} \text{ is lower semicontinuous at } 0 \text{ and exact at } 0 \text{ and the} \\ \text{function } f^* \square g^* \text{ is lower semicontinuous and exact at } 0. \end{array} \right.$$

7.5 Dealing with the Maximal Monotonicity of Bifunctions via Representative Functions

The study of the maximal monotonicity of bifunctions began with the seminal paper [12], followed by works like [116, 135, 160], and in all of them the investigations were based on the theory of equilibrium problems. However, motivated by the recent results on maximally monotone operators, obtained almost exclusively by means of representative functions, we involved the latter in the new approach of the maximal monotonicity of bifunctions proposed in [33]. In this way we succeeded in extending some statements from the literature and, moreover, in proving some recent conjectures. This section is dedicated to presenting these results, but before stating them some preliminaries on monotone bifunctions are necessary.

7.5.1 Monotone Bifunctions

We begin with some preliminaries on bifunctions, following [116,135]. Take further X to be a normed space. Let the nonempty set $C \subseteq X$. A function $F : C \times C \to \mathbb{R}$ is called *bifunction*. The bifunction F is called *monotone* if $F(x,y) + F(y,x) \leq 0$ for all $x, y \in C$. To the bifunction F one can attach the *diagonal subdifferential operators* $A^F : X \rightrightarrows X^*$ and ${}^F\!A : X \rightrightarrows X^*$ defined by

$$A^F(x) = \begin{cases} \{x^* \in X^* : F(x,y) - F(x,x) \geq \langle x^*, y - x\rangle \ \forall y \in C\}, & \text{if } x \in C, \\ \emptyset, & \text{otherwise}, \end{cases}$$

and, respectively,

$${}^F\!A(x) = \begin{cases} \{x^* \in X^* : F(x,x) - F(y,x) \geq \langle x^*, y - x\rangle \ \forall y \in C\}, & \text{if } x \in C, \\ \emptyset, & \text{otherwise}. \end{cases}$$

When $F(x,x) = 0$ for all $x \in C$ and F (respectively $-F$) is monotone, then A^F (${}^F\!A$) is a monotone operator. When F is monotone and $F(x,x) = 0$ for all $x \in C$ one has $G\big(A^F\big) \subseteq G\big({}^F\!A\big)$.

The monotone bifunction F fulfilling $F(x,x) = 0$ for all $x \in C$ is said to be *maximally monotone* if A^F is maximally monotone and, respectively, *BO-maximally monotone* (where BO stands for Blum-Oettli, as this type of monotone bifunction was introduced in [12]) when for every $(x,x^*) \in C \times X^*$ it holds

$$F(y,x) + \langle x^*, y - x\rangle \leq 0 \ \forall y \in C \ \Rightarrow F(x,y) \geq \langle x^*, y - x\rangle \ \forall y \in C.$$

When F is monotone and $F(x,x) = 0$ for all $x \in C$, its BO-maximal monotonicity is equivalent to ${}^F\!A = A^F$. Any maximally monotone bifunction is BO-maximally monotone, but the opposite implication is not always valid, as the situation in [116, Example 2.2] shows.

In order not to overcomplicate the presentation, when $x \in C$ we denote by a slight abuse of notation by $F(x,\cdot) + \delta_C$ the function defined on X with extended real values which is equal to $F(x,\cdot)$ on C and takes the value $+\infty$ otherwise. Analogously, when $y \in C$ we denote by $-F(\cdot,y) + \delta_C$ the function defined on X with extended real values which is equal to $-F(\cdot,y)$ on C and takes the value $+\infty$ otherwise. Hence, when $F(x,x) = 0$ for all $x \in C$, one can write $A^F(x) = \partial(F(x,\cdot) + \delta_C)(x)$ and ${}^F\!A(x) = \partial(-F(\cdot,x) + \delta_C)(x)$ for all $x \in X$. Note that A^F and ${}^F\!A$ are not subdifferentials of functions, being at each point the subdifferential of another function.

We close this preliminary subsection by presenting a statement which holds in a more general framework than originally considered in [12, Lemma 3], followed by a consequence needed later in our investigations.

Lemma 7.7. *Let F and G be two bifunctions defined on the nonempty and convex set $C \subseteq X$, satisfying $F(x,x) = G(x,x) = 0$ for all $x \in C$, such that F is monotone, $F(x,\cdot)$ and $G(x,\cdot)$ are convex for all $x \in C$ and $F(\cdot,y)$ is upper hemicontinuous for all $y \in C$. Then the following statements are equivalent*

(i) $\bar{x} \in C$ and $F(y,\bar{x}) \le G(\bar{x},y)$ for all $y \in C$;
(ii) $\bar{x} \in C$ and $0 \le F(\bar{x},y) + G(\bar{x},y)$ for all $y \in C$.

Remark 7.22. The monotonicity of F is required only for proving the implication "$(ii) \Rightarrow (i)$" in Lemma 7.7, which actually holds even if the convexity and topological hypotheses are removed.

Lemma 7.8. *Let F be a bifunction defined on the nonempty and convex set $C \subseteq X$, satisfying $F(x,x) = 0$ for all $x \in C$. If $F(x,\cdot)$ is convex for all $x \in C$ and $F(\cdot,y)$ is upper hemicontinuous for all $y \in C$, then $G({}^F\!A) \subseteq G(A^F)$.*

Proof. Let $(x,x^*) \in G({}^F\!A)$. Then $x \in C$ and $F(y,x) \le \langle x^*, x-y\rangle$ for all $y \in C$. By Lemma 7.7$(i) \Rightarrow (ii)$ one gets $0 \le F(x,y) + \langle x^*, x-y\rangle$ for all $y \in C$, thus $(x,x^*) \in G(A^F)$. □

Remark 7.23. If in addition to the assumptions of Lemma 7.8 F is taken moreover monotone, one also gets that F is BO-maximally monotone.

7.5.2 *Maximal Monotone Bifunctions*

Let $F : C \times C \to \mathbb{R}$ be a bifunction, where $C \subseteq X$ is nonempty. In order to deal with its maximal monotonicity, we attach to F the functions $h_F, g_F : X \times X^* \to \overline{\mathbb{R}}$, defined at $(x,x^*) \in X \times X^*$ by

$$h_F(x,x^*) = \sup_{y \in C} \{\langle x^*, y\rangle - F(x,y)\} + \delta_C(x) = (F(x,\cdot) + \delta_C)^*(x^*) + \delta_C(x)$$

and

$$g_F(x,x^*) = \sup_{y\in C}\{\langle x^*,y\rangle + F(y,x)\} + \delta_C(x) = (-F(\cdot,x)+\delta_C)^*(x^*) + \delta_C(x).$$

Regarding their conjugates, for $(x^*,x) \in X^* \times X$ one has

$$h_F^*(x^*,x) = \sup_{y\in C}\{\langle x^*,y\rangle + (F(y,\cdot)+\delta_C)^{**}(x)\}$$

and

$$g_F^*(x^*,x) = \sup_{y\in C}\{\langle x^*,y\rangle + (-F(\cdot,y)+\delta_C)^{**}(x)\}.$$

Other properties of these functions are given in the following statements, whose proofs are trivial hence skipped.

Proposition 7.1. *(a) For all $(x,x^*) \in X \times X^*$, it holds $g_F(x,x^*) \geq h_F^*(x^*,x)$.*
(b) If $F(x,x) = 0$ for all $x \in C$, then $h_F \geq c$ and $g_F \geq c$.
(c) If F is monotone, then $h_F(x,x^*) \geq g_F(x,x^*)$ and $\overline{\text{co}}h_F(x,x^*) \geq h_F^*(x^*,x)$ for all $(x,x^*) \in X \times X^*$.

Remark 7.24. If $F(x,x) = 0$ for all $x \in C$, one has that $h_F(x,x^*) = c(x,x^*)$ if and only if $(x,x^*) \in G(A^F)$ and, respectively, $g_F(x,x^*) = c(x,x^*)$ if and only if $(x,x^*) \in G({}^F\!A)$. However, g_F and h_F are in general neither convex nor lower semicontinuous, therefore they are not always representative functions for A^F in case this is monotone. Note also that in [2] a function that slightly extends g_F is called the *Fitzpatrick transform* of the monotone bifunction F.

In the next statements we provide sufficient conditions for the maximal monotonicity of A^F. We begin with an assertion where F is not even asked to be monotone.

Theorem 7.11. *Let C be convex and closed and F be fulfilling $F(x,x) = 0$ for all $x \in C$. If $\text{sqri}\, C \neq \emptyset$, $F(x,\cdot)$ is convex and lower semicontinuous for all $x \in C$ and $F(\cdot,y)$ concave and upper semicontinuous for all $y \in C$, then A^F is maximally monotone and $A^F = {}^F\!A$.*

Proof. The convexity and topological assumptions on C and $F(x,\cdot)$, for $x \in C$, yield that the function $F(x,\cdot) + \delta_C$ is proper, convex and lower semicontinuous whenever $x \in C$. Then $(F(x,\cdot)+\delta_C)^{**}(z) = F(x,z)+\delta_C(z)$ whenever $x \in C$ and $z \in X$, consequently, via Proposition 7.1, $h_F^{*\top} = g_F \geq c$ on $X \times X^*$. Analogously, the convexity and topological assumptions on C and $-F(\cdot,y)$, $y \in C$, imply $h_F = g_F^{*\top} \geq c$ on $X \times X^*$. Obviously, h_F and g_F are in this case convex functions, whose properness follows immediately, too.

One gets $\text{Pr}_X(\text{dom}\, h_F) \subseteq \text{Pr}_X(\text{dom}\, g_F) \subseteq C$. Taking an $x \in C$, since $F(x,\cdot) + \delta_C$ is proper, convex and lower semicontinuous, its conjugate is proper (cf. [221, Theorem 2.3.3]), so there exists an $x^* \in X^*$ such that $(F(x,\cdot) + \delta_C)^*(x^*) < +\infty$. Consequently, $h_F(x,x^*) < +\infty$, i.e. $C \subseteq \mathrm{Pr}_X(\mathrm{dom}\, h_F)$. Therefore $\mathrm{Pr}_X(\mathrm{dom}\, h_F) = \mathrm{Pr}_X(\mathrm{dom}\, g_F) = C$. We are now ready to apply Lemma 7.5 for h_F and g_F, obtaining that the operators (identified through their graphs)

$$\begin{aligned}&\{(x,x^*)\in X \times X^* : h_F^*(x^*,x) = c(x,x^*)\}\\&\quad= \{(x,x^*)\in X \times X^* : g_F(x,x^*) = c(x,x^*)\},\end{aligned}$$

which is actually $G\left({}^F\!A\right)$, and

$$\begin{aligned}&\{(x,x^*)\in X \times X^* : g_F^*(x^*,x) = c(x,x^*)\}\\&\quad= \{(x,x^*)\in X \times X^* : h_F(x,x^*) = c(x,x^*)\},\end{aligned}$$

that is $G\left(A^F\right)$, are maximally monotone.

Using Lemma 7.8, it follows $G\left({}^F\!A\right) \subseteq G\left(A^F\right)$, consequently, $A^F = {}^F\!A$, since both are maximally monotone operators. □

Remark 7.25. If X is reflexive, the hypothesis $\mathrm{sqri}\, C \neq \emptyset$ is no longer needed in Theorem 7.11, since one can use in its proof in this case Lemma 7.4 instead of Lemma 7.5.

If $C = X$ the condition $\mathrm{sqri}\, C \neq \emptyset$ is automatically satisfied and Theorem 7.11 yields the following statement, noting that the lower/upper semicontinuity of a real valued convex/concave function on the entire space is equivalent to its continuity (cf. [221, Proposition 2.1.6]).

Corollary 7.14. *Let $F(x,x) = 0$ for all $x \in X$, $F(x,\cdot)$ be convex and continuous for all $x \in X$ and $F(\cdot,y)$ concave and continuous for all $y \in X$. Then A^F is maximally monotone and $A^F = {}^F\!A$.*

Remark 7.26. In Theorem 7.12 we prove one of the conjectures formulated at the end of [135], actually slightly weakening its hypotheses since instead of taking F continuous we ask it to be continuous in each of its variables. If X is reflexive, Theorem 7.12 slightly improves [135, Theorem 3.6(i)], by bringing the mentioned weakening of its hypotheses.

Taking F to be monotone, here are some hypotheses that guarantee its maximality even in the absence of convexity assumptions in its first variable.

Theorem 7.12. *Let C be convex and closed and F be monotone and fulfilling $F(x,x) = 0$ for all $x \in C$. If $\mathrm{sqri}\, C \neq \emptyset$, $F(x,\cdot)$ is convex and lower semicontinuous for all $x \in C$ and $F(\cdot,y)$ upper hemicontinuous for all $y \in C$, then F is maximally monotone.*

Proof. The convexity and topological assumptions on C and $F(x,\cdot)$, for $x \in C$, yield that the function $F(x,\cdot) + \delta_C$ is proper, convex and lower semicontinuous whenever $x \in C$. Then $(F(x,\cdot) + \delta_C)^{**}(z) = F(x,z) + \delta_C(z)$ whenever $x \in C$ and $z \in X$, hence $h_F^*(x^*,x) = g_F(x,x^*)$ for all $(x,x^*) \in X \times X^*$. Consequently, via Proposition 7.1 and taking into consideration the properties of the conjugate function, one has

$$h_F(x,x^*) \geq \overline{\text{co}} h_F(x,x^*) \geq h_F^*(x^*,x) \geq c(x,x^*) \ \forall (x,x^*) \in X \times X^*. \tag{7.5.11}$$

Assuming that h_F were improper leads to a contradiction with (7.5.11), consequently h_F, $\overline{\text{co}} h_F$ and h_F^* are all proper. Like in the proof of Theorem 7.11 one can show that $\text{Pr}_X(\text{dom}\, h_F) = C$. Then

$$\text{Pr}_X(\text{dom}\, h_F) \subseteq \text{Pr}_X(\text{dom}\, \overline{\text{co}} h_F) \subseteq \overline{\text{co}} \text{Pr}_X(\text{dom}\, h_F) \tag{7.5.12}$$

and, since C is convex and closed, we get $\text{Pr}_X\left(\text{dom}\left(\overline{\text{co}} h_F\right)\right) = C$.

In the following we show that

$$\begin{aligned} G(A^F) &= \left\{(x,x^*) \in X \times X^* : \overline{\text{co}} h_F(x,x^*) = c(x,x^*)\right\} \\ &= \left\{(x,x^*) \in X \times X^* : h_F^*(x^*,x) = c(x,x^*)\right\}. \end{aligned} \tag{7.5.13}$$

If $(x,x^*) \in G(A^F)$, (7.5.11) yields $h_F^*(x^*,x) = c(x,x^*)$.

Let now $(x,x^*) \in X \times X^*$ for which $h_F^*(x^*,x) = c(x,x^*)$. Then $(x,x^*) \in G({}^F\!A)$, so Lemma 7.8 yields $(x,x^*) \in G(A^F)$. This implies that $\overline{\text{co}} h_F(x,x^*) = c(x,x^*)$ holds if and only if $(x,x^*) \in G(A^F)$. Applying Lemma 7.5 for $\overline{\text{co}} h_F$, it follows that A^F is maximally monotone, i.e. F is maximally monotone, too. □

Remark 7.27. In Theorem 7.12 we provide a positive answer to the conjecture formulated at the end of [136]. When the space X is reflexive, the regularity condition $\text{sqri}\, C \neq \emptyset$ is no longer necessary in the hypotheses of Theorem 7.12 and this statement rediscovers [116, Proposition 3.1], by means of representative functions, employing tools of convex analysis and without renorming the space X.

Corollary 7.15. *Let X be reflexive, C be convex and closed and F be monotone and fulfilling $F(x,x) = 0$ for all $x \in C$. If $F(x,\cdot)$ is convex and lower semicontinuous for all $x \in C$ and $F(\cdot,y)$ upper hemicontinuous for all $y \in C$, then F is maximally monotone.*

Proof. Things work in the lines of the proof of Theorem 7.12, noticing that (7.5.11) and (7.5.13) are fulfilled. Then we apply Lemma 7.4. □

When $C = X$ we obtain from Theorem 7.12 the following statement.

Corollary 7.16. *Let F be monotone and fulfilling $F(x,x) = 0$ for all $x \in X$. If $F(x,\cdot)$ is convex and continuous for all $x \in X$ and $F(\cdot,y)$ upper hemicontinuous for all $y \in X$, then F is maximally monotone.*

Remark 7.28. In [135, Theorem 3.6(ii)] the same conclusion as in Corollary 7.16 is obtained when X is reflexive for a monotone bifunction F that fulfills $F(x,x) = 0$ for all $x \in X$, by assuming $F(x,\cdot)$ only convex for all $x \in X$ and $F(\cdot, y)$ continuous for all $y \in X$. However, we doubt that this result holds without any topological assumption on the functions $F(x,\cdot)$, $x \in X$, since in its proof is used [135, Theorem 3.4(ii)], whose hypotheses should contain also the lower semicontinuity of $F(x,\cdot)$ for all $x \in X$. A similar comment can be made also for [135, Theorem 3.6(iii)] and for the conjectures extending the two mentioned statements to nonreflexive spaces given at the end of [135].

Whenever a monotone bifunction F fulfills $F(x,x) = 0$ for all $x \in C$ is BO-maximally monotone, one has $A^F = {}^F\!A$, so Lemma 7.7 is not longer needed in the proof of Theorem 7.12. Hence we rediscover, in the reflexive case, and extend, when X is a general Banach space, [160, Proposition 3.2], as follows.

Corollary 7.17. *Let C be convex and closed with* $\operatorname{sqri} C \neq \emptyset$ *and F be BO-maximally monotone. If $F(x,\cdot)$ is convex and lower semicontinuous for all $x \in C$, then F is maximally monotone.*

Corollary 7.18. *Let X be reflexive, C convex and closed and F be BO-maximally monotone. If $F(x,\cdot)$ is convex and lower semicontinuous for all $x \in C$, then F is maximally monotone.*

When $C = X$ one can formulate another maximality criterium for a monotone bifunction, extending [116, Proposition 3.5] to general Banach spaces.

Theorem 7.13. *Let F be monotone and fulfilling $F(x,x) = 0$ for all $x \in X$. If $D(A^F) = X$ and $F(\cdot, y)$ is upper hemicontinuous for all $y \in X$, then F is maximally monotone.*

Proof. As $D(A^F) = X$, for all $x \in X$ one has $\partial F(x,\cdot)(x) \neq \emptyset$, which yields $\overline{\text{co}} F(x,\cdot)\ (x) = F(x,x) = 0$. On the other hand, for all $x \in X$ it holds $X = \operatorname{dom} F(x,\cdot) \subseteq \operatorname{dom} \overline{\text{co}} F(x,\cdot)$, which implies $\operatorname{dom} \overline{\text{co}} F(x,\cdot) = X$ and via [221, Proposition 2.2.5], as $\overline{\text{co}} F(x,\cdot)(x) = 0$, also the properness of $\overline{\text{co}} F(x,\cdot)$. Then, for any $(x,x^*) \in X \times X^*$, one has

$$h_F^*(x^*,x) = \sup_{y\in X} \big\{\langle x^*, y\rangle + (F(y,\cdot))^{**}(x)\big\} = \sup_{y\in X} \big\{\langle x^*, y\rangle + \overline{\text{co}} F(y,\cdot)(x)\big\} \geq$$

$$\langle x^*, x\rangle + \overline{\text{co}} F(x,\cdot)(x) = \langle x^*, x\rangle,$$

consequently, $h_F \geq \overline{\text{co}} h_F \geq h_F^{*\top} \geq c$ on $X \times X^*$. As $D(A^F) = X$, $\Pr_X(\operatorname{dom} h_F) = X$, using (7.5.12) it follows $\Pr_X(\operatorname{dom} \overline{\text{co}} h_F) = X$. Applying Lemma 7.5 for $\overline{\text{co}} h_F$, the operator having the graph $\{(x,x^*) \in X \times X^* : h_F^*(x^*,x) = c(x,x^*)\}$ turns out to be maximally monotone. This graph includes $G(A^F)$. To show that the opposite inclusion holds, too, let $(x,x^*) \in X \times X^*$ for which $h_F^*(x^*,x) = c(x,x^*)$. Then $h_F^*(x^*,x) \leq c(x,x^*)$, so for all $y \in X$ it holds $\overline{\text{co}} F(y,\cdot)(x) \leq \langle x^*, x - y\rangle$. This means nothing but $(x,x^*) \in G({}^H\!A)$, where the bifunction $H : X \times X \to \mathbb{R}$ is defined by $H(x, y) := \overline{\text{co}}F(x, \cdot)(y)$. It follows immediately that $H(z, z) = 0$ for all $z \in X$. As $H(z, \cdot) = \overline{\text{co}}F(z, \cdot)$ is convex for all $z \in X$ and for all $y \in X$ one can verify that $H(\cdot, y)$ is upper hemicontinuous, Lemma 7.8 yields $(x, x^*) \in G(A^H)$. This means that for all $y \in X$ one has $\overline{\text{co}}F(x, \cdot)(y) \geq \langle x^*, y - x\rangle$, followed by $F(x, y) \geq \langle x^*, y - x\rangle$. Thus $(x, x^*) \in G(A^F)$, therefore (7.5.13) holds. Consequently, F is maximally monotone. □

Remark 7.29. One can see in the proofs of Theorems 7.11–7.13 that not only $\overline{\text{co}}h_F$ (which coincides with h_F under the hypotheses of the first of them), but also the restriction to $X \times X^*$ of $h_F^{*\top}$ are representative functions of the maximally monotone operator A^F.

In Theorems 7.11–7.13 we have shown with the help of the theory of representative functions that under some hypotheses A^F is maximally monotone. Now let us show that the representative functions of it identified there are actually representative to A^F whenever it is maximally monotone.

Theorem 7.14. *Let F be maximally monotone. Then $\overline{\text{co}}h_F$ and the restriction to $X \times X^*$ of $h_F^{*\top}$ are representative functions of A^F.*

Proof. The maximal monotonicity of F implies via Lemma 7.3 that

$$\begin{aligned} G(A^F) &= \{(x, x^*) \in X \times X^* : \psi_{A^F}(x, x^*) = c(x, x^*)\} \\ &= \{(x, x^*) \in X \times X^* : \varphi_{A^F}(x, x^*) = c(x, x^*)\}. \end{aligned}$$

On the other hand, the way h_F is constructed implies $(c + \delta_{A^F})(x, x^*) \geq h_F(x, x^*)$ for all $(x, x^*) \in X \times X^*$, which yields

$$h_F^*(x^*, x) \geq (c + \delta_{A^F})^*(x^*, x) = \psi_{A^F}^*(x^*, x) = \varphi_{A^F}(x, x^*) \; \forall (x, x^*) \in X \times X^*.$$

Since the monotonicity of F implies, via Proposition 7.2, $h_F(x, x^*) \geq \overline{\text{co}}h_F\,(x, x^*) \geq h_F^*(x^*, x)$ for all $(x, x^*) \in X \times X^*$, it follows immediately that for all $(x, x^*) \in X \times X^*$ it holds

$$\psi_{A^F}(x, x^*) \geq \overline{\text{co}}h_F(x, x^*) \geq h_F^*(x^*, x) \geq \varphi_{A^F}(x, x^*) \geq c(x, x^*).$$

Consequently,

$$\begin{aligned} G(A^F) &= \{(x, x^*) \in X \times X^* : \overline{\text{co}}h_F(x, x^*) = c(x, x^*)\} \\ &= \{(x, x^*) \in X \times X^* : h_F^*(x^*, x) = c(x, x^*)\}, \end{aligned}$$

which implies that $\overline{\text{co}}h_F$ and $h_F^{*\top}$ restricted to $X \times X^*$ are representative functions of A^F. □

Remark 7.30. One can easily see that, when F is maximally monotone with $F(x,x) = 0$ for all $x \in C$, then $\overline{\text{co}}g_F$ and the restriction to $X \times X^*$ of $g_F^{*\top}$ are representative functions of A^F, too.

Remark 7.31. In the lines of the proof of Theorem 7.14, one can show that if $T : X \rightrightarrows X^*$ is a maximally monotone operator and $h : X \times X^* \to \overline{\mathbb{R}}$ is a function fulfilling $h(x,x^*) \geq h^*(x^*,x)$ for all $(x,x^*) \in X \times X^*$ and $h(x,x^*) \leq c(x,x^*)$ whenever $(x,x^*) \in G(T)$, then $\overline{\text{co}}h_F$ and the restriction to $X \times X^*$ of $h_F^{*\top}$ are representative functions of T.

7.5.3 The Sum of Two Monotone Bifunctions

One of the most dealt with questions regarding maximally monotone operators is what guarantees that the sum of two of them remains maximally monotone. This issue was extended for maximally monotone bifunctions in [116], by means of equilibrium problems. We provide another answer in this matter, preceded by a preliminary result.

Proposition 7.2. *Let F and G be monotone bifunctions defined on a nonempty set $C \subseteq X$. Then $A^F(x) + A^G(x) \subseteq A^{F+G}(x)$ for all $x \in X$ and $F + G$ is monotone.*

Proof. Let $x \in X$, $y^* \in A^F(x)$ and $z^* \in A^G(x)$. Then $x \in C$ and for all $y \in C$ one has $F(x,y) \geq \langle y^*, y-x\rangle$ and $G(x,y) \geq \langle z^*, y-x\rangle$. Adding these inequalities, one gets $F(x,y)+G(x,y) \geq \langle y^*+z^*, y-x\rangle$ for all $y \in C$, i.e. $y^*+z^* \in A^{F+G}(x)$.

Analogously, writing what the monotonicity of F and G means and adding the obtained inequalities one gets that $F + G$ is monotone. □

For the following statement we need to introduce the bivariate infimal convolution of two functions defined on a cartesian product of sets. Let A and B be two nonempty sets. When $f, g : A \times B \to \overline{\mathbb{R}}$ are proper, their *bivariate infimal convolution* is the function $f \Box_2 g : A \times B \to \overline{\mathbb{R}}$, $f \Box_2 g(a,b) = \inf\{f(a,c) + g(a,b-c) : c \in B\}$.

Theorem 7.15. *Let X be reflexive and F and G two maximally monotone bifunctions defined on a nonempty set $C \subseteq X$ with f_F and f_G their corresponding representative functions. If $0 \in \text{sqri}\big(D(A^F) - D(A^G)\big)$ (or, equivalently, $0 \in \text{sqri}\big(\text{Pr}_X(\text{dom } f_F) - \text{Pr}_X(\text{dom } f_G)\big)$), then $F + G$ is maximally monotone, $A^F + A^G = A^{F+G}$ and $f_F \Box_2 f_G$ is a representative function of A^{F+G}.*

Proof. By [172, Corollary 3.6] we obtain that the hypotheses yield the maximal monotonicity of $A^F + A^G$, to which $f_F \Box_2 f_G$ is a representative function. Then Proposition 7.2 implies that $A^F(x) + A^G(x) = A^{F+G}(x)$ for all $x \in X$. Consequently, $F + G$ is maximally monotone and $f_F \Box_2 f_G$ is a representative function of A^{F+G}, too. □

Remark 7.32. Note that under the hypotheses of Theorem 7.15 also the function $(f_F \Box_2 f_G)^{*\top}$ is a representative function of A^{F+G}. If one takes $f_F := \overline{\text{co}} h_F$ and $f_G := \overline{\text{co}} h_G$, then it holds

$$(f_F \Box_2 f_G)^*(x^*, x) = \sup_{y \in C} \left\{ \langle x^*, y \rangle + (F(y, \cdot) + \delta_C)^{**}(x) + (G(y, \cdot) + \delta_C)^{**}(x) \right\}$$

and this is less than $h^*_{F+G}(x^*, x)$ for all $(x, x^*) \in X \times X^*$. Thus the just identified representative function of A^{F+G} is smaller than the ones obtained for it via Theorem 7.14.

Remark 7.33. If both F and G satisfy the hypotheses of one of Theorems 7.11–7.12, Corollary 7.15 or, when $C = X$, Theorem 7.15, then $F + G$ fulfills them, too, and this has as consequence its maximal monotonicity.

Now let us present a situation, different from the one displayed in Theorem 7.15, when the inclusion proven in Proposition 7.2 turns out to be actually an equality. Note that the space X needs not be reflexive for this statement.

Proposition 7.3. *Let F and G be monotone bifunctions defined on the convex and closed set C fulfilling $F(x, x) = G(x, x) = 0$ for all $x \in C$, such that for all $x \in C$ the functions $F(x, \cdot)$ and $G(x, \cdot)$ are convex and lower semicontinuous. If $0 \in \text{sqri}(C - C)$, then $A^F + A^G = A^{F+G}$.*

Proof. Let $x \in C$. One has $\text{dom}(F(x, \cdot) + \delta_C) = \text{dom}(G(x, \cdot) + \delta_C) = \text{dom}((F + G)(x, \cdot) + \delta_C) = C$. By definition, $A^F(x) = \partial(F(x, \cdot) + \delta_C)(x)$. Note also that $(F(x, \cdot) + \delta_C) + (G(x, \cdot) + \delta_C) = (F + G)(x, \cdot) + \delta_C$. By [221, Theorem 2.8.7], the hypotheses imply

$$\partial(F(x, \cdot) + \delta_C)(x) + \partial(G(x, \cdot) + \delta_C)(x) = \partial(F(x, \cdot) + G(x, \cdot) + \delta_C)(x).$$

Consequently, $A^F(x) + A^G(x) = A^{F+G}(x)$ and since $x \in C$ was arbitrarily chosen, the conclusion follows. □

Remark 7.34. Note that the hypotheses of Proposition 7.3 ensure that $\overline{\text{co}} h_{F+G}(x, x^*) \geq h^*_{F+G}(x^*, x) \geq c(x, x^*)$ for all $(x, x^*) \in X \times X^*$. Unfortunately, this is not enough in order to guarantee the maximality of $F + G$, which would follow for instance provided the BO-maximal monotonicity of this bifunction. However, checking also Remark 7.33, this additional assumption would make, at least in the reflexive case, the condition $0 \in \text{sqri}(C - C)$ redundant. Therefore, it remains as an open question what should one add to the hypotheses of Proposition 7.3 in order to obtain the maximality of $F + G$ under no stronger hypotheses than the ones in Remark 7.33.

References

1. Aliprantis, C.D., Florenzano, M., Martins-da-Rocha, V.F., Tourky, R.: Equilibrium analysis in financial markets with countably many securities. J. Math. Econ. **40**, 683–699 (2004)
2. Alizadeh, M.H., Hadjisavvas, N.: On the Fitzpatrick transform of a monotone bifunction. Optimization **62**, 693–701 (2013)
3. Anbo, Y.: Nonstandard arguments and the characterization of independence in generic structures. RIMS Kôkyûroku **1646**, 4–17 (2009)
4. Attouch, H., Baillon, J.-B., Théra, M.: Variational sum of monotone operators. J. Convex Anal. **1**, 1–29 (1994)
5. Attouch, H., Théra, M.: A general duality principle for the sum of two operators. J. Convex Anal. **3**, 1–24 (1996)
6. Bao, T.Q., Mordukhovich, B.S.: Relative Pareto minimizers for multiobjective problems: existence and optimality conditions. Math. Program. Ser. A. **122**, 301–347 (2010)
7. Bao, T.Q., Mordukhovich, B.S.: Extended Pareto optimality in multiobjective problems. In: Ansari, Q.H., Yao, J.-C. (eds.) Recent Developments in Vector Optimization, pp. 467–515. Springer, Berlin/Heidelberg (2012)
8. Bartz, S., Bauschke, H.H., Borwein, J.M., Reich, S., Wang, X.: Fitzpatrick functions, cyclic monotonicity and Rockafellar's antiderivative. Nonlinear Anal. Theory Methods Appl. **66**, 1198–1223 (2007)
9. Bauschke, H.H., Moffat, S.M., Wang, X.: Near equality, near convexity, sums of maximally monotone operators, and averages of firmly nonexpansive mappings. Math. Program. **139**, 55–70 (2013)
10. Bauschke, H.H., Wang, X., Yao, L.: Rectangularity and paramonotonicity of maximally monotone operators. Optimization **63**, 487–504 (2014)
11. Benson, H.P.: An improved definition of proper efficiency for vector maximization with respect to cones. J. Math. Anal. Appl. **71**, 232–241 (1979)
12. Blum, E., Oettli, W.: From optimization and variational inequalities to equilibrium problems. Math. Stud. **63**, 123–145 (1994)
13. Boncea, H.V., Grad, S.-M.: Characterizations of ε-duality gap statements for composed optimization problems. Nonlinear Anal. Theory Methods Appl. **92**, 96–107 (2013)
14. Boncea, H.V., Grad, S.-M.: Characterizations of ε-duality gap statements for constrained optimization problems. Cent. Eur. J. Math. **11**, 2020–2033, (2013)
15. Bonnel, H., Iusem, A.N., Svaiter, B.F.: Proximal methods in vector optimization. SIAM J. Optim. **15**, 953–970 (2005)
16. Borwein, J.M.: Proper efficient points for maximizations with respect to cones. SIAM J. Control Optim. **15**, 57–63 (1977)

S.-M. Grad, *Vector Optimization and Monotone Operators via Convex Duality*, Vector Optimization, DOI 10.1007/978-3-319-08900-3

17. Borwein, J.M.: The geometry of Pareto efficiency over cones. Math. Operationsforsch. Stat. Ser. Optim. **11**, 235–248 (1980)
18. Borwein, J.M.: On the existence of Pareto efficient points. Math. Oper. Res. **8**, 64–73 (1983)
19. Borwein, J.M.: Maximal monotonicity via convex analysis. J. Convex Anal. **13**, 561–586 (2006)
20. Borwein, J.M., Lewis, A.S.: Partially finite convex programming, Part I: quasi relative interiors and duality theory. Math. Program. Ser. B **57**, 15–48 (1992)
21. Boţ, R.I.: Conjugate Duality in Convex Optimization. Lecture Notes in Economics and Mathematical Systems, vol. 637. Springer, Berlin/Heidelberg (2010)
22. Boţ, R.I., Csetnek, E.R.: Error bound results for convex inequality systems via conjugate duality. Top **20**, 296–309 (2012)
23. Boţ, R.I., Csetnek, E.R.: Regularity conditions via generalized interiority notions in convex optimization: new achievements and their relation to some classical statements. Optimization **61**, 35–65 (2012)
24. Boţ, R.I., Csetnek, E.R., Moldovan, A.: Revisiting some duality theorems via the quasirelative interior in convex optimization. J. Optim. Theory Appl. **139**, 67–84 (2008)
25. Boţ, R.I., Csetnek, E.R., Wanka, G.: Regularity conditions via quasi-relative interior in convex programming. SIAM J. Optim. **19**, 217–233 (2008)
26. Boţ, R.I., Dumitru, A., Wanka, G.: A new Fenchel dual problem in vector optimization. Proc. Indian Acad. Sci. Math. Sci. **119**, 251–265 (2009)
27. Boţ, R.I., Grad, S.-M.: Regularity conditions for formulae of biconjugate functions. Taiwan. J. Math. **12**, 1921–1942 (2008)
28. Boţ, R.I., Grad, S.-M.: Lower semicontinuous type regularity conditions for subdifferential calculus. Optim. Methods Softw. **25**, 37–48 (2010)
29. Boţ, R.I., Grad, S.-M.: Wolfe duality and Mond-Weir duality via perturbations. Nonlinear Anal. Theory Methods Appl. **73**, 374–384 (2010)
30. Boţ, R.I., Grad, S.-M.: Closedness type regularity conditions for surjectivity results involving the sum of two maximal monotone operators. Cent. Eur. J. Math. **9**, 162–172 (2011)
31. Boţ, R.I., Grad, S.-M.: Duality for vector optimization problems via a general scalarization. Optimization **60**, 1269–1290 (2011)
32. Boţ, R.I., Grad, S.-M.: Extending the classical vector Wolfe and Mond-Weir duality concepts via perturbations. J. Nonlinear Convex Anal. **12**, 81–101 (2011)
33. Boţ, R.I., Grad, S.-M.: Approaching the maximal monotonicity of bifunctions via representative functions. J. Convex Anal. **19**, 713–724 (2012)
34. Boţ, R.I., Grad, S.-M.: On linear vector optimization duality in infinite-dimensional spaces. Numer. Algebra Control Optim. **1**, 407–415 (2011)
35. Boţ, R.I., Grad, S.-M., Wanka, G.: Brézis-Haraux-type approximation of the range of a monotone operator composed with a linear mapping. In: Kása, Z., Kassay, G., Kolumbán, J. (eds.) Proceedings of the International Conference in Memoriam Gyula Farkas, Cluj-Napoca, pp. 36–49. Cluj University Press, Cluj-Napoca (2006)
36. Boţ, R.I., Grad, S.-M., Wanka, G.: Fenchel-Lagrange duality versus geometric duality in convex optimization. J. Optim. Theory Appl. **129**, 33–54 (2006)
37. Boţ, R.I., Grad, S.-M., Wanka, G.: A general approach for studying duality in multiobjective optimization. Math. Methods Oper. Res. **65**, 417–444 (2007)
38. Boţ, R.I., Grad, S.-M., Wanka, G.: A new regularity condition for subdifferential calculus and Fenchel duality in infinite dimensional spaces. Applications for maximal monotone operators. In: Castellani, G. (ed.) Seminario Mario Volpato, vol. 3, pp. 16–30. Ca'Foscari University of Venice, Venice (2007)
39. Boţ, R.I., Grad, S.-M., Wanka, G.: Almost convex functions: conjugacy and duality. In: Konnov, I.V., Luc, D.T., Rubinov, A.M. (eds.) Generalized Convexity and Related Topics. Lecture Notes in Economics and Mathematical Systems, vol. 583, pp. 101–114. Springer, Berlin (2007)
40. Boţ, R.I., Grad, S.-M., Wanka, G.: Brézis-Haraux-type approximation in nonreflexive Banach spaces. In: Allevi, E., Bertocchi, M., Gnudi, A., Konnov, I.V. (eds.) Nonlinear Analysis with

Applications in Economics, Energy and Transportation, pp. 155–170. Bergamo University Press, Bergamo (2007)
41. Boţ, R.I., Grad, S.-M., Wanka, G.: Fenchel's duality theorem for nearly convex functions. J. Optim. Theory Appl. **132**, 509–515 (2007)
42. Boţ, R.I., Grad, S.-M., Wanka, G.: Maximal monotonicity for the precomposition with a linear operator. SIAM J. Optim. **17**, 1239–1252 (2007)
43. Boţ, R.I., Grad, S.-M., Wanka, G.: New constraint qualification and conjugate duality for composed convex optimization problems. J. Optim. Theory Appl. **135**, 241–255 (2007)
44. Boţ, R.I., Grad, S.-M., Wanka, G.: Weaker constraint qualifications in maximal monotonicity. Numer. Funct. Anal. Optim. **28**, 27–41 (2007)
45. Boţ, R.I., Grad, S.-M., Wanka, G.: A new constraint qualification for the formula of the subdifferential of composed convex functions in infinite dimensional spaces. Math. Nachr. **281**, 1088–1107 (2008)
46. Boţ, R.I., Grad, S.-M., Wanka, G.: New regularity conditions for strong and total Fenchel-Lagrange duality in infinite dimensional spaces. Nonlinear Anal. Theory Methods Appl. **69**, 323–336 (2008)
47. Boţ, R.I., Grad, S.-M., Wanka, G.: On strong and total Lagrange duality for convex optimization problems. J. Math. Anal. Appl. **337**, 1315–1325 (2008)
48. Boţ, R.I., Grad, S.-M., Wanka, G.: Duality in Vector Optimization. Springer, Berlin/Heidelberg (2009)
49. Boţ, R.I., Grad, S.-M., Wanka, G.: Generalized Moreau-Rockafellar results for composed convex functions. Optimization **58**, 917–933 (2009)
50. Boţ, R.I., Grad, S.-M., Wanka, G.: New regularity conditions for Lagrange and Fenchel-Lagrange duality in infinite dimensional spaces. Math. Inequal. Appl. **12**, 171–189 (2009)
51. Boţ, R.I., Grad, S.-M., Wanka, G.: Classical linear vector optimization duality revisited. Optim. Lett. **6**, 199–210 (2012)
52. Boţ, R.I., Kassay, G., Wanka, G.: Strong duality for generalized convex optimization problems. J. Optim. Theory Appl. **127**, 45–70 (2005)
53. Boţ, R.I., Kassay, G., Wanka, G.: Duality for almost convex optimization problems via the perturbation approach. J. Glob. Optim. **42**, 385–399 (2009)
54. Boţ, R.I., Wanka, G.: An analysis of some dual problems in multiobjective optimization (I). Optimization **53**, 281–300 (2004)
55. Boţ, R.I., Wanka, G.: An analysis of some dual problems in multiobjective optimization (II). Optimization **53**, 301–324 (2004)
56. Boţ, R.I., Wanka, G.: A weaker regularity condition for subdifferential calculus and Fenchel duality in infinite dimensional spaces. Nonlinear Anal. Theory Methods Appl. **64**, 2787–2804 (2006)
57. Boţ, R.I., Wanka, G.: An alternative formulation for a new closed cone constraint qualification. Nonlinear Anal. Theory Methods Appl. **64**, 1367–1381 (2006)
58. Breckner, W., Kolumbán, J.: Dualität bei Optimierungsaugaben in Topologischen Vektorräumen. Mathematica **10**, 229–244 (1968)
59. Breckner, W., Kolumbán, J.: Konvexe Optimierungsaufgaben in Topologischen Vektorräumen. Math. Scand. **25**, 227–247 (1969)
60. Brézis, H., Haraux, A.: Image d'une somme d'opérateurs monotones et applications. Isr. J. Math. **23**, 165–186 (1976)
61. Brumelle, S.: Duality for multiple objective convex programs. Math. Oper. Res. **6**, 159–172 (1981)
62. Bueno, O., Martínez-Legaz, J.-E., Svaiter, B.F.: On the monotone polar and representable closures of monotone operators. J. Convex Anal. **21**, 495–505 (2014)
63. Burachik, R.S., Jeyakumar, V.: A dual condition for the convex subdifferential sum formula with applications. J. Convex Anal. **12**, 279–290 (2005)
64. Burachik, R.S., Jeyakumar, V., Wu, Z.-Y.: Necessary and sufficient conditions for stable conjugate duality. Nonlinear Anal. Theory Methods Appl. **64**, 1998–2006 (2006)

65. Burachik, R.S., Svaiter, B.F.: Maximal monotone operators, convex functions and a special family of enlargements. Set-Valued Anal. **10**, 297–316 (2002)
66. Cambini, R., Carosi, L.: Duality in multiobjective optimization problems with set constraints. In: Eberhard, A., Hadjisavvas, N., Luc, D.T. (eds.) Generalized Convexity, Generalized Monotonicity and Applications, Proceedings of the 7th International Symposium on Generalized Convexity and Generalized Monotonicity held in Hanoi, 27–31 Aug 2002. Nonconvex Optimization and Its Applications, vol. 77, pp. 131–146. Springer, New York (2005)
67. Cammaroto, F., Di Bella, B.: Separation theorem based on the quasirelative interior and application to duality theory. J. Optim. Theory Appl. **125**, 223–229 (2005)
68. Cammaroto, F., Di Bella, B.: On a separation theorem involving the quasi-relative interior. Proc. Edinb. Math. Soc. II. Ser. **50**, 605–610 (2007)
69. Carrizosa, E., Fliege, J.: Generalized goal programming: polynomial methods and applications. Math. Program. **93**, 281–303 (2002)
70. Chbani, Z., Riahi, H.: The range of sums of monotone operators and applications to Hammerstein inclusions and nonlinear complementarity problems. In: Benkirane, A., Grossez, P. (eds.) Nonlinear Partial Differential Equations, Fés, 1994. Pitman Research Notes in Mathematics Series, vol. 343, pp. 61–72. Longman, Harlow (1996)
71. Chien, T.Q.: Nondifferentiable and quasidifferentiable duality in vector optimization theory. Kybernetika **21**, 298–312 (1985)
72. Chu, L.-J.: On Brézis-Haraux approximation with applications. Far East J. Math. Sci. **4**, 425–442 (1996)
73. Chu, L.-J.: On the sum of monotone operators. Mich. Math. J. **43**, 273–289 (1996)
74. Coulibaly, A., Crouzeix, J.-P.: Condition numbers and error bounds in convex programming. Math. Program. Ser. B **116**, 79–113 (2009)
75. Crouzeix, J.-P., Ocaña Anaya, E.: Maximality is nothing but continuity. J. Convex Anal. **17**, 521–534 (2010)
76. Daniele, P., Giuffré, S.: General infinite dimensional duality theory and applications to evolutionary network equilibrium problems. Optim. Lett. **1**, 227–243 (2007)
77. Daniele, P., Giuffré, S., Idone, G., Maugeri, A.: Infinite dimensional duality and applications. Math. Ann. **339**, 221–239 (2007)
78. Dauer, J.P., Saleh, O.A.: A characterization of proper minimal points as solutions of sublinear optimization problems. J. Math. Anal. Appl. **178**, 227–246 (1993)
79. Debreu, G.: Theory of Value. Wiley, New York (1959)
80. Durea, M., Dutta, J., Tammer, C.: Lagrange multipliers for ε-Pareto solutions in vector optimization with nonsolid cones in Banach spaces. J. Optim. Theory Appl. **145**, 196–211 (2010)
81. Durea, M., Tammer, C.: Fuzzy necessary optimality conditions for vector optimization problems. Optimization **58**, 449–467 (2009)
82. Egudo, R.R.: Proper efficiency and multiobjective duality in nonlinear programming. J. Inf. Optim. Sci. **8**, 155–166 (1987)
83. Egudo, R.R.: Efficiency and generalized convex duality for multiobjective programs. J. Math. Anal. Appl. **138**, 84–94 (1989)
84. Egudo, R.R., Weir, T., Mond, B.: Duality without constraint qualification for multiobjective programming. J. Aust. Math. Soc. Ser. B **33**, 531–544 (1992)
85. Eichfelder, G., Jahn, J.: Set-semidefinite optimization. J. Convex Anal. **15**, 767–801 (2008)
86. Fitzpatrick, S.P.: Representing monotone operators by convex functions. In: Fitzpatrick, S.P., Giles, J.R. (eds.) Workshop/Miniconference on Functional Analysis and Optimization. Proceedings of the Centre for Mathematical Analysis, vol. 20, pp. 59–65. Australian National University, Canberra (1988)
87. Fliege, J.: Approximation Techniques for the Set of Efficient Points. Habilitation Thesis, Faculty of Mathematics, University of Dortmund (2001)
88. Fliege, J., Heseler, A.: Constructing approximations to the efficient set of convex quadratic multiobjective problems. Reports on applied mathematics, 211, Faculty of Mathematics, University of Dortmund (2002)

89. Flores-Bazán, F., Flores-Bazán, F., Laengle, S.: Characterizing efficiency on infinite-dimensional commodity spaces with ordering cones having possibly empty interior. J. Optim. Theory Appl. (2014). doi:10.1007/s10957-014-0558-y
90. Flores-Bazán, F., Flores-Bazán, F., Vera, C.: Gordan-type alternative theorems and vector optimization revisited. In: Ansari, Q.H., Yao, J.-C. (eds.) Recent Developments in Vector Optimization, pp. 29–59. Springer, Berlin/Heidelberg (2012)
91. Flores-Bazán, F., Hadjisavvas, N., Vera, C.: An optimal alternative theorem and applications. J. Glob. Optim. **37**, 229–243 (2007)
92. Flores-Bazán, F., Hernández, E.: A unified vector optimization problem: complete scalarizations and applications. Optimization **60**, 1399–1419 (2011)
93. Flores-Bazán, F., Hernández, E.: Optimality conditions for a unified vector optimization problem with not necessarily preordering relations. J. Glob. Optim. **56**, 299–315 (2013)
94. Focke, J: Vektormaximumproblem und parametrische Optimierung. Math. Operationsforsch. Stat. **4**, 365–369 (1973)
95. Friedman, H.M.: A way out. In: Link, G. (ed.) One Hundred Years of Russell's Paradox, pp. 49–86. de Gruyter, Berlin/New York (2004)
96. Gale, D., Kuhn, H.W., Tucker, A.W.: Linear programming and the theory of games. In: Koopmans, T.C. (ed.) Activity Analysis of Production and Allocation, pp. 317–329. Wiley, New York (1951)
97. Gerstewitz, C.: Nichtkonvexe Dualität in der Vektoroptimierung. Wiss. Z. Tech. Hochsch. Carl Schorlemmer Leuna-Merseburg **25**, 357–364 (1983)
98. Gerstewitz, C., Iwanow, E.: Dualität für nichtkonvexe Vektoroptimierungsprobleme. Wiss. Z. Tech. Hochsch. Ilmenau **31**, 61–81 (1985)
99. Gerth, C., Weidner, P.: Nonconvex separation theorems and some applications in vector optimization. J. Optim. Theory Appl. **67**, 297–320 (1990)
100. Gong, X.-H.: Optimality conditions for Henig and globally proper efficient solutions with ordering cone has empty interior. J. Math. Anal. Appl. **307**, 12–31 (2005)
101. Göpfert, A.: Multicriterial duality, examples and advances. In: Fandel, G., Grauer, M., Kurzhanski, A., Wierzbicki, A.P. (eds.) Large-Scale Modelling and Interactive Decision Analysis, Proceedings of the International Workshop Held in Eisenach, 18–21 Nov 1985. Lecture Notes in Economics and Mathematical Systems, vol. 273, pp. 52–58. Springer, Berlin (1986)
102. Göpfert, A., Gerth, C.: Über die Skalarisierung und Dualisierung von Vektoroptimierungsproblemen. Z. Anal. Anwend. **5**, 377–384 (1986)
103. Göpfert, A., Nehse, R.: Vektoroptimierung: Theorie, Verfahren und Anwendungen. Teubner, Leipzig (1990)
104. Gossez, J.-P.: Opérateurs monotones non linéaires dans les espaces de Banach non réflexifs. J. Math. Anal. Appl. **34**, 371–395 (1971)
105. Grad, A.: Generalized Duality and Optimality Conditions. Editura Mega, Cluj-Napoca (2010)
106. Grad, A.: Quasi interior-type optimality conditions in set-valued duality. J. Nonlinear Convex Anal. **14**, 301–317 (2013)
107. Grad, S.-M.: Recent Advances in Vector Optimization and Set-Valued Analysis via Convex Duality. Habilitation Thesis, Faculty of Mathematics, Chemnitz University of Technology (2014)
108. Grad, S.-M., Pop, E.L.: Alternative generalized Wolfe type and Mond-Weir type vector duality. J. Nonlinear Convex Anal. **15**, 867–884 (2014)
109. Grad, S.-M., Pop, E.L.: Characterizing relatively minimal elements via linear scalarization. In: Huisman, D., Louwerse, I., Wagelmans, A.P.M. (eds.) Operations Research Proceedings 2013. Springer, Berlin/Heidelberg, 153–159 (2014)
110. Grad, S.-M., Pop, E.L.: Vector duality for convex vector optimization problems by means of the quasi interior of the ordering cone. Optimization **63**, 21–37 (2014)
111. Graña Drummond, L.M., Iusem, A.N.: First order conditions for ideal minimization of matrix-valued problems. J. Convex Anal. **10**, 1–19 (2003)

112. Graña Drummond, L.M., Iusem, A.N., Svaiter, B.F.: On first order optimality conditions for vector optimization. Acta Math. Appl. Sin. Engl. Ser. **19**, 371–386 (2003)
113. Guerraggio, A., Molho, E., Zaffaroni, A.: On the notion of proper efficiency in vector optimization. J. Optim. Theory Appl. **82**, 1–21 (1994)
114. Gutiérrez, C., Jiménez, B., Novo, V.: On approximate solutions in vector optimization problems via scalarization. Comput. Optim. Appl. **35**, 305–324 (2006)
115. Ha, T.X.D.: Optimality conditions for various efficient solutions involving coderivatives: from set-valued optimization problems to set-valued equilibrium problems. Nonlinear Anal. Theory Methods Appl. **75**, 1305–1323 (2012)
116. Hadjisavvas, N., Khatibzadeh, H.: Maximal monotonicity of bifunctions. Optimization **59**, 147–160 (2010)
117. Hamel, A.H., Heyde, F., Löhne, A., Tammer, C., Winkler, K.: Closing the duality gap in linear vector optimization. J. Convex Anal. **11**, 163–178 (2004)
118. Hartley, R.: On cone-efficiency, cone-convexity and cone-compactness. SIAM J. Appl. Math. **34**, 211–222 (1978)
119. Helbig, S.: Parametric optimization with a bottleneck objective and vector optimization. In: Jahn, J., Krabs, W. (eds.) Recent Advances and Historical Development of Vector Optimization, pp. 146–159. Springer, Berlin (1987)
120. Henig, M.I.: Proper efficiency with respect to cones. J. Optim. Theory Appl. **36**, 387–407 (1982)
121. Heyde, F., Löhne, A.: Geometric duality in multiple objective linear programming. SIAM J. Optim. **19**, 836–845 (2008)
122. Heyde, F., Löhne, A., Tammer, C.: Set-valued duality theory for multiple objective linear programs and application to mathematical finance. Math. Methods Oper. Res. **69**, 159–179 (2009)
123. Heyde, F., Löhne, A., Tammer, C.: The attainment of the solution of the dual program in vertices for vectorial linear programs. In: Barichard, V., Ehrgott, M., Gandibleux, X., T'kindt, M.V. (eds.) Multiobjective Programming and Goal Programming. Lecture Notes in Economics and Mathematical Systems, vol. 618, pp. 13–24. Springer, Berlin/Heidelberg (2009)
124. Hiriart-Urruty, J.-B.: New concepts in nondifferentiable programming. Bull. Soc. Math. Fr. Suppl. Mém. **60**, 57–85 (1979)
125. Hiriart-Urruty, J.-B.: Tangent cones, generalized gradients and mathematical programming in Banach spaces. Math. Oper. Res. **4**, 79–97 (1979)
126. Hiriart-Urruty, J.-B.: ε-subdifferential calculus. In: Aubin, J.-P., Vinter, R.B. (eds.) Convex Analysis and Optimization. Pitman Research Notes in Mathematics Series, vol. 57, pp. 43–92. Pitman, Boston (1982)
127. Hiriart-Urruty, J.-B., Lemaréchal, C.: Convex Analysis and Minimization Algorithms I: Fundamentals. Grundlehren der Mathematischen Wissenschaften, vol. 305. Springer, Berlin (1993)
128. Hiriart-Urruty, J.-B., Lemaréchal, C.: Convex Analysis and Minimization Algorithms II: Advanced Theory and Bundle Methods. Grundlehren der Mathematischen Wissenschaften, vol. 306. Springer, Berlin (1993)
129. Huong, N.T.T., Yen, N.D.: The Pascoletti-Serafini scalarization scheme and linear vector optimization. J. Optim. Theory Appl. **162**, 559–576 (2014)
130. Hurwicz, L: Programming in linear spaces. In: Arrow, K.J., Hurwicz, L., Uzawa, H. (eds.) Studies in Linear and Non-Linear Programming, pp. 38–102. Stanford University Press, Stanford (1958)
131. Isermann, H.: Proper efficiency and the linear vector maximum problem. Oper. Res. **22**, 189–191 (1974)
132. Isermann, H.: Duality in multiple objective linear programming. In: Zionts, S. (ed.) Multiple Criteria Problem Solving. Lecture Notes in Economics and Mathematical Systems, vol. 155, pp. 274–285. Springer, Berlin (1978)

133. Isermann, H.: On some relations between a dual pair of multiple objective linear programs. Z. Oper. Res. Ser. A **22**, A33–A41 (1978)
134. Islam, M.A.: Sufficiency and duality in nondifferentiable multiobjective programming. Pure Appl. Math. Sci. **39**, 31–39 (1994)
135. Iusem, A.N.: On the maximal monotonicity of diagonal subdifferential operators. J. Convex Anal. **18**, 489–503 (2011)
136. Iusem, A.N., Svaiter, B.F.: On diagonal subdifferential operators in nonreflexive Banach spaces. Set-Valued Var. Anal. **20**, 1–14 (2012)
137. Iwanow, E.H., Nehse, R.: Some results on dual vector optimization problems. Optimization **16**, 505–517 (1985)
138. Jahn, J.: Duality in vector optimization. Math. Program. **25**, 343–353 (1983)
139. Jahn, J.: Scalarization in vector optimization. Math. Program. **29**, 203–218 (1984)
140. Jahn, J.: Vector Optimization – Theory, Applications, and Extensions. Springer, Berlin/Heidelberg (2004)
141. Jeyakumar, V., Li, G.Y.: New dual constraint qualifications characterizing zero duality gaps of convex programs and semidefinite programs. Nonlinear Anal. Theory Methods Appl. **71**, 2239–2249 (2009)
142. Jeyakumar, V., Li, G.Y.: Stable zero duality gaps in convex programming: complete dual characterizations with applications to semidefinite programs. J. Math. Anal. Appl. **360**, 156–167 (2009)
143. Kaliszewski, I.: Norm scalarization and proper efficiency in vector optimization. Found. Control Eng. **11**, 117–131 (1986)
144. Kaliszewski, I.: Generating nested subsets of efficient solutions. In: Jahn, J., Krabs, W. (eds.) Recent Advances and Historical Development of Vector Optimization, pp. 173–182. Springer, Berlin (1987)
145. Kanniappan, P.: Duality theorems for convex programming without constraint qualification. J. Aust. Math. Soc. Ser. A **36**, 252–266 (1984)
146. Khanh, P.Q.: Optimality conditions via norm scalarization in vector optimization. SIAM J. Control Optim. **31**, 646–658 (1993)
147. Kim, G.S., Lee, G.M.: On ε-approximate solutions for convex semidefinite optimization problems. Taiwan. J. Math. **11**, 765–784 (2007)
148. Klee, V.L.: Convex sets in linear spaces. Duke Univ. Math. Ser. **16**, 443–466 (1948)
149. Kolumbán, J.: Dualität bei Optimierungsaufgaben. In: Alexits, G., Stechkin, S.B. (eds.) Proceedings of the Conference on the Constructive Theory of Functions (Approximation Theory), Budapest, 1969, pp. 261–265. Akadémiai Kiadó, Budapest (1972)
150. Kornbluth, J.S.H.: Duality, indifference and sensitivity analysis in multiple objective linear programming. Oper. Res. Q. **25**, 599–614 (1974)
151. Kusraev, A.G., Kutateladze, S.S.: Subdifferentials: Theory and Applications. Mathematics and Its Applications, vol. 323. Kluwer, Dordrecht (1995)
152. Kuwano, I., Tanaka, T., Yamada, S.: Several nonlinear scalarization methods for set-valued maps. RIMS Kôkyûroku **1643**, 75–86 (2009)
153. Lampe, U.: Dualität und eigentliche Effizienz in der Vektoroptimierung. Humboldt-Univ. Berl., Fachbereich Math., Semin.ber. **37**, 45–54 (1981)
154. Liu, J., Song, W.: On proper efficiencies in locally convex spaces – a survey. Acta Math. Vietnam. **26**, 301–312 (2001)
155. Luc, D.T.: Theory of Vector Optimization. Lecture Notes in Economics and Mathematical Systems, vol. 319. Springer, Berlin/Heidelberg (1989)
156. Luc, D.T.: On duality in multiple objective linear programming. Eur. J. Oper. Res. **210**, 158–168 (2011)
157. Luc, D.T., Phong, T.Q., Volle, M.: A new duality approach to solving concave vector maximization problems. J. Glob. Optim. **36**, 401–423 (2006)
158. Lowe, T.J., Thisse, J.-F., Ward, J.E., Wendell, R.E.: On efficient solutions to multiple objective mathematical problems. Manag. Sci. **30**, 13–46 (1984)
159. Mangasarian, O.L.: Nonlinear Programming. McGraw-Hill, New York (1969)

160. Mansour, M.A., Chbani, Z., Riahi, H.: Recession bifunction and solvability of noncoercive equilibrium problems. Commun. Appl. Anal. **7**, 369–377 (2003)
161. Marques Alves, M., Svaiter, B.F.: Maximal monotonicity, conjugation and the duality product in non-reflexive Banach spaces. J. Convex Anal. **17**, 553–563 (2010)
162. Marques Alves, M., Svaiter, B.F.: On the surjectivity properties of perturbations of maximal monotone operators in non-reflexive Banach spaces. J. Convex Anal. **18**, 209–226 (2011)
163. Martínez-Legaz, J.-E.: Some generalizations of Rockafellar's surjectivity theorem. Pac. J. Optim. **4**, 527–535 (2008)
164. Mbunga, P.: Structural stability of vector optimization problems. In: Pardalos, P.M., Tseveendorj, I., Enkhbat, R. (eds.) Optimization and Optimal Control, Ulaanbaatar, 2002. Series on Computers and Operations Research, vol. 1, pp. 175–183. World Scientific, River Edge (2003)
165. Miettinen, K., Mäkelä, M.M.: On scalarizing functions in multiobjective optimization. OR Spectr. **24**, 193–213 (2002)
166. Miglierina, E., Molho, E.: Scalarization and stability in vector optimization. J. Optim. Theory Appl. **114**, 657–670 (2002)
167. Miglierina, E., Molho, E., Rocca, M.: Well-posedness and scalarization in vector optimization. J. Optim. Theory Appl. **126**, 391–409 (2004)
168. Mitani, K., Nakayama, H.: A multiobjective diet planning support system using the satisficing trade-off method. J. Multi-Criteria Decis. Anal. **6**, 131–139 (1997)
169. Mond, B., Weir, T.: Generalized concavity and duality. In: Schaible, S., Zemba, W.T. (eds.) Generalized Concavity in Optimization and Economics. Proceedings of the NATO Advanced Study Institute, University of British Columbia, Vancouver, 1980, pp. 263–279. Academic, New York (1981)
170. Nakayama, H.: Some remarks on dualization in vector optimization. J. Multi-Criteria Decis. Anal. **5**, 218–255 (1996)
171. Pennanen, T.: On the range of monotone composite mappings. J. Nonlinear Convex Anal. **2**, 193–202 (2001)
172. Penot, J.-P., Zălinescu, C.: Some problems about the representation of monotone operators by convex functions. ANZIAM J. **47**, 1–20 (2005)
173. Phelps, R.R.: Lectures on maximal monotone operators. Extr. Math. **12**, 193–230 (1997)
174. Precupanu, T.: Closedness conditions for the optimality of a family of nonconvex optimization problems. Math. Operationsforsch. Stat. Ser. Optim. **15**, 339–346 (1984)
175. Qiu, J.H., Hao, Y.: Scalarization of Henig properly efficient points in locally convex spaces. J. Optim. Theory Appl. **147**, 71–92 (2010)
176. Riahi, H.: On the range of the sum of monotone operators in general Banach spaces. Proc. Am. Math. Soc. **124**, 3333–3338 (1996)
177. Rocco, M., Martínez-Legaz, J.-E.: On surjectivity results for maximal monotone operators of type (D). J. Convex Anal. **18**, 545–576 (2011)
178. Rockafellar, R.T.: Convex Analysis. Princeton University Press, Princeton (1970)
179. Rockafellar, R.T.: On the maximal monotonicity of subdifferential mappings. Pac. J. Math. **33**, 209–216 (1970)
180. Rockafellar, R.T.: On the maximality of sums of nonlinear monotone operators. Trans. Am. Math. Soc. **149**, 75–88 (1970)
181. Rödder, W.: A generalized saddlepoint theory; its application to duality theory for linear vector optimum problems. Eur. J. Oper. Res. **1**, 55–59 (1977)
182. Rubinov, A.M.: Sublinear operator and theirs applications. Russ. Math. Surv. **32**, 113–175 (1977)
183. Rubinov, A.M., Gasimov, R.N.: Scalarization and nonlinear scalar duality for vector optimization with preferences that are not necessarily a pre-order relation. J. Glob. Optim. **29**, 455–477 (2004)
184. Rubinov, A.M., Glover, B.M.: Quasiconvexity via two step functions. In: Crouzeix, J.-P., Martínez Legaz, J.E., Volle, M. (eds.) Generalized Convexity, Generalized Monotonicity: Recent Results. Nonconvex Optimization and Its Applications, vol. 27, pp. 159–183. Kluwer, Dordrecht (1998)

185. Schandl, B., Klamroth, K., Wiecek, M.M.: Norm-based approximation in multicriteria programming. Comput. Math. Appl. **44**, 925–942 (2002)
186. Schechter, M.: A subgradient duality theorem. J. Math. Anal. Appl. **61**, 850–855 (1977)
187. Schönefeld, P.: Some duality theorems for the non-linear vector maximum problem. Unternehmensforsch. **14**, 51–63 (1970)
188. Shimizu, A., Nishizawa, S., Tanaka, T.: On nonlinear scalarization methods in set-valued optimization. RIMS Kôkyûroku **1415**, 20–28 (2005)
189. Simons, S.: The range of a monotone operator. J. Math. Anal. Appl. **199**, 176–201 (1996)
190. Simons, S.: From Hahn-Banach to Monotonicity. Lecture Notes in Mathematics, vol. 1693. Springer, Berlin (2008)
191. Tammer, C.: A variational principle and applications for vectorial control approximation problems. Preprint 96–09, Reports on Optimization and Stochastics, Martin-Luther University Halle-Wittenberg (1996)
192. Tammer, C., Göpfert, A.: Theory of vector optimization. In: Ehrgott, M., Gandibleux, X. (eds.) Multiple Criteria Optimization: State of the Art – Annotated Bibliographic Surveys. International Series in Operations Research & Management Science, vol. 52, pp. 1–70. Kluwer, Boston (2002)
193. Tammer, C., Göpfert, A., Riahi, H., Zălinescu, C.: Variational Methods in Partially Ordered Spaces. CMS Books in Mathematics/Ouvrages de Mathématiques de la SMC, vol. 17. Springer, New York (2003)
194. Tammer, C., Winkler, K.: A new scalarization approach and applications in multicriteria d.c. optimization. J. Nonlinear Convex Anal. **4**, 365–380 (2003)
195. Tanaka, T., Kuroiwa, D.: Some general conditions assuring $\operatorname{int} A + B = \operatorname{int}(A + B)$. Appl. Math. Lett. **6**, 51–53 (1993)
196. Tanaka, T., Kuroiwa, D.: The convexity of A and B assures $\operatorname{int} A + B = \operatorname{int}(A + B)$. Appl. Math. Lett. **6**, 83–86 (1993)
197. Tanino, T., Kuk, H.: Nonlinear multiobjective programming. In: Ehrgott, M., Gandibleux, X. (eds.) Multiple Criteria Optimization: State of the Art – Annotated Bibliographic Surveys. International Series in Operations Research & Management Science, vol. 52, pp. 71–128. Kluwer, Boston (2002)
198. Tasset, T.N.: Lagrange Multipliers for Set-Valued Functions When Ordering Cones Have Empty Interior. PhD Thesis, University of Colorado (2010)
199. Tidball, M.M., Pourtallier, O., Altman, E.: Approximations in dynamic zero-sum games. SIAM J. Optim. **35**, 2101–2117 (2006)
200. Wanka, G., Boţ, R.I.: A new duality approach for multiobjective convex optimization problems. J. Nonlinear Convex Anal. **3**, 41–57 (2002)
201. Wanka, G., Boţ, R.I., Grad, S.-M.: Multiobjective duality for convex semidefinite programming problems. Z. Anal. Anwend. **22**, 711–728 (2003)
202. Wanka, G., Boţ, R.I., Vargyas, E.T.: Duality for location problems with unbounded unit balls. Eur. J. Oper. Res. **179**, 1252–1265 (2007)
203. Weidner, P.: An approach to different scalarizations in vector optimization. Wiss. Z. Tech. Hochsch. Ilmenau **36**, 103–110 (1990)
204. Weidner, P.: The influence of proper efficiency on optimal solutions of scalarizing problems in multicriteria optimization. OR Spektrum **16**, 255–260 (1994)
205. Weir, T.: Proper efficiency and duality for vector valued optimization problems. J. Aust. Math. Soc. Ser. A **43**, 21–34 (1987)
206. Weir, T.: A note on invex functions and duality in multiple objective optimization. Opsearch **25**, 98–104 (1988)
207. Weir, T.: On efficiency, proper efficiency and duality in multiobjective programming. Asia-Pac. J. Oper. Res. **7**, 46–54 (1990)
208. Weir, T., Mond, B.: Duality for generalized convex programming without a constraint qualification. Util. Math. **31**, 233–242 (1987)
209. Weir, T., Mond, B.: Generalised convexity and duality in multiple objective programming. Bull. Aust. Math. Soc. **39**, 287–299 (1989)

210. Weir, T., Mond, B.: Multiple objective programming duality without a constraint qualification. Util. Math. **39**, 41–55 (1991)
211. Weir, T., Mond, B., Craven, B.D.: On duality for weakly minimized vector valued optimization problems. Optimization **17**, 711–721 (1986)
212. Weir, T., Mond, B., Craven, B.D.: Weak minimization and duality. Numer. Funct. Anal. Optim. **9**, 181–192 (1987)
213. Wierzbicki, A.P.: Basic properties of scalarizing functionals for multiobjective optimization. Math. Operationsforsch. Stat. Ser. Optim. **8**, 55–60 (1977)
214. Winkler, K.: Skalarisierung mehrkriterieller Optimierungsprobleme mittels schiefer Normen. In: Habenicht, W., Scheubrein, B., Scheubein, R. (eds.) Multi-Criteria- und Fuzzy-Systeme in Theorie und Praxis, pp. 173–190. Deutscher Universitäts-Verlag, Wiesbaden (2003)
215. Wolfe, P.: A duality theorem for non-linear programming. Q. Appl. Math. **19**, 239–244 (1961)
216. Zaffaroni, A.: Degrees of efficiency and degrees of minimality. SIAM J. Control Optim. **42**, 1071–1086 (2003)
217. Zeidler, E.: Nonlinear Functional Analysis and Applications II-B: Nonlinear Monotone Operators. Springer, New York (1990)
218. Zheng, X.Y.: Scalarization of Henig proper efficient points in a normed space. J. Optim. Theory Appl. **105**, 233–247 (2000)
219. Zhou, Z.A., Yang, X.M.: Optimality conditions of generalized subconvexlike set-valued optimization problems based on the quasi-relative interior. J. Optim. Theory Appl. **150**, 327–340 (2011)
220. Zălinescu, C.: Stability for a class of nonlinear optimization problems and applications. In: Clarke, F.H., Demyanov, V.F., Giannessi, F. (eds.) Nonsmooth Optimization and Related Topics, Erice 1988, pp. 437–458. Plenum, New York (1988)
221. Zălinescu, C.: Convex Analysis in General Vector Spaces. World Scientific, River Edge (2002)
222. Zălinescu, C.: A new convexity property for monotone operators. J. Convex Anal. **13**, 883–887 (2006)
223. Zălinescu, C.: On three open problems related to quasi relative interior J. Convex Anal. **22** (2015, in press)

Index

S.-M. Grad, *Vector Optimization and Monotone Operators via Convex Duality*,
Vector Optimization, DOI 10.1007/978-3-319-08900-3

MIX
Papier aus verantwortungsvollen Quellen
Paper from responsible sources
FSC® C105338
FSC
www.fsc.org

If you have any concerns about our products,
you can contact us on
ProductSafety@springernature.com

In case Publisher is established outside the EU,
the EU authorized representative is:
Springer Nature Customer Service Center GmbH
Europaplatz 3, 69115 Heidelberg, Germany

Printed by Libri Plureos GmbH
in Hamburg, Germany